THERMAL
PHYSICS

THERMAL PHYSICS

PHILIP M. MORSE

Professor of Physics, Massachusetts Institute of Technology

W. A. BENJAMIN, INC.
New York · Amsterdam 1965

THERMAL PHYSICS

Library of Congress Catalog Card Number 63–22799
Manufactured in the United States of America

*The final manuscript was received June 24, 1963, and this volume
was published January 6, 1964. Second printing with corrections, January 1965.*
*The publisher is pleased to acknowledge the assistance of Sophie
Adler, who designed the book; and William Prokos, who
designed the dust jacket*

W. A. BENJAMIN, INC.
New York, New York

PREFACE

This book, in its preliminary form, has been used as a text in the subject "Statistical Mechanics and Thermodynamics," a one-semester course required of seniors majoring in physics at the Massachusetts Institute of Technology and also taken by mathematics and electrical engineering majors. Prerequisites for the subject are introductory calculus, introductory mechanics, and a course in atomic physics that includes the ideas of quantization and atomic structure. The material in it and in two other senior subjects, classical mechanics and electromagnetic theory, are required of entering graduate students before they can progress to strictly graduate subjects in physics.

The material in this revised and considerably expanded edition has been chosen to prepare the student to go on into fields of modern statistical physics, such as low-temperature physics, solid-state physics, and plasma physics. Thus the illustrative material includes discussions of the properties of liquid helium, superconductors, and paramagnetic substances. To emphasize the fact that thermodynamics is more than a study of the relationships between pressure, volume, and temperature, the thermomagnetic properties of materials that obey Curie's law are discussed in some detail, instead of the usual examples taken from physical chemistry.

These illustrative topics are each taken up several times in the course of the text, to illustrate new concepts or techniques as they are encountered.

The text follows roughly the historical development of the subject, beginning with thermodynamics, going on to kinetic theory, and then to classical and quantum statistical mechanics. Emphasis is laid on the fact that the subject is still developing, is not yet in its final, moribund form. Mention is made of basic changes that have had to be made in the theory to accommodate new experimental data, and it is suggested that other changes may have to be made in the future. Thus the presentation carries forward the point of view enunciated in the Preface to the Physical Science Study Committee text, "that physics is a developing subject, and that this development is the imaginative work of men and women like the student himself".

The topics covered in the present expanded version are rather more than can be taught in one semester to the average physics major. To reduce the amount of assigned material, the latter portions of Chapters 7, 8, 13, 14, 15, 21, 22, 25, or 26 and all of Chapters 10 or 27, or both, may be omitted. In the one-semester subject given at M.I.T., illustrative material from four or five of these more specialized topics is discussed in the lectures, which are three hours per week, and one or two others are taken up in the problem sections, which are one hour per week. It has been found that the more advanced students will read, and absorb, most of the material in the present book.

The problems, listed at the ends of chapters, have been chosen to illuminate basic concepts rather than for drill in techniques. Usually only two or three such problems are assigned each week; these are discussed in some detail in the problem sections. At times the instructors have assigned or have discussed additional drill problems specifically written to improve the student's facility in handling some techniques.

Because the subject of thermal physics impinges on so many parts of physics, the duplication of symbolism is unavoidable if one tries to conform to standard notation in each of the fields touched on. Some ameliorization has been achieved by use of different type fonts, such as German script for magnetization, magnetic and electric intensity, etc. It is hoped that confusion will be minimized with the help of the Glossary of Terms, in which is listed the place or places in the text where the symbol is defined.

Since this is a text, not a treatise, no references are given to original papers on various subjects. The bibliography lists other texts which cover some aspect in more detail, or from a different point of view.

In developing the present volume the author has had valuable assistance from many of his colleagues, notably from Professors L. Tisza, F. Villars, and L. C. Bradley. They are to be thanked for numerous improvements and corrections; they should not be blamed for the shortcomings that remain.

PHILIP M. MORSE

Cambridge, Massachusetts
September 1963

CONTENTS

I THERMODYNAMICS

CHAPTER

INTRODUCTION

The subject matter of this book, thermodynamics, kinetic theory, and statistical mechanics, constitutes the portion of physics having to do with heat. Thus it may be called *thermal physics*. Since heat and other related properties, such as pressure and temperature, are characteristics of aggregates of matter, the subject constitutes a part of the physics of matter in bulk. Its theoretical models, devised to correspond with the observed behavior of matter in bulk (and to predict other unobserved behavior), rely on the mathematics of probability and statistics; thus the subject may alternatively be called *statistical physics*.

Historical

The part of the subject called thermodynamics has had a long and controversial history. It began at the start of the industrial revolution, when it became important to understand the relation between heat and chemical transformations and the conversion of heat into mechanical energy. At that time the atomic nature of matter was not yet understood and the mathematical model, developed to represent the relation between thermal and mechanical behavior, had to be put together with the guidance of the crude experiments that could then be made. Many initial mistakes were made and the theory had to be drastically revised several times. This theory, now called thermodynamics, concerns itself solely with the macroscopic properties of aggregated matter, such as temperature and pressure, and their interrelations, without reference to the underlying atomic structure. The general pattern of these

3

interrelations is summarized in the laws of thermodynamics, from which one can predict the complete thermal behavior of any substance, given a relatively few empirical relationships, obtained by macroscopic measurements made on the substance in question.

In the latter half of the nineteenth century, when the atomic nature of matter began to be understood, efforts were made to learn how the macroscopic properties of matter, dealt with by thermodynamics, could depend on the assumed behavior of constituent atoms. The first successes of this work were concerned with gases, where the interactions between the atomic components are minimal. The results provide a means of expressing the pressure, temperature, and other macroscopic properties of the gas in terms of average values of properties of the molecules, such as their kinetic energy. This part of the subject came to be called *kinetic theory*.

In the meantime a much more ambitious effort was begun by Gibbs in this country, and by Boltzmann and others in Europe, to provide a statistical correspondence between the atomic substructure of any piece of matter and its macroscopic behavior. Gibbs called this theory *statistical mechanics*. Despite the fragmentary knowledge of atomic physics at the time, statistical mechanics was surprisingly successful from the first. Since then, of course, increased atomic knowledge has enabled us to clarify its basic principles and extend its techniques. It now provides us with a means of understanding the laws of thermodynamics and of predicting the various relations between thermodynamic variables, hitherto obtained empirically.

Thermodynamics and Statistical Mechanics

Thus thermodynamics and statistical mechanics are mutually complementary. For example, if the functional relationship between the pressure of a gas, its temperature, and the volume it occupies is known, and if the dependence of the heat capacity of the gas on its temperature and pressure has been determined, then thermodynamics can predict how the temperature and pressure are related when the gas is isolated thermally, or how much heat it will liberate when compressed at constant temperature. Statistical mechanics, on the other hand, seeks to derive the functional relation between pressure, volume, and temperature, and also the behavior of the heat capacity, in terms of the properties of the molecules that make up the gas.

In this volume we shall first take up thermodynamics, because it is more obviously related to the gross physical properties we wish to study. But we shall continue to refer back to the underlying microstructure, by now well understood, to remind ourselves that the thermodynamic variables are just another manifestation of atomic behavior. In fact, because it does not make use of atomic concepts, thermodynamics is a rather abstract subject, employing sophisticated concepts, which have many logical interconnections; it is not easy to understand one part until one understands the whole. In such a case it is better pedagogy to depart from strict logical presentation. Hence several derivations and definitions will be given in steps, first presented in simple form and, only after other concepts have been introduced, later re-enunciated in final, accurate form.

Part of the difficulty comes from the fact, more apparent now than earlier, that the thermodynamic quantities such as temperature and pressure are aggregate effects of related atomic properties. In thermodynamics we assume, with considerable empirical justification, that whenever a given amount of gas, in a container of given volume, is brought to a given temperature, its pressure and other thermodynamic properties will take on specific values, no matter what has been done to the gas previously. By this we do not mean that when the gas is brought back to the same temperature each molecule of the gas returns to the same position and velocity it had previously. All we mean is that the average effects of all the atoms return to their original values, that even if a particular molecule does not return to its previous position or velocity, its place will have been taken by another, so that the aggregate effect is the same.

To thus assume that the state of a given collection of atoms can be at all adequately determined by specifying the values of a small number of macroscopic variables, such as temperature and pressure, would at first seem to be an unworkable oversimplification. Even if there were a large number of different configurations of atomic positions and motions which resulted in the same measurement of temperature, for example, there is no a priori reason that all, or even most, of these same configurations would produce the same pressure. What must happen (and what innumerable experiments show does happen) is that a large number of these configurations do produce the same pressure and that thermodynamics has a method of distinguishing this subset of configurations from others, which do not produce the same pressure. The distinguishing feature of the favored subset is embodied in the concept of the equilibrium state.

Equilibrium States

A detailed specification of the position, velocity, and quantum state of each atom in a given system is called a *microstate* of the system. The definition is useful conceptually, not experimentally, for we cannot determine by observation just what microstate a system is in at some instant, and we would not need to do so even if we could. As we said before, many different microstates will produce the same macroscopic effects; all we need to do is to find a method of confining our attention to that set of microstates which exhibits the simple relations between macroscopic variables, with which thermodynamics concerns itself.

Consider for a moment all those microstates of a particular gas for which the total kinetic energy of all the molecules is equal to some value U. Some of the microstates will correspond to the gas being in a state of turbulence, some parts of the gas having net momentum in one direction, some in another. But a very large number of microstates will correspond to a fairly uniform distribution of molecular kinetic energies and directions of motion, over all regions occupied by the gas. In these states, which we shall call the equilibrium microstates, we shall find that the temperature and pressure are fairly uniform throughout the gas. It is a fact, verified by many experiments, that if a gas is started in a microstate corresponding to turbulence, it will sooner or later reach one of the equilibrium microstates, in which temperature and pressure are uniform. From then on, although the system will change from microstate to microstate as the molecules move about and collide, it will confine itself to equilibrium microstates. To put it in other language, although the gas may start in a state of turbulence, if it is left alone long enough internal friction will bring it to that state of thermodynamic quiescence we call *equilibrium*, where it will remain.

Classical mechanics, as developed by Newton, Lagrange, and Hamilton, is able to predict the behavior of systems with a few elements, or with many elements arranged in some very symmetrical pattern. The two-body problem of celestial mechanics is not difficult to work out in complete detail, but the three-body problem has not yet been solved in general. The motions of many molecules can be worked out only if they are held in some regular pattern by elastic forces, as in a crystal; otherwise the solution is beyond our largest electronic computers. Thermodynamics, on the other hand, deals with the opposite kind of problem, with aggregates of molecules having completely random distributions of positions and velocities. It can work out the behavior of only those states of maximum disorder, which we call the *states of thermodynamic equilibrium*.

We must wait until we take up statistical mechanics to set forth a quantitative measure of the disorder of a system. But even the crudest observations demonstrate that it is a universal property of systems with many elements to become more and more disordered as time goes on, until they reach equilibrium. We can start a card game with a new deck, with all the cards in regular order; but after a few shuffles the hands which are dealt have roughly equal numbers of the different suits; it may take several years of playing poker before one is dealt a royal flush. Likewise, if we introduce 10^{20} helium atoms into a closed container it will not be long before each unit volume inside the container has a more or less equal number of atoms, exerting more or less the same pressure and having roughly the same temperature. The chance that these 10^{20} atoms will "unshuffle" themselves—that they all will spontaneously crowd into one half of the contained volume, for instance—is so small as to be truly negligible. All complex systems we encounter have this tendency to increase their disorder to its maximum, this trend toward thermodynamic equilibrium.

Classical thermodynamics only deals with *equilibrium states* of a system, each of which corresponds to a set of indistinguishable microstates, indistinguishable because the temperature, the pressure, and all the other applicable thermodynamic variables have the same values for each microstate of the set. These equilibrium states are reached by letting the system settle down long enough so that quantities such as temperature and pressure become uniform throughout, so that the system has a chance to forget its past history, so to speak.

Quantities, such as pressure and temperature, which return to the same values whenever the system returns to the same equilibrium state, are called *state variables*. A thermodynamic state of a given system is thus completely defined by specifying the values of a relatively few state variables (which then become the *independent variables*), whereupon the values of all the other applicable state variables (the *dependent variables*) are uniquely determined. Dependent state variable U is thus specified as a function $U(x, y, ..., z)$ of the independent variables $x, ..., z$, where it is tacitly understood that the functional relationship only holds for equilibrium states.

It should be emphasized that the fact that there are equilibrium states, to which matter in bulk tends to approach spontaneously if left to itself, and the fact that there are thermo-dynamic variables which are uniquely specified by the equilibrium state (independent of the past history of the system) are *not* conclusions deduced logically from some philosophical first princi-ples. They are conclusions ineluctably drawn from more than two centuries of experiments.

CHAPTER

2

HEAT, TEMPERATURE, AND PRESSURE

In our introductory remarks we have used the words temperature and pressure without definition, because we could be sure the reader had encountered them earlier. Before long, however, these quantities must be defined, by describing how they are measured and also by indicating how they are related to each other and to all the other quantities that enter into the theoretical construct we call thermodynamics. As mentioned earlier, this construct is so tightly knit that an adequate definition of temperature involves other concepts and quantities, themselves defined in terms of temperature. Thus a stepwise procedure is required.

Temperature

The first step in a definition of temperature, for example, is to refer to our natural perception of heat and cold, noting that the property seems to be one-dimensional, in that we can arrange objects in a one-dimensional sequence, from coldest to hottest. Next we note that many materials, notably gases and some liquids, perceptibly expand when heated; so we can devise and arbitrarily calibrate a thermometer. The usual calibration, corresponding to the centigrade scale, sets 0° at the temperature of melting ice and 100° at the temperature of boiling water, and makes the intermediate scale proportional to the expansion of mercury within this range. We use such a thermometer to provide a preliminary definition of temperature, T, until we have learned enough thermodynamics to understand a better one.

We measure the temperature of a body by setting our thermometer in "thermal contact" with the body and letting it "come to the temperature of the body." It is useful to be explicit about the meaning of the two phrases in quotation marks. Bodies are in thermal contact when the temperature of one can affect that of the other. We know, from long experience, that two bodies in thermal contact tend to come to the same temperature; the colder one gets warmer, the hotter one gets cooler. This is, of course, a simple example of the spontaneous process of coming to thermal equilibrium. When equilibrium is attained, the thermometer will read a temperature intermediate between the original temperatures of body and thermometer.

A series of experiments with bodies of different sizes (experiments we all have made, in various unscientific ways, in our youth) will serve to show that when a large body is brought into thermal contact with a small body, the equilibrium temperature of the two is nearer the original temperature of the larger body. We don't try to cool off a hot brick by pouring a spoonful of water on it, we dump the brick in a pailful of water. In other words, in addition to temperature, a body possesses the property of *thermal capacity* for something we call *heat*, and a pailful of water has a greater capacity than does a spoonful. We need a large amount of this heat to bring the pailful up to a given temperature, just as we need to pump a lot of gas into a large tank before we can attain a given pressure.

When a large, hot body is placed in thermal contact with a small, cold one, only a small amount of the heat in the large body is needed to bring the small one up to the equilibrium temperature. And thus, if we use a thermometer of considerably smaller heat capacity than that of the body, whose temperature is to be measured, the equilibrium temperature of thermometer and body is fairly close to the initial temperature of the body. In this rather pedantic way we have thus made explicit what we instinctively do when we use a thermometer.

It was natural for the early investigators of thermal phenomena to think of heat as a fluid, something like a gas or like the "electric fluid" which was being studied at about the same time. Temperature was then thought of as analogous to a pressure, being proportional to the quantity of heat fluid contained by the body and inversely proportional to the heat capacity of the body. To put it quantitatively, the addition of an amount dQ of heat to a body will raise its temperature by an amount $dT = (1/C)\,dQ$, where C is the heat capacity of the body (C may be a function of T, of course). The tendency of bodies to reach thermal equilibrium is thus equivalent to the tendency of heat to flow from a region of higher temperature to one of lower.

That this picture of heat as a fluid is more or less self-consistent was shown by the fact that heat can be "produced" in several ways; by sending an electric current through the body, if it is metallic, or by producing a chemical reaction in it, such as the burning of a mixture of fuel and air. A given current, or the burning of a given amount of fuel, produces a small increase of temperature in a large body or a large increase in a small body, so it is consistent to consider that the burning of a given amount of fuel produces a given amount of heat. Within certain limits, to be discussed later, this picture of heat, produced by chemical reactions, related to temperature by the equation $dQ = C\,dT$, is a satisfactory model for representing thermal behavior. We can arbitrarily define the unit of heat, the kilogram-calorie, as being the amount which will raise the temperature of a kilogram of water from 4° to 5° centigrade, according to our mercury thermometer.

Having gone this far, the temptation was great to give the entity, heat, other properties of a fluid, such as inertia and velocity. The rate of flow **J** of heat, by conduction across an interface, per second per unit area, turns out to be proportional to the temperature gradient across the interface, $\mathbf{J} = -K\,\mathrm{grad}\,T$, where K is the thermal conductivity of the material. We might expect **J** to be the product of the density of the heat fluid times its velocity of flow. And, if the fluid has inertia, we might expect the flow to "overshoot" at times, leading to thermal oscillations above and below equilibrium, analogous to the acoustic vibrations which occur when a gas is suddenly introduced into a previously evacuated vessel.

It was natural for the early investigators to develop this concept of heat as a fluid and temperature as its pressure. They naturally went on to assume that the amount of heat contained by a system in equilibrium would be a state variable of the system. Thus, for a gas in equilibrium, they assumed that the heat Q it contains would be a definite function of the independent state variables of the gas, its temperature T and the amount of gas in the vessel, or its pressure, for example. In particular, if Q were a state variable, then, after withdrawal of heat from a body, the only way the body could be returned to its original equilibrium state would be to return the same amount of heat fluid it had originally had, either by appropriate chemical reaction or by thermal conduction across the body's outer surface.

Heat and Energy

Unfortunately for this simple theory, heat does not turn out to have the properties of a fluid, such as inertia and velocity. Furthermore it does not turn out to be a state variable. Count Rumford demonstrated that it was possible to withdraw an unlimited amount of heat from a system, provided only that mechanical energy were fed into the system in amounts proportional to the heat withdrawn. His experiments demonstrated that we could change the state of a system by withdrawing heat, to warm up an initially cold body for example, and then could bring it back to its initial state by adding, not heat, but *mechanical energy*. This could be done in many ways; the mechanical energy could be used to rub two parts of the system together or could actuate any other frictional device to dissipate mechanical energy and generate heat. Passage of an electric current I through a resistance R is a good way of generating heat from electromagnetic energy; the electric energy lost is $I^2 R$, and this turns out to be proportional to the amount of heat generated. Careful measurements indicate that 4182 joules of energy must be supplied to generate 1 kilogram-calorie of heat; so we can measure heat in energy units, joules, rather than in the specialized units of calories. We shall do so in this volume.

Heat thus cannot be a state variable, for we can add energy to a body in the form of heat and then take it away in the form of mechanical energy, bringing the body back to its initial equilibrium state at the end of such a cycle of operation. If heat were a state variable, as much would have to be given off during the cycle as was absorbed for Q to come back to its original value at the end of the cycle. But if heat can be changed to work, the net amount of heat added during the cycle may be positive, zero, or negative, depending on the net amount of work done by or on the body during the cycle. The quantity which is conserved, and which thus is the state variable, is U, the *energy* possessed by the body, which can be drawn off either as heat or as mechanical work, depending on the circumstances. As we shall see more clearly later, heat represents that energy content of the body which is added or removed in disorganized form; work is the energy added or removed in organized form; within certain limits disorganization can be changed to organization and vice versa.

We can increase the energy of a body by elevating it, doing work against the force of gravity. This increase is in potential energy, which is immediately available again as work. The temperature of the body is not changed by the elevation, so its heat content is not changed. We can translate the potential energy into organized kinetic energy by dropping the body; this also makes no change in its heat content. But if the body, in its fall, hits the ground the organized motion of the body is changed to disorganized internal vibrations; its temperature rises; heat has been produced. In a sense, the reason that classical thermodynamics is usually limited to a study of equilibrium states is because an equilibrium state is the state in which heat energy can easily and unmistakably be distinguished from mechanical energy. Before equilibrium is reached, sound waves or turbulence may be present, and it is difficult to decide when such motion ceases to be "mechanical" and becomes "thermal."

However this talk of disorganization and of molecular motions is outside the realm of thermodynamics, which does not concern itself with atomic details. Thermodynamics takes note of the empirical facts of thermal equilibrium and the relations between state variables, building up from there, without asking why or how. For this reason, perhaps, it may appear to modern physicists as being unduly abstract or somehow unrealistic. From the standpoint of atomic theory the measurable quantities we call pressure and temperature are gross manifestations of atomic motions, with the variability of individual motions averaged out. In situations of maximum disorder these averages turn out to bear simple relations to each other. From the standpoint of thermodynamics, pressure and temperature are quantities to be measured in specified ways; if the measurements are carried out under equilibrium conditions we find experimentally that they are functionally related.

Strictly speaking, we could discuss thermodynamics without once mentioning atoms or molecules. But physics is not just a disparate set of mathematical models, each of which interrelates a separate set of empirical facts. We hope that physical theory is interconnected and that the ideas of atomic theory will cast light on, will make more understandable, the concepts of thermodynamics, even though they may logically be superfluous to these concepts. So, in the first third of this book, we shall occasionally turn from thermodynamics to atomic theory, to maintain our awareness of the interconnection between the behavior of atoms in bulk and the empirical equations and abstract laws of thermodynamics.

The details of this interconnection are the task of statistical mechanics, the subject of the last third of this book, but it is useful to anticipate important ideas, rather than withholding them until they are logically necessary.

We have already indicated how thermodynamics defines temperature, in terms of a thermometer and the procedure of achieving thermal equilibrium. A like definition of pressure is as simple. We can use the flexure of a diaphragm or the balancing of a column of liquid as a means of measuring pressure difference; we need also to specify that we wait for pressure equilibrium to be reached between the measuring instrument and the material measured. Pressure is a mechanical quantity, representative of a great number of internal stresses which can be imposed on a body, such as tensions or torques or shears, changes in any of which represent an addition or subtraction of mechanical energy to the body. Pressure is more usually encountered in thermodynamic problems; it is the only stress that a gas can sustain in equilibrium. The usual units of pressure are *newtons per square meter*, although the *atmosphere* (about 10^5 newtons per square meter) is also used.

Pressure and Atomic Motion

The pressure exerted by a gas on its container walls is a very good example of a mechanical quantity which is the resultant of the random motions of the gas molecules and which nonetheless is a remarkably stable function of a relatively small number of state variables. To illustrate this point we shall digress for a few pages into a discussion of kinetic theory. The pressure P on a container wall is the force exerted by the gas, normal to an area dA of the wall, divided by dA. This force is caused by the collisions of the gas molecules against the area dA. Each collision delivers a small amount of momentum to dA; the amount of momentum delivered per second is the force $P\,dA$.

Let us assume a very simplified model of a gas, one consisting of N similar atoms, each of mass m and of "negligible" dimensions, with negligible interactions between them so that the sole energy of the ith atom is its kinetic energy of translation,

$$\tfrac{1}{2}m(v_{ix}^2 + v_{iy}^2 + v_{iz}^2)$$

The gas is confined in a container of internal volume V, the walls of which are perfect reflectors for incident gas atoms. By "negligible dimensions" we mean the atoms are very small compared to the mean distance of separation, so collisions are very rare and most of the time each atom is in free motion. We also mean that we do not need to consider the effects of atomic rotation. We shall call this simple model a *perfect gas of point atoms*.

Next we must ask what distribution of velocities and positions of the N atoms in volume V corresponds to a state of equilibrium. As the atoms rebound from the walls and from each other, they cannot lose energy, for the collisions are elastic. In collisions between the atoms what energy one loses the other gains. The total energy of translational motion of all the atoms,

$$U = \tfrac{1}{2}m \sum_{i=1}^{N} (v_{ix}^2 + v_{iy}^2 + v_{iz}^2) = \tfrac{1}{2}m \sum_i v_i^2 = N\langle K.E.\rangle_{tran} \quad (2\text{-}1)$$

is constant. The last part of this set of equations defines the average kinetic energy $\langle K.E.\rangle_{tran}$ of translation of an atom of the gas as being U divided by the number of atoms N. (The angular brackets $\langle \, \rangle$ will symbolize average values.)

As the gas settles down to equilibrium, U does not change but the randomness of the atomic motion increases. At equilibrium the atoms will be uniformly distributed throughout the container, with a density (N/V) atoms per unit volume; their velocities will also be randomly distributed, as many moving in one direction as in another. Some atoms are going slowly and some rapidly, of course, but at equilibrium the total x component of atomic momentum, Σmv_{ix}, is zero; similarly with the total y and z components of momentum. The total x component of the kinetic energy, $\tfrac{1}{2}\Sigma mv_{ix}^2$, is not zero, however. At equilibrium it is equal to the total y component and to the total z component, each of which is equal to one-third of the total kinetic energy, according to Eq. (2-1),

$$\sum_i \tfrac{1}{2}mv_{ix}^2 = \tfrac{1}{3}N\langle K.E.\rangle_{tran} \quad (2\text{-}2)$$

At equilibrium all directions of motions are equally likely.

Pressure in a Perfect Gas

Next we ask how many atoms strike the area dA of container wall per second. For simplicity we orient the axes so that the positive x axis points normally into dA (Figure 2-1). Consider first all those atoms in V which have their x component of velocity, v_{ix}, equal to some value v_x (v_x must be positive if the atom is to hit dA). All

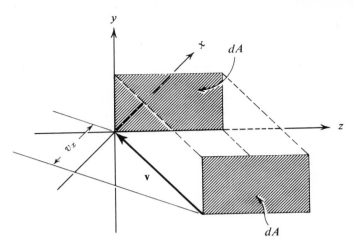

FIGURE 2-1
Impact of atoms, with velocity **v**, *on area element dA in the yz plane.*

these kinds of atoms, which are a distance v_x from the wall or closer, will hit the wall in the next second, and a fraction proportional to dA of those will hit dA in the next second. In fact a fraction $(v_x/V) \, dA$ of all the atoms in V which have x component of velocity equal to v_x will hit dA per second. Each of these atoms, as it strikes dA during the second, rebounds with an x component $-v_x$, so each of these atoms imparts·a momentum $2mv_x$ to dA. Thus the momentum imparted per second by the atoms with x velocity equal to v_x is $2mv_x(v_x/V) \, dA$ times the total number of atoms in V having x velocity equal to v_x. And therefore the total momentum imparted to dA per second is the sum of $(2mv_{ix}^2/V) \, dA$ for each atom in V that has a positive value of v_{ix}.

Since half the atoms have a positive value of v_{ix}, the total momentum imparted to dA per second is

$$\frac{1}{2} \sum_{i=1}^{N} (2mv_{ix}^2/V) \, dA = \tfrac{2}{3}(N/V)\langle \text{K.E.}\rangle_{\text{tran}} \, dA = \tfrac{2}{3}(U/V) \, dA \quad (2\text{-}3)$$

where we have used Eq. (2-2) to express the result in terms of the mean atomic kinetic energy, defined in Eq. (2-1). Since this is equal to the force $P \, dA$ on dA, we finally arrive at an equation giving the pressure P of a perfect gas of point atoms in a volume V, in terms of the mean kinetic energy $\langle \text{K.E.}\rangle_{\text{tran}}$ of translation per atom or in terms of the total energy content of the gas per unit volume

(U/V) (total as long as the only energy is kinetic energy of translation, that is):

$$P = \tfrac{2}{3}(U/V) \qquad \text{or} \qquad PV = \tfrac{2}{3}U = \tfrac{2}{3}N\langle \text{K.E.}\rangle_{\text{tran}} \qquad (2\text{-}4)$$

This is a very interesting result, for it demonstrates the great stability of the relationships between aggregate quantities such as P and U for systems in equilibrium. The relationship of Eq. (2-4) holds no matter what the distribution in speed the atoms have as long as their total energy is U, as long as the atoms are uniformly distributed in space, and as long as all directions of motion are equally likely (i.e., as long as the gas is in equilibrium). Subject to these provisos, every atom could have kinetic energy $\langle \text{K.E.}\rangle_{\text{tran}}$, or half of them could have kinetic energy $\tfrac{1}{2}\langle \text{K.E.}\rangle_{\text{tran}}$ and the other half energy $\tfrac{3}{2}\langle \text{K.E.}\rangle_{\text{tran}}$, or any other distribution having an average value $\langle \text{K.E.}\rangle_{\text{tran}}$. As long as it is uniform in space and isotropic in direction the relation between P, V, and $N\langle \text{K.E.}\rangle_{\text{tran}}$ is that given in Eq. (2-4). Even the proportionality constant is fixed; PV is not just proportional to $N\langle \text{K.E.}\rangle_{\text{tran}}$—the factor is $\tfrac{2}{3}$, no matter what the velocity distribution is. In fact the denominator of the fraction is just the number of dimensions of space, as a review of the derivation of Eq. (2-2) will demonstrate.

From our earlier discussion we may suspect that $\langle \text{K.E.}\rangle_{\text{tran}}$ is a function of the gas temperature T; if T is increased, the kinetic energy of the gas atoms should increase. We shall see what this relationship is in the next chapter.

CHAPTER

3

STATE VARIABLES AND EQUATIONS OF STATE

To recapitulate, when a thermodynamic system is in a certain equilibrium state, certain aggregate properties of the system, such as pressure and temperature, called *state variables*, have specific values, determined only by the state and not by the previous history of the system. Alternately, specifying the values of a certain number of state variables specifies the state of the system; the number of variables required to specify the state uniquely depends on the system and on its constraints. For example, if the system is a definite number of like molecules in gaseous form within a container, then only two variables are needed to specify the state—either the pressure of the gas and the volume of the container, or the pressure and temperature of the gas, or else the volume and temperature. If the system is a mixture of two gases (such as hydrogen and oxygen) which react chemically to form a third product (such as water vapor), the relative abundance of two of the three possible molecular types must be specified, in addition to the total pressure and volume (or P and T, or T and V), to determine the state. If the gas is paramagnetic, and we wish to investigate its thermomagnetic properties, then the strength $\mathfrak{H}$ of the applied magnetic field (or else the magnetic polarization of the gas) must also be specified.

Extensive and Intensive Variables

One state variable is simply the amount of each chemical substance present in the system. The convenient unit for this variable is the *mole*; 1 mole of a substance, which has molecular weight M, is M kilograms of the substance (1 mole of hydrogen gas is 2 kg of H_2, 1 mole of oxygen is 32 kg of O_2). By definition, each mole contains the same number of molecules, $N_0 = 6 \times 10^{26}$, called *Avogadro's number*. In many respects a mole of gas behaves the same, no matter what its composition. When the thermodynamic system is made up of a single substance, then the number of moles present (which we shall denote by the letter n) is constant. But if the system is a chemically reacting mixture the n's for the different substances may change.

State variables are of two sorts, one sort directly proportional to n, the other not. For example, suppose we have two equal amounts of the same kind of gas, each of n moles and each in equilibrium at the same temperature T in containers of equal volume V. We then connect the containers so the two samples of gas can mix. The combined system now has $2n$ moles of gas in a volume $2V$, and the total internal energy of the system is twice the internal energy U of each original part. But the common temperature T and pressure P of the mixed gas have the same values they had in the original separated states. Variables of the former type, proportional to n (such as U and V), are called *extensive* variables; those of the latter type (such as T and P) are called *intensive* variables. At thermodynamic equilibrium the intensive variables have uniform values throughout the system.

A basic state variable for all thermodynamic systems (almost by definition) is its temperature T, which is an intensive variable. At present we have agreed to measure its value by a thermometer; a better definition will be given later. Related to the temperature is the heat capacity of the system, the heat required to raise the system 1 degree in temperature. Because heat is not a state variable, the amount of heat required depends on the way the heat is added. For example, the amount of heat required to raise T by 1 degree, when the volume occupied by the system is kept constant, is called the *heat capacity* at *constant volume* and is denoted by C_v. The heat required to raise T by 1 degree when the pressure is kept constant is called the *heat capacity* at *constant pressure* and is denoted C_p. A system at constant pressure expands when heated, thus doing work, so C_p is greater than C_v. We shall prove this later.

These heat capacities are state variables, in fact they are extensive variables; their units are joules per degree. The capacities per mole of substance, $c_v = (C_v/n)$ and $c_p = (C_p/n)$, are called *specific heats*, at constant volume or pressure, respectively. They

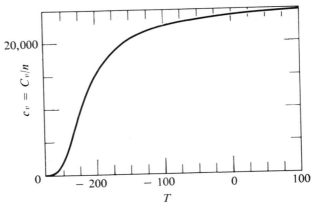

FIGURE 3-1

Specific heat of a solid, in units of joules per degree per mole, as function of temperature in degrees centigrade.

have been measured, for many materials, and a number of interesting regularities have emerged. For example, the specific heat at constant volume, c_v, for any monatomic gas is roughly equal to 12,000 joules per degree centigrade per mole, independent of T and P over a wide range of states, whereas c_v for diatomic gases is roughly 20,000 joules per degree centigrade per mole, with a few exceptions. A typical plot of c_v for a solid is given in Figure 3-1, showing that c_v is independent of T for solids only when T is large.

Pairs of Mechanical Variables

Other state variables are of a mechanical, rather than thermal, type. For example, there is the pressure P (in newtons per square meter), an intensive variable appropriate for fluids, although applicable also for solids that are uniformly compressed (in general, in solids, one needs a tensor to describe the stress). Related to P is the extensive variable V (in cubic meters), the volume occupied by the system. The pair define a mechanical energy; work $P \, dV$ (in joules) is done by the system on the container walls if its volume is increased by dV when it is in equilibrium at pressure P. The pair P and V are the most familiar of the mechanical state variables. For a bar of material, the change in dimensions may be simple stretching, in which case the extensive variable would be the length L of the bar, the intensive variable would be the tension J, and the work done on the bar by stretching it an additional amount dL would be $J\,dL$.

Or possibly the material may be polarized by a magnetic field. The intensive variable here is the impressed magnetic intensity $\mathfrak{H}$ (in ampere turns per meter) and the extensive variable is the total magnetization $\mathfrak{M}$ of the material. Reference to a text on electromagnetic theory will remind us that, related to $\mathfrak{H}$, there is the magnetic induction field $\mathfrak{B}$ (in webers per square meter). For a material of magnetic susceptibility χ, a magnetic field causes a polarization $\mathfrak{P}$ of the material, which is related to $\mathfrak{H}$ and $\mathfrak{B}$ through the equation $\mathfrak{B} = \mu_0 \mathfrak{H}(1 + \chi) = \mu_0(\mathfrak{H} + \mathfrak{P})$, where μ_0 is the permeability of vacuum, $4\pi \times 10^{-7}$ henrys per meter. Diamagnetic materials have no permanent magnets attached to their molecules; the polarization is caused by currents induced by the magnetic field and the susceptibility χ is negative and quite small. Paramagnetic materials contain permanent molecular magnets which tend to line up with the field, so their susceptibility is positive and can be quite large. Since the susceptibility of paramagnetic materials is strongly temperature-dependent, they can be used to convert magnetic energy into heat and vice versa, particularly at very low temperatures.

Referring to the atomic picture, the polarization of a paramagnetic material is the average effect of the alignment of atomic magnets in the direction of the applied field $\mathfrak{H}$. At thermal equilibrium, in the absence of a field, these magnets are pointed at random, so their combined effects cancel out. An applied field $\mathfrak{H}$ tends to line some of them in the direction of the field, producing a net magnetic moment $\mathfrak{P}$ per unit volume of the material. Measurements show [see also Eq. (13-17)] that $\mathfrak{P}$ is proportional to $\mathfrak{H}$, at least for small intensities; the ratio between $\mathfrak{P}$ and $\mathfrak{H}$, the susceptibility χ, is independent of $\mathfrak{H}$ for small values of $\mathfrak{H}$. On the other hand, if $\mathfrak{H}$ is large enough to force all the atomic magnets to line up, in spite of thermal agitation, then the polarization $\mathfrak{P}$ will have its maximum value, equal to the magnitude of the atomic moment times the number of moments per unit volume, independent of $\mathfrak{H}$, so that χ is inversely proportional to $\mathfrak{H}$ when $\mathfrak{H}$ is large.

The energy density of the magnetic field is

$$\tfrac{1}{2}\mathfrak{H}\mathfrak{B} = \tfrac{1}{2}\mu_0 \mathfrak{H}(\mathfrak{H} + \mathfrak{P}).$$

The first term of this, $\frac{1}{2}\mu_0\mathfrak{H}^2$, is the energy density of the applied field in vacuum, and is independent of the presence of the paramagnetic substance. The second term, $\frac{1}{2}\mu_0\mathfrak{H}\mathfrak{P}$, is the energy density which the applied field has devoted to the lining up of the atomic moments. Therefore, for a volume V of the substance, the applied magnetic field $\mathfrak{H}$ contributes an amount $\frac{1}{2}\mu_0 V\mathfrak{H}\mathfrak{P}$ to the internal energy of the substance. Therefore we define the total *magnetization* of the body as being the quantity $\mathfrak{M} = \mu_0 V\mathfrak{P}$ (in weber-meters); for paramagnetic materials $\mathfrak{M}$ would equal $\mu_0 V\chi\mathfrak{H}$. Then the magnetic work done on the body in magnetic field $\mathfrak{H}$, when its magnetization is increased by $d\mathfrak{M}$, would be $\mathfrak{H}\,d\mathfrak{M}$, the integral of which, for $\mathfrak{M} = \mu_0 V\chi\mathfrak{H}$, becomes $\frac{1}{2}\mu_0 V\mathfrak{H}\mathfrak{P}$, as desired. Magnetization $\mathfrak{M}$ is thus the extensive variable related to $\mathfrak{H}$. A similar pair of state variables can be devised for dielectrics and the electric field.

As mentioned earlier, n itself, the number of moles of each chemical substance involved, is an extensive variable. The related intensive variable, usually called μ, is an energy per mole of the substance itself. The meaning of μ becomes clearer if we compare a mixture of 2 moles of hydrogen and 1 mole of oxygen with 2 moles of steam. The mixture of H_2 and O_2 can change into the H_2O, without the addition of any more matter, but with the emission of a lot of heat. In other words, each mole of hydrogen and oxygen carried with them potential energy of chemical combination, which was released in the reaction, and our internal energy inventory must keep track of this fact; the heat is part of the released energy.

Thus when an amount dn_i of substance i is introduced into the system, an amount $\mu_i\,dn_i$ of chemical potential energy is added; μ_i is appropriately called the *chemical potential* of material i. It depends on temperature and pressure, as well as on the chemical properties of the substance involved. As with most energy scales, only energy differences are measurable, so we can be arbitrary about zero levels. What is usually done is to say that the chemical potential of chemical elements in their usual form (molecular or atomic) at standard pressure and temperature are all zero. Then, if two elements produce heat when they combine, the chemical potential of the compound is negative; if energy is required to get them to combine, the μ of the compound is positive. Since 115.6 kilogram-calories of heat are evolved in going from 2 moles of H_2 and 1 mole of O_2 at standard pressure and temperature to 2 moles of H_2O at standard pressure and temperature, we can say that the μ for H_2O at standard conditions is $-242,000$ joules per mole. But this will be discussed in Chapter 10.

Thus thermodynamics deals with a whole series of pairs of state variables, each pair related to a particular way of modifying the gross structure of the system under study. One variable of the ith pair (call it Y_i) is an intensive variable, the other (call it X_i) is extensive. The units of the variables are so chosen that the energy added to or subtracted from the system, when variable X_i is changed by an amount dX_i, when the system is kept in equilibrium, is $Y_i\, dX_i$ in joules. There are, potentially, a large number of such pairs, but we need only concern ourselves with a few at a time, those which are involved in the experiment under study. Thus we can omit the magnetic terms when no magnetic field is involved, and we need not consider longitudinal stress and strain when we study a gas. Pressure P and its extensive counterpart V, the volume occupied by the material, are nearly always involved, however. And, whenever material is added or taken away, n and its intensive counterpart μ, the chemical potential, must be included. The intensive variable T, the temperature, is nearly always involved. We are not ready to discuss its extensive counterpart; it will be taken up in Chapter 6.

As we mentioned earlier, we need to determine experimentally a certain minimum number of relationships between the state variables of a system before the theoretical machinery of thermodynamics can "take hold" to predict the system's other thermal properties. One of these relationships is the dependence of one of the heat capacities, either C_v or C_p (or C_L or C_J, or $C_{\mathfrak{H}}$ or $C_{\mathfrak{M}}$ depending on the mechanical or electromagnetic variables of interest) on T and on P or V (or L or J, or on $\mathfrak{H}$ or $\mathfrak{M}$, as the case may be). We shall show later that, if C_v is measured, C_p can be computed (and similarly for the other pairs of heat capacities); thus only one heat capacity needs to be measured as a function of the independent variables of the system.

The Equation of State of a Gas

Another necessary empirical relationship is the relation between the pair of mechanical variables P and V (or J and L, or $\mathfrak{H}$ and $\mathfrak{M}$) and the temperature T for the system under study. Such a relationship, expressed as a mathematical equation, is called an *equation of state*. There must be an equation of state known for each pair of mechanical variables of interest. We shall write down some of those of general interest, which will be used often later.

Of course to write down a quantitative relationship between P and V and T, for example, is to include in the relation the possible idiosyncrasies of the temperature scale we have chosen. So our discussion of equations of state will automatically take us one step further toward a final determination of temperature scale.

The equation of state first to be experimentally determined was the one between P and V and T for a gas, and this equation served for some time as a basis for a temperature scale. Boyle first demonstrated that as a gas was compressed into a smaller volume its pressure increased and that, over a wide range of pressures the pressure P is inversely proportional to the volume V occupied by the gas. In other words, to a good approximation the product PV is equal to a constant, proportional to n, the amount of gas involved, but independent of P and V. Charles next showed that this constant

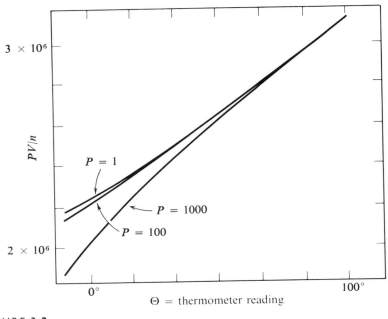

FIGURE 3-2

Equation of state for a gas, P in newtons per square meter and V in cubic meters; therefore (PV/n) is in joules per mole; Θ is temperature read on a mercury thermometer.

does depend on temperature T. Over fairly wide ranges of T he showed that a plot of (PV/n) against the temperature, as measured by his thermometer, was nearly a straight line.

To distinguish between temperature scales at this point, let us call Θ the temperature as read by the mercury thermometer we use, instead of T. If it is calibrated as described on page 9, Θ is zero at the temperature of melting ice and is 100 at the temperature of boiling water. If we now hold P constant and measure V for different values of Θ for n moles of some gas, we obtain a set of curves of the sort shown in Figure 3-2, curves of (PV/n) against Θ for different values of P. We note that the curves for P small are nearly straight lines and are nearly independent of P. Similar measurements for other gases result in curves which are nearly identical with those drawn for small P; they differ only for P large. In other words, for all gases which condense into liquids well below $0°C$ (H_2, O_2, He, etc.) the curves of (PV/n) against Θ for P small are practically identical, being almost a straight line going through 2.271×10^6 at $\Theta = 0°$ and going through 3.102×10^6 at $\Theta = 100°$.

Thus the quantity (PV/n) for P small, which we can call $(PV/n)_0$, being nearly independent of the gas used, is a more universal temperature indicator than is the mercury thermometer which we have been talking about using. Instead of saying that $(PV/n)_0$ is a function of Θ, we can turn around and calibrate the mercury thermometer in terms of a gas thermometer scale. The quantity

$$\tau = (1/8314)\,[(PV/n)_0 - 2.271 \times 10^6]$$

is zero at $\Theta = 0°$ and 100 at $\Theta = 100°$ and automatically makes $(PV/n)_0$ a straight line between these points. Thus the expansion of a very rarified gas (called an *ideal gas*) is the basis of our next temperature scale, which is the centigrade scale in present use. In its terms the equation of state for a rarified gas is

$$PV \simeq nR(\tau + \tau_0) \qquad (P \to 0)$$

where $R = 8314$ and $\tau_0 = (2.271 \times 10^6/8314) = 273.2$.

We have thus devised our temperature scale so that PV for rarified gases is accurately a linear function of temperature. We might as well go farther and move the temperature origin so as to remove the additive constant τ_0. We define the *absolute temperature* as being $(\tau + \tau_0) = [(PV/n)_0/8314] = T$, so that ice melts at $T = 273.2°$ and water boils at $T = 373.2°$ and $(PV/n)_0$ is, by definition, proportional to T. This completes our next step in the definition of temperature. For the next few chapters we shall be using the T scale, measured by a gas thermometer calibrated as described, in *degrees Kelvin* (thus ice melts at 273°K).

For T in degrees Kelvin the equation of state for a rarified gas is

$$PV = nRT \tag{3-1}$$

where $R = 8314$ joules per mole per degree Kelvin is called the *gas constant*. This is called the *ideal gas law*. It is only an approximation to the equation of state of an actual gas. Figure 3-3 shows curves displaying the departure from the ideal gas law of the equations of

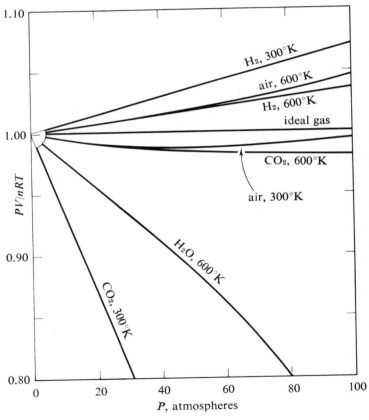

FIGURE 3-3

Equations of state for a few actual gases.

state of a few gases. We see that, except for gases near their temperature of condensation, such as CO_2 at 300° and H_2O at 600° the ideal gas law is correct to within a few per cent over a wide range of pressures and temperatures. Having thus used the equation of state for a rarified gas to determine a temperature scale T, we shall use this scale in all the other equations of state.

We are now in a position to illustrate how kinetic theory can supplement an empirical thermodynamic formula with a physical model. In the previous chapter we calculated the pressure exerted by a gas of N point particles confined at equilibrium in a container of volume V. This should be a good model of an ideal gas. As a matter of fact, Eq. (2-4) has a form remarkably like that of Eq. (3-1). All we need to do to make the equations identical is to set

$$nRT = \tfrac{2}{3}U = \tfrac{2}{3}N\langle\text{K.E.}\rangle_{\text{tran}}$$

The juxtaposition is most suggestive. We have already pointed out that N, the number of molecules, is equal to nN_0, where N_0, Avogadro's number, is equal to 6×10^{26} for any substance. Thus we reach the remarkably simple result, that $RT = \tfrac{2}{3}N_0\langle\text{K.E.}\rangle_{\text{tran}}$ for any perfect gas. For this model, therefore, the average kinetic energy per molecule is proportional to the temperature, and the proportionality constant $\tfrac{3}{2}(R/N_0)$ is independent of the molecular mass. The ratio $(R/N_0) = 1.4 \times 10^{-23}$ joules per degree Kelvin is called k, the *Boltzmann constant*.

Thus the model suggests that for those gases which obey the ideal gas law fairly accurately, the average kinetic energy of molecular translation is directly proportional to the absolute temperature,

$$\langle\text{K.E.}\rangle_{\text{tran}} = \tfrac{3}{2}kT \qquad \text{perfect gas} \qquad (3\text{-}2)$$

independent of the molecular mass. Only the kinetic energy of translation enters into this formula; our model of point atoms assumed their rotational kinetic energy was negligible. We might expect that this would be true for actual monatomic gases, like helium and argon, and that for these gases the total internal energy is

$$U = N\langle\text{K.E.}\rangle_{\text{tran}} = \tfrac{3}{2}NkT = \tfrac{3}{2}nRT \qquad (3\text{-}3)$$

Measurement shows this to be nearly correct [see discussion of Eqs. (6-11) and (22-5)]. For polyatomic gases U is greater, corresponding to the additional kinetic energy of rotation (the additional term does not enter into the equation for P, however). We shall return to this point, to enlarge on and to modify it, as we learn more. In accord with usage, we define an *ideal gas* to be one obeying the equation of state (3-1), a *perfect gas* to be one which, in addition, has its internal energy U a function of T only, as in Eq. (3-3). We shall see in Chapter 6 that for a gas to be perfect it must be ideal and vice versa. Thus the adjectives are interchangeable.

Other Equations of State

Of course the equation of state for an actual gas is not as simple as Eq. (3-1). We could, instead of transforming our measurements into an equation, simply present the relationship between P, V, and T in the form of a table of numbers or a set of curves as in Figure 3-3. But, as we shall see, the thermal behavior of bodies usually is expressed in terms of the first and second derivatives of the equation of state, and taking derivatives of a table of numbers is tedious and subject to error. It is often better to fit an analytic formula to the data, so we can differentiate it more easily.

A formula that fits the empirical behavior of many gases, over a wider range of T and P than does Eq. (3-1), is the *van der Waals* approximation,

$$(V - nb)\left(P + \frac{an^2}{V^2}\right) = nRT \qquad (3\text{-}4)$$

For large-enough values of V this approaches the ideal gas law. Typical curves for P against V for different values of T are shown in Figure 3-4. For temperatures smaller than $(8a/27bR)$ there is a range of P and V for which a given value of pressure corresponds to three different values of V. This represents (as we shall see later) the transition from gas to liquid. Thus the van der Waals formula covers approximately both the gaseous and liquid phases, although the accuracy of the formula for the liquid phase is not very good.

The atomic picture gives some reasons why a first approximation to the equation of state of an actual gas would have the van der Waals form. In the first place the individual molecules occupy some volume, each of them. So the free space for the molecules to wander around in is V minus the volume occupied by the molecules, which is the nb term. In the second place molecules of an actual gas attract each other slightly, not enough to cause them to condense into a liquid or a solid at room temperatures and above, but enough to reduce the momentum with which they try to leave the others and hit the wall of the container. Therefore the pressure

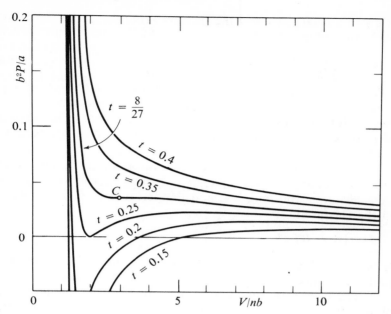

FIGURE 3-4

The van der Waals equation of state. Plots of b^2P/a versus V/nb for different values of $t = RbT/a$. Point C is the critical point (see Figure 9-3).

P should be reduced somewhat below the amount $nRT/(V - nb)$ which it would have if there were no intermolecular attraction. This is the meaning of the (an^2/V^2) term, which gets larger the smaller V is, the closer the molecules are together.

It is also possible to express the equation of state as a power series in (n/V),

$$P = (nRT/V)\left[1 + \frac{n}{V}B(T) + \left(\frac{n}{V} \right)^2 C(T) + \cdots \right] \qquad (3\text{-}5)$$

This form is called the *virial equation* and the functions $B(T)$, $C(T)$, etc., are called virial coefficients. Values of these coefficients and their derivatives can then be tabulated or plotted for the substance under study.

Corresponding equations of state can be devised for solids. A simple one, which is satisfactory for temperatures and pressures that are not too large (below the melting point for T, up to several hundred atmospheres for P), is

$$V = V_0(1 + \beta T - \kappa P) \tag{3-6}$$

Both β, which is called the *thermal expansion coefficient*, and κ, called the *compressibility*, are small quantities, of the order of 10^{-6} for metals, for example. They are not quite constant; they vary as T and P are changed, although the variation for most solids is not large for the usual range of T and P.

The other pairs of mechanical variables also have their equations of state. For example, in a stretched rod the relation between tension J and length L and temperature T is, for stretches within the elastic limit,

$$J = (A + BT)(L - L_0) \tag{3-7}$$

where A, B, and L_0 are constants (approximately); B is negative for many substances but positive for a few, such as rubber.

There also is a magnetic equation of state for the paramagnetic substances discussed in the previous chapter. This relates the applied field $\mathfrak{H}$, the magnetization $\mathfrak{M} = \mu_0 V \mathfrak{P}$, and the absolute temperature T. A fairly good approximation, for low enough fields so the substance is not near saturation, is *Curie's law*,

$$\mathfrak{M} = (nD\mathfrak{H}/T) \tag{3-8}$$

The constant D, called Curie's constant, is proportional to the magnetic susceptibility of the substance. Later in this book [see Eqs. (13-22) and (18-24)] we shall show how this equation of state can be predicted from the atomic properties of a paramagnetic substance.

One fortunate characteristic of most equations of state should be pointed out; they usually involve only one pair of state variables at a time. The relation between P, V, and T for a paramagnetic gas is independent of $\mathfrak{H}$ and $\mathfrak{M}$; conversely the constant D in Curie's law (3-8) is independent of P or V. This separation of the state variables is fortunate; otherwise we could not neglect most of them when we make any thermodynamic calculation or measurement. This virtually complete separation is not true in the case of the chemical potential μ; it is a function of P (or alternatively of V) as well as of T.

We thus see that a thermodynamic state of a system can be completely specified by giving the value of the absolute temperature

T, plus the values of just one of each pair of variables which happen to be involved in the problem. Thus the state of a simple gas is defined by specifying T and either P or V, so that either T and P or T and V (or even P and V) can be chosen for the independent variables. For a paramagnetic gas, three independent variables are needed to specify the state, either T, P, and $\mathfrak{H}$ or T, V, and $\mathfrak{M}$ or any other of eight different combinations of three of the five variables T, P, V, $\mathfrak{H}$, $\mathfrak{M}$.

Of course in every case we must specify n, if the system consists of a single substance, or the number of moles n_i of each substance, if the system is heterogeneous. A value of the chemical potential μ (or a set of values of the μ_i's) could be specified instead of n (or the n_i's).

Partial Derivatives

In all these equations there is a relationship between at least three variables. We can choose any pair of them to be the independent variables; the other one is then a dependent variable. We shall often wish to compute the rate of change of a dependent variable with respect to one of the independent variables, holding the other constant. This rate, called a partial derivative, is discussed at length in courses in advanced calculus. In thermodynamics, since we are all the time changing from one pair of independent variables to another, we find it advisable to label each partial by *both* independent variables, the one varied and the one held constant. The partial $(\partial P/\partial V)_T$, for example, is the rate of change of P with respect to V, when T is held constant; V and T are the independent variables in this case, and P is expressed explicitly as a function of V and T before performing the differentiation.

There are a number of relationships between partial derivatives that we shall find useful. If z and u are dependent variables, functions of x and y, then, by manipulation of the basic equations,

$$dz = (\partial z/\partial x)_y\, dx + (\partial z/\partial y)_x\, dy \qquad \text{and}$$
$$(\partial z/\partial x)_y = (\partial z/\partial u)_y(\partial u/\partial x)_y$$

we can obtain

$$\left(\frac{\partial z}{\partial x}\right)_y = \frac{(\partial u/\partial x)_y}{(\partial u/\partial z)_y} = \frac{1}{(\partial x/\partial z)_y}; \qquad \left(\frac{\partial x}{\partial y}\right)_z = -\frac{(\partial z/\partial y)_x}{(\partial z/\partial x)_y} \qquad (3\text{-}9)$$

The last equation can be interpreted as follows: On the left we express x as a function of y and z, on the right z is expressed as a function of x and y before differentiating and the ratio is then reconverted to be a function of y and z to effect the equation. Each partial is itself a function of the independent variables and thus may also be differentiated. Since the order of differentiation is immaterial, we have the useful relationship

$$\left[\frac{\partial}{\partial x}\left(\frac{\partial z}{\partial y}\right)_x\right]_y = \left[\frac{\partial}{\partial y}\left(\frac{\partial z}{\partial x}\right)_y\right]_x \tag{3-10}$$

As an example of the use of these formulas, we can find the partial $(\partial V/\partial T)_P$, as function of P and V or of T and V, for the van der Waals formula (3-4):

$$\left(\frac{\partial V}{\partial T}\right)_P = -\frac{(\partial P/\partial T)_v}{(\partial P/\partial V)_T} = \frac{nR/(V - nb)}{[nRT/(V - nb)^2] - (2an^2/V^3)}$$

$$= \frac{R(V - nb)V^3}{RTV^3 - 2an(V - nb)^2}$$

We shall often be given the relevant partial derivatives of a state function and be required to compute the function itself by integration. If $(\partial z/\partial x)_y = f(x)$ and $(\partial z/\partial y)_x = g(y)$ this is straightforward; we integrate each partial separately and add

$$z = \int f(x)\,dx + \int g(y)\,dy$$

But if either partial depends on the other independent variable it is not quite so simple. For example, if $(\partial z/\partial x)_y = f(x) + ay$ and $(\partial z/\partial y)_x = ax$, then the integral is

$$z = \int f(x)\,dx + axy \qquad [\text{not} \int f(x)\,dx + 2axy]$$

as may be seen by taking partials of z. The cross term appears in both partials and we include it only once in the integral. To thus coalesce two terms of the integral, the two terms must of course be equal. This seems to be assuming more of a relationship between $(\partial z/\partial x)_y$ and $(\partial z/\partial y)_x$ than we have any right to do, until we remember that they *are* related, according to Eq. (3-10), in just the right way so the cross terms can be coalesced. When this is so, the differential $dz = (\partial z/\partial x)_y\,dx + (\partial z/\partial y)_x\,dy$ is a *perfect differential*, which can be integrated in the manner just illustrated to obtain z, a function of x and y, the integrated value coming out the same no matter what

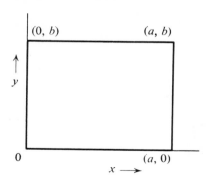

FIGURE **3-5**

Integration in the x, y plane.

path in the x, y plane we choose to perform the integration along, as long as the terminal points of the path are unchanged (Figure 3-5).

A differential $dz = f(x, y)\, dx + g(x, y)\, dy$, where $(\partial f/\partial y)_x$ is *not* equal to $(\partial g/\partial x)_y$, results in an integral which depends on the path of integration as well as the end points, is called an *imperfect differential*, and is distinguished by the bar through the d. The integral of the perfect differential $dz = y\, dx + x\, dy$, from $(0, 0)$ to (a, b) over the path from $(0, 0)$ to $(0, b)$ to (a, b) is

$$0 \cdot \int_0^b dy + b \cdot \int_0^a dx = ab,$$

which equals that for the path $(0, 0)$, $(a, 0)$, (a, b),

$$0 \cdot \int_0^a dx + a \cdot \int_0^b dy = ab.$$

On the other hand, the integral from $(0, 0)$ to (a, b) of the imperfect integral $dz = y\, dx - x\, dy$ is ab over the first route and $-ab$ over the second. Such a differential cannot be integrated to produce a state function $z(x, y)$. However, we can multiply the imperfect differential dz by an appropriate function of x and y (in this case $1/y^2$), which will turn it into a perfect differential, du; in this case

$$(dz/y^2) = du = (1/y)\, dx - (x/y^2)\, dy \qquad \text{and} \qquad u = (x/y)$$

The factor that converts an imperfect differential into a perfect one is called an *integrating factor*. One always exists (although it may be hard to find) for differentials of two independent variables. For more than two independent variables there are imperfect differentials for which no integrating factor exists.

THE FIRST LAW OF THERMODYNAMICS

The laws of thermodynamics are deceptively simple statements, each implying many properties of thermal behavior which are not at all self-evident and which were agreed to only after a great deal of argument and a lot of careful experimentation. As with any physical theory these laws concentrate a large number of concepts into a small number of words or into a simple-looking equation. To understand one of them we must first carefully define each term in its statement; then we must take it apart, to understand the various concepts it embodies.

Many writers treat the laws of thermodynamics as they would a complete set of logically sufficient axioms, from which the rest of thermodynamics can be derived, analogous to a sequence of theorems in mathematics. Perhaps it is better to consider a discussion of the consequences of each law as a sort of guided tour of the thermal idiosyncrasies of matter in bulk, each of them being examples, of which the law is a generalization.

Some writers, to ensure completeness of their set of axioms, set down a *zero law of thermodynamics*; that two bodies, each in thermodynamic equilibrium with a third system, are in equilibrium with each other. Others prefer to leave "self-evident" facts unstated.

Implications of the First Law

The first law of thermodynamics is just the statement that energy is conserved, plus the added comment that heat is a form

of energy. One sometimes gets the feeling that the law of conservation of energy is just a series of definitions of what we mean
by energy. At first energy was thought of as purely mechanical,
force times distance, both of which could be felt and measured by
bodily action. Newton's laws implied that this quantity was conserved. But we soon found situations where mechanical energy was
not conserved, as for example when we do mechanical work in
turning the shaft of an electric generator. So we say that energy is
not lost, it has simply changed from mechanical to electromagnetic
energy. Also, as mentioned in the previous chapter, we find that
mechanical energy disappears when we do work against friction; so
we say the energy has been changed into heat, which is a form
of energy. It might seem that, as fast as we find a situation where
energy, as we know it, is not conserved, we invent a new form
of energy to conserve the law of conservation of energy.

But the first law of thermodynamics is more than just a
definition. To say that heat is a form of energy implies properties
of heat which are not obvious and which will require several chapters
of discussion to elucidate.

Both the first and second laws of thermodynamics are most
simply stated in differential form. We are not often interested in
the total quantity of energy possessed by a body, even if we could
define or measure it. What is important is the relationship between
the amounts of different sorts of energy which are added to or taken
away from the body. The first law says that there is a generalized
store of energy, possessed by a thermodynamic system, called its
internal energy U, which can be changed by adding or subtracting
energy of any form, and that the algebraic sum of all these added
or subtracted amounts is equal to the net change, dU, of the internal
energy of the system. Put another way, it states that U *is a state
variable* of the system, that dU is a perfect differential, that when
the system is in equilibrium in a given state, its internal energy
always has the same value, no matter how the state was reached.
It is not difficult to see that U is an extensive variable; if a system in
equilibrium is divided into two equal parts, each part has half the
internal energy of the original whole.

Work and Internal Energy

The internal energy U can be changed by having the system
do work dW against some externally applied force, or by having
this force do work $-dW$ on the system. For example, if the system
is confined under uniform pressure, an increase in volume would

mean that the system did work $dW = P\,dV$; if the system is under tension J, it would require work $-dW = J\,dL$ to be done on the system to increase its length dL. Similarly an increase in magnetization $d\mathfrak{M}$ in the presence of a field $\mathfrak{H}$ will increase U by $\mathfrak{H}\,d\mathfrak{M}$. Or, if dn moles of a substance with chemical potential μ is added, U would increase by $\mu\,dn$. In all these cases work dW is being done in an organized way by the system and U is increased by $-dW$. Note our convention, a positive dW is work done *by* the system, a negative value represents work done *on* the system, so that the change in U is opposite in sign to dW.

Note also that we have been using the symbol of the imperfect differential for dW, implying that the amount of work done by the system depends on the path (i.e., on how it is done). For example, the work done by an ideal gas in going from state 1 of Figure 4-1 to state 2 differs whether we go via path a or path b. Along path $1a$, V does not change, so no work is done by or on the gas, although the temperature changes from $T_1 = (P_1 V_1/nR)$ to $T_a = (P_2 V_1/nR)$. Along path $a2$, P does not change, so that the work done by the gas in going along the whole of path $1a2$ is $\Delta W_a = P_2(V_2 - V_1)$. Similarly, the work done by the gas in going along path $1b2$ is $\Delta W_b = P_1(V_2 - V_1)$, differing from ΔW_a by the factor P_1. This same sort of argument can be used to show that work done by the system, in consequence of a variation of any of the mechanical variables that describe its state, cannot be a state variable.

Something more should be said about the meaning of the diagram of Figure 4-1. In our calculations we tacitly assumed that at each point along each path the system was in an equilibrium state, for which the equation of state $PV = nRT$ held. But for a system to be in equilibrium, so that P and T have any meaning, it must be allowed to settle down, so that P and V (and therefore T) are assumed constant. How, then, can we talk about going along a path, about changing P and V, and at the same time assume that the system successively occupies equilibrium states as the change is made? Certainly if the change is made rapidly, sound waves and turbulence will be created and the equation of state will no longer hold. What we must do is to make the change slowly and in small steps, going from 1 to i and waiting till the system settles down, then going slowly to j, and so on. Only in the limit of many steps and slow change can we be sure that the system is never far from equilibrium and that the actual work done will approach the value computed from the equation of state. In thermodynamics we have to limit our calculations to such slow, stepwise changes (called *quasistatic* processes) in order to have our formulas hold during the change. This may seem to be an intolerable limitation on the kinds of processes thermodynamics can deal with; we shall see later that the limitation is not as severe as it may seem.

Heat and Internal Energy

If the only way to change the system's energy is to perform work on it or have it do work, then the picture would be simple. Not only dU but also dW would be a perfect differential; whatever work was performed on the system could eventually be recovered as mechanical (or electrical or magnetic) energy. This was the original theory of thermodynamic systems; work was work and heat was heat. The introduction of heat dQ served to raise the temperature of the body (indeed the rise in temperature was the usual way in which the heat added could be measured, as was pointed out at the beginning of Chapter 2), and when the body was brought back to its initial temperature it would have given up the same heat that had been given it earlier. We could thus talk about an internal energy of the system, which was the net balance of work done on or by the system, and we could talk about the heat possessed by the body, the net balance of heat intake and output, measured by the body's temperature.

It was quite a shock to find that this model of matter in bulk was inconsistent with observation. A body's temperature could be changed by doing work on it; a body could take in heat (from a furnace, say) and produce mechanical work. It was realized that we cannot talk about the heat "contained" by the system, nor about the mechanical energy it contains. It possesses just one pool of contained energy, which we call its internal energy U, contributed to by input of *both* mechanical work and also of heat, which can be withdrawn either as mechanical energy or as heat. Any change in U, dU, is the difference between the heat added, dQ, and the work done by the system, dW, during any process,

$$dU = dQ - dW$$
$$= dQ - P\,dV + J\,dL + \mathfrak{H}\,d\mathfrak{M} + \mu\,dn + \cdots \qquad (4\text{-}1)$$

where dU is a perfect differential and dQ and dW are imperfect ones. Note the convention used here; dQ is the heat *added* to the system, dW is the work *done* by the system.

This set of equations is the *first law of thermodynamics*. It states that mechanical work and heat are two forms of energy and

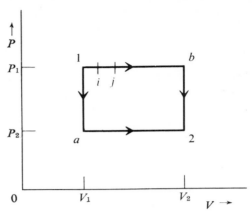

FIGURE **4-1**
Plots of quasistatic processes on the PV plane.

must be lumped together when we compute the change in internal energy of the system. It was not obvious to physicists of the early nineteenth century. To have experiments show that heat could be changed into work *ad libitum*, that neither dQ nor dW were perfect differentials, seemed at the time to introduce confusion into a previously simple, symmetric theory.

There were some compensations. Gone was the troublesome question of how to measure the heat "contained" by the system. The question has no meaning; there is no state variable Q, there is only internal energy U to measure. Also the amount of heat dQ added could sometimes be most easily measured by measuring dU and dW and computing $dQ = dU + dW$. An accurate measurement of the amount of heat added is even now difficult to make in many cases (the heat produced by passage of electric current through a resistance is relatively easy to measure, but the direct measurement of heat produced in a chemical reaction is still not easy).

Of course, compensations or not, Eq. (4-1) was the one that corresponded with experiment, so it was the one to use, and people had to persuade themselves that the new theory was really more "simple" and "obvious" than the old one. By now this revision of simplicity has been achieved; the idea of heat as a separate substance appears to us "illogical."

Just as with work, the total amount of heat added or withdrawn from a system depends on the process, on the path in the

P, V plane of Figure 4-1, for example. Of course the process must be that slow, stepwise kind, called quasistatic, if we are to use our thermodynamic formulas to calculate its change. To go from 1 to a in Figure 4-1 we must remove enough heat from the gas, keeping its volume constant meanwhile, to lower its temperature from $T_1 = (P_1V_1/nR)$ to $T_a = (P_2V_1/nR)$. We could do this relatively quickly (but not quasistatically) by placing the gas in thermal contact with a constant-temperature heat source at temperature T_a. Such a source, sometimes called a *heat reservoir*, is supposed to have such a large heat capacity that the amount of heat contributed by the gas will not change its temperature. In this case the gas would not be in thermal equilibrium until it settled down once more into equilibrium at $T = T_a$. To carry out a quasistatic process, for which we could use our formulas to compute the heat added, we should have to place the gas first into contact with a heat reservoir at temperature $T_1 - dT$, allowing it to come to equilibrium, then place it in contact with a reservoir at $T_1 - 2dT$, and so on.

To be sure, if the gas is a perfect gas of point atoms, we already know that $U = \frac{3}{2}PV$, so that $U_a - U_1 = -\frac{3}{2}V_1(P_1 - P_2)$, whether the system passes through intermediate equilibrium states or not, as long as states 1 and a are equilibrium states. Then since in this case $dW = 0$, we can immediately find dQ. But if we did not know the formula for U, but only knew the heat capacity of the gas at constant volume, we should be required (conceptually) to limit the process of going from 1 to a to a quasistatic one, in order to use C_v to compute the heat added. For a quasistatic process, for a perfect gas of point atoms where $C_v = \frac{3}{2}nR$, the heat added to the gas between 1 and a is

$$Q_{1a} = \int C_v \, dT = (\tfrac{3}{2}nR) \int (V_1 \, dP/nR) = -\tfrac{3}{2}V_1(P_1 - P_2)$$

checking with the value calculated from the change in U.

Quasistatic Processes

In going from a to 2 the same problem arises. We can imagine that the gas container is provided with a piston, which can be moved to change the volume V occupied by the gas. We could place the gas in thermal contact with a heat reservoir at temperature $T_2 = (P_2V_2/nR)$ and also move the piston so the volume changes rapidly from V_1 to V_2 and then wait until the gas settles down to equilibrium. In this case we can be sure that the internal energy U will end up having the value $\frac{3}{2}nRT_2 = \frac{3}{2}P_2V_2$, but we cannot use thermodynamics to compute how much work was done during the process or how much heat was absorbed from the heat reservoir. If, for example, instead of moving a piston, we turned a stopcock and let the gas expand freely into a previously evacuated volume $(V_2 - V_1)$, the gas would do no work while expanding. Whereas, if we moved the piston very slowly, useful work would be done and more heat would have to be taken from the reservoir in order to end up with the same value of U at state 2. In the case of free expansion, the energy not given up as useful work would go into turbulence and sound energy, which would then degenerate into heat and less would be taken from the reservoir by the time the system settled down to state 2.

If we did not know how U depends on P and T, but only knew the value of the heat capacity at constant pressure (which we shall show later equals $\frac{5}{2}nR$ for a perfect gas of point atoms) we should have to devise a quasistatic process, going from a to 2, for which to compute ΔQ_{a2} and ΔW_{a2} and thence, by Eq. (4-1), to obtain ΔU. For example, we can attach the piston to a device (such as a spring) which will maintain a constant pressure P_2 on the gas no matter what position the piston takes up (such a device could be called a constant-pressure work source, or a *work reservoir*). We then place the gas in contact with a heat reservoir at temperature $T_a + dT$, wait until the gas comes to equilibrium at slightly greater volume, place it in contact with another reservoir at temperature $T_a + 2dT$, and so on. The work done in this quasistatic process at constant pressure is, as we said earlier,

$$\Delta W_{a2} = P_2(V_2 - V_1).$$

The heat donated by the heat reservoir (if $C_p = \frac{5}{2}nR$) is

$$\Delta Q_{a2} = \tfrac{5}{2}nR \int dT = \tfrac{5}{2}P_2(V_2 - V_1)$$

and the difference is $\Delta U = \Delta Q_{a2} - \Delta W_{a2} = \frac{3}{2}P_2(V_2 - V_1)$, as it must be.

Thus thermodynamic computations, using an appropriate quasistatic process, can predict the change in internal energy U (or in any other state variable) for any process, fast or slow, which begins and ends in an equilibrium state. But these calculations cannot predict the amount of intake of heat or the production of work during the process unless the process differs only slightly from the quasistatic one used in the calculations. It behooves us to avoid incomplete differentials, such as dW and dQ, and to express the thermodynamic changes in a system during a process in terms of state variables, which can be computed for any equilibrium state, no matter how the system actually arrived at the state.

Heat Capacities

To integrate U for a simple system, where

$$dU = dQ - P \, dV \tag{4-2}$$

we need to work out some relationships between the heat capacities and the partial derivatives of U. For example, if T and V are chosen to be the independent variables, the heat absorbed in a quasistatic process is

$$dQ = dU + P \, dV = (\partial U/\partial T)_V \, dT + [(\partial U/\partial V)_T + P] \, dV \tag{4-3}$$

Since C_v is defined as the heat absorbed per unit increase in T, when $dV = 0$, we see that

$$C_v = (\partial U/\partial T)_V \tag{4-4}$$

so that Eq. (4-3) can be written

$$dQ = C_v \, dT + [(\partial U/\partial V)_T + P] \, dV \tag{4-5}$$

If T and V are varied so that P remains constant, then when T changes by dT, V will change by $(\partial V/\partial T)_P \, dT$ and the amount of heat absorbed is

$$dQ = C_P \, dT = C_v \, dT + [(\partial U/\partial V)_T + P](\partial V/\partial T)_P \, dT$$

or

$$C_P = C_v + (\partial V/\partial T)_P[(\partial U/\partial V)_T + P] \tag{4-6}$$

In our earlier discussion we stated that for a perfect gas of point atoms $C_v = \frac{3}{2}nR$ and $C_P = \frac{5}{2}nR$; we can now justify our statements. From Eq. (3-3) we know that for such a gas $U = \frac{3}{2}nRT$, so Eq. (4-4) gives us C_v immediately. It also shows that, for this gas $(\partial U/\partial V)_T = 0$, so that, from Eq. (4-6),

$$C_v = \tfrac{3}{2}nR; \qquad C_P = C_v + \left(\frac{nRT}{V}\right)\frac{V}{T} = \tfrac{5}{2}nR \tag{4-7}$$

for a perfect gas of point atoms.

A similar set of relationships can be derived for other pairs of mechanical variables. For example, for paramagnetic materials, the specific heats for constant $\mathfrak{M}$ and for constant $\mathfrak{H}$ are obtained from Eq. (4-1) (assuming that V, L, and n are constant):

$$dQ = (\partial U/\partial T)_{\mathfrak{M}}\, dT + [(\partial U/\partial \mathfrak{M})_T - \mathfrak{H}]\, d\mathfrak{M}$$

from which we can obtain

$$C_{\mathfrak{M}} = \left(\frac{\partial U}{\partial T}\right)_{\mathfrak{M}}; \qquad C_{\mathfrak{H}} = C_{\mathfrak{M}} - \left(\frac{\partial \mathfrak{M}}{\partial T}\right)_{\mathfrak{H}}\left[\mathfrak{H} - \left(\frac{\partial U}{\partial \mathfrak{M}}\right)_T\right] \qquad (4\text{-}8)$$

For a material obeying Curie's law $\mathfrak{M} = (nD\mathfrak{H}/T)$, it again turns out that $(\partial U/\partial \mathfrak{M})_T = 0$, analogous to the perfect gas, so that

$$C_{\mathfrak{H}} = C_{\mathfrak{M}} + (nD\mathfrak{H}^2/T^2) = C_{\mathfrak{M}} + (1/nD)\mathfrak{M}^2 \qquad (4\text{-}9)$$

Strictly speaking, $C_{\mathfrak{M}}$ should be written $C_{V\mathfrak{M}L\cdots}$, but since we are usually concerned with one pair of variables at a time, no ambiguity arises if we omit all but the variables of immediate interest

Isothermal and Adiabatic Processes

Other quasistatic processes can be devised beside those at constant volume and at constant pressure. For example, the system may be placed in thermal contact with a heat reservoir and the mechanical variables may be varied slowly enough so that the temperature of the system remains constant during the process. This is called an *isothermal* process. A heat capacity for this process does not exist (formally speaking, C_T is infinite). However it is important to be able to calculate the relationship between the heat dQ absorbed from the reservoir and the work dW done by the system while it proceeds.

For the perfect gas, where $(\partial U/\partial V)_T = 0$, and for para-magnetic materials, where $(\partial U/\partial \mathfrak{M})_T = 0$, and for other systems where U turns out to be a function of T alone, the heat absorbed from the reservoir during the isothermal process exactly equals the work done by the system. Such systems are perfect isothermal energy transformers, changing work into heat or vice versa without holding out any of it along the way. The transformation cannot continue indefinitely, however, for physical limits of volume or elastic breakdown or magnetic saturation or the like will intervene.

For less simple substances the heat absorbed in an element of an isothermal process is

$$dQ = dU + dW = \left[\left(\frac{\partial U}{\partial V}\right)_{T\mathfrak{M}\cdots} + P\right] dV +$$

$$+ \left[\left(\frac{\partial U}{\partial \mathfrak{M}} \right)_{TV \ldots} - H \right] d\mathfrak{M}$$

$$+ \sum_i \left[\left(\frac{\partial U}{\partial n_i} \right)_{TV \ldots} - \mu_i \right] dn_i + \cdots \qquad (4\text{-}10)$$

differing from the work done by the amount by which U increases as V or M or n_i is changed isothermally. We remind ourselves that $\mu_i \, dn_i$ is the chemical energy introduced into the system when dn_i moles of substance i is introduced or created in the system, and thus that $-\mu_i \, dn_i$ is the chemical analogue of work done.

Another quasistatic process can be carried out with the system isolated thermally, so that dQ is zero. This is called an *adiabatic* process; for it the heat capacity of the system is zero. The relationship between the variables can be obtained from Eq. (4-1) by setting $dQ = 0$. For example, for a system with V and T as independent variables, using Eqs. (4-5) and (4-6), the change of T with V in an adiabatic process is

$$C_v \, dT = - \frac{C_P - C_v}{(\partial V/\partial T)_P} \, dV \quad \text{or} \quad \left(\frac{\partial T}{\partial V} \right)_s = - (\gamma - 1) \left(\frac{\partial T}{\partial V} \right)_P$$

$$(4\text{-}11)$$

where $\gamma = (C_P/C_v)$ is a state variable. The reason for using the subscript s to denote an adiabatic process will be elucidated in Chapter 6. We see that when γ is constant (as it is for a perfect gas) the adiabatic change of T with V is proportional to the change of T with V at constant pressure.

For the perfect gas of point atoms, where $C_P = \frac{5}{2}nR$ and $C_v = \frac{3}{2}nR$, $\gamma = \frac{5}{3}$ and $(\partial T/\partial V)_P = (P/nR) = (T/V)$. In this case the relation between T and V for an adiabatic expansion is

$$(dT/T) + (\gamma - 1)(dV/V) = 0 \quad \text{or}$$

$$TV^{\gamma - 1} = (PV^{\gamma}/nR) = \text{const.} \qquad (4\text{-}12)$$

For diatomic gases $C_v = \frac{5}{2}nR$, $C_P = \frac{7}{2}nR$, and $\gamma = \frac{7}{5}$; for polyatomic gases $\gamma = \frac{4}{3}$ [see Eq. (22-5)]. Compressing a gas adiabatically increases its temperature, because $\gamma > 1$ pressure increases more rapidly, with change of volume, in an adiabatic compression than in an isothermal compression, where (PV/nR) is constant.

Sound waves in a gas involve changes in pressure which are so rapid that there is no time for heat to flow from one part of the gas to another, so the expansions and compressions are adiabatic.

The equation of motion of the gas states that the time rate of change of the momentum $\rho\mathbf{u}$ of a unit volume of the gas is equal to the negative gradient of pressure,

$$\rho(\partial\mathbf{u}/\partial t) = -\operatorname{grad} P \qquad (4\text{-}13)$$

If the velocity of motion $\mathbf{u}$ has a divergence at some point, then there will be a net outflow of gas at this point, and the rate of change of density $(\partial\rho/\partial t)$ at that point will be equal to $-\operatorname{div}(\rho\mathbf{u})$. In sound waves the density change is not great, so we can set $\operatorname{div}(\rho\mathbf{u})$ roughly equal to $\rho\operatorname{div}(\mathbf{u})$, and we thus have the equation

$$\operatorname{div}\mathbf{u} = -(1/\rho)(\partial\rho/\partial t) = -(1/\rho)(\partial\rho/\partial P)(\partial P/\partial t) \qquad (4\text{-}14)$$

where the last expression puts it in terms of pressure change, instead of density.

The density is, by definition, the reciprocal of the volume V_u occupied by a unit mass of the gas, so that the factor multiplying $(\partial P/\partial t)$ is equal to the compressibility κ of the gas, as defined in Problem 3-1;

$$(1/\rho)(\partial\rho/\partial P) = V_u\left(\frac{\partial}{\partial P}\frac{1}{V_u}\right) = -\frac{1}{V}\left(\frac{\partial V}{\partial P}\right) = \kappa$$

For acoustic waves, the partial is for an adiabatic process, not for an isothermal one, so the subscripts should be S instead of T.

Taking the divergence of the vector Eq. (4-13) and substituting for $\operatorname{div}\mathbf{u}$ from Eq. (4-14), we obtain the wave equation for sound waves

$$\nabla^2 P = (\partial\rho/\partial P)_s(\partial^2 P/\partial t^2) = \rho\kappa_s(\partial^2 P/\partial t^2) \qquad (4\text{-}15)$$

with the wave velocity equal to $\sqrt{(1/\rho\kappa_s)} = \sqrt{[(\partial P/\partial\rho)_s]}$. If the expansions in the sound wave were isothermal, we would use the ideal gas law, $P = (nRT/V) = (RT\rho/M)$, where $(1/M)$, the reciprocal of the molecular weight of the gas, is the number of moles per unit mass, and ρ is the number of units of mass per unit volume of the gas. In this case the velocity of sound would be $\sqrt{(RT/M)} = \sqrt{(P/\rho)}$, which does not agree with the measured velocity.

For adiabatic expansion, according to Eq. (4-12),

$$P = P_0(V_0/V)^\gamma = P_0(\rho/\rho_0)^\gamma, \qquad (4\text{-}16)$$

where P_0, ρ_0 are the pressure and density in the absence of sound, so the wave velocity should be $[(\gamma P_0/\rho_0)(\rho/\rho_0)^{\gamma-1}]^{1/2} \simeq \sqrt{(\gamma P_0/\rho_0)}$ if ρ never differs much from ρ_0. This velocity, which is $\sqrt{\gamma}$ times the isothermal velocity, is the value actually measured.

We note also that whenever there are acoustic waves of pressure in the gas, there are accompanying waves of temperature variation. As the pressure is increased adiabatically, the temperature increases, in accordance with the equation $T = \text{const.}\ P^{\gamma-1}$, which can be obtained from Eq. (4-12).

Adiabatic processes can involve other pairs of state variables than P and V. For example, for a paramagnetic material that obeys Curie's law and happens to have $C_\mathfrak{M}$ independent of T and $\mathfrak{M}$, the relation between T and $\mathfrak{M}$ during adiabatic magnetization is [see Eqs. (4-8) and (4-9)]

$$C_\mathfrak{M}\, dT = \mathfrak{H}\, d\mathfrak{M} \qquad \text{or} \qquad C_\mathfrak{M}\, dT = \frac{\mathfrak{M}T}{nD}\, d\mathfrak{M} \qquad \text{or}$$

$$T = T_0 \exp(\mathfrak{M}^2/2nDC_\mathfrak{M}) \tag{4-17}$$

When $\mathfrak{M} = 0$, the atomic magnets, responsible for the paramagnetic properties of the material, are rotating at random with thermal motion; impressing a magnetic field on the material tends to line up the magnets and reduce their thermal motion and so to "squeeze out" their heat energy, which must go into translational energy of the atoms (increased temperature) since heat is not removed in an adiabatic process. Reciprocally, if a paramagnetic material is magnetized, brought down to as low a temperature as possible and then demagnetized, the material's temperature will be still further reduced. By this process of adiabatic demagnetization, paramagnetic materials have been cooled from about $1\,°\text{K}$ to less than $0.0001\,°\text{K}$, the closest to absolute zero that has been attained. This process is but one example of the various thermal idiosyncrasies of matter in bulk, which are embodied in the general statement that heat is a form of energy. See also Eq. (8-29).

5

THE SECOND LAW OF THERMODYNAMICS

The laws of thermodynamics have a negative quality which distinguishes them from most other laws of physics, which makes direct, positive, experimental proof quite difficult. The first law may be phrased as saying that energy cannot be destroyed. This sort of negative is much harder to demonstrate than is the positive statement that gravitational force varies inversely as the square of the distance and directly as the product of the two attracting masses. In a sense we have to draw on the experience of all of physics to support the first law; we have never found an example which contradicts it and our use of it enables us to understand and correlate more and more new data.

Implications of the Second Law

The second law of thermodynamics also has this negative quality, and the lack of direct verification is even more noticeable than with the first law (where we at least have the experiments of Rumford as a direct demonstration of the equivalence of heat and energy). One way of phrasing the second law is that the spontaneous tendency of a system to go toward thermodynamic equilibrium cannot be reversed without at the same time changing some organized energy, work, into disorganized energy, heat. No single experiment will convince one of the validity of such a negative statement. All we can say is that the theory based on it, namely thermodynamics, has been, and still is, successful in interpreting and predicting all thermal phenomena so far.

So we cannot schedule our discussion of the second law as we do with other theories in physics, starting with a few crucial experiments and building the theory up in an obviously inductive manner from the empirical facts. Here, after defining terms, we immediately state the second law, not once but in many different-sounding ways (we have already started), and only then can we work out the various consequences, which can be verified experimentally. Here again the presentation resembles the deductive flow of mathematics, from axioms to consequent theorems, whereas the actual path of justification is the reverse, going from myriads of experiments to the unifying law.

Several questions must have occurred to the reader of the previous chapter. We showed that each term, in the sum of the differentials which makes up the mechanical work dW done by the system, is a product of an intensive variable (such as P) and a differential of a related extensive variable (such as V). We did not express the heat differential dQ as a similar product. Why not? If T is the intensive variable, what is the extensive variable, whose differential multiplies T to equal dQ? A related question is: dQ is an incomplete differential; can we find an integrating factor Φ (see page 33) so that $\Phi\, dQ$ is a complete differential? Another question is whether there is a quantitative measure of the difference between the process of going from equilibrium state 1 of Figure 4-1 to equilibrium state a by quasistatic steps and the process of going from 1 to a by nonequilibrium steps, as discussed on page 39.

Finally, since we have already stated that the internal energy of a body can be extracted either as heat or as mechanical energy, we could ask whether it is possible to convert all of a body's internal energy into mechanical energy, whether it is possible to carry out the inverse of Rumford's experiment by completely converting heat into mechanical energy in a continuous manner. We shall find that an investigation of this question will lead to answers to all the questions raised in the previous paragraph, as well as provide an introduction to the ramifications of the second law. The rest of the book is, in a sense, a further discussion of these ramifications.

It is not difficult to show that the second law, in the form we have already phrased it, denies the possibility of devising an instrumentality which would induce heat to flow spontaneously from a colder to a warmer body. We can, of course, produce such a flow by expending work, and we shall shortly see how this can be done by quasistatic procedures. But to induce such a flow without simultaneously changing some work into heat would be to discover an exception to the spontaneous trend to thermodynamic equilibrium we have talked about, and the second law says this is impossible.

The first law does not forbid such up-temperature flow, for heat at a lower temperature is the same amount of energy as the same amount of heat at a higher temperature. No energy would be destroyed or created if heat flowed from a colder to a hotter body, but the second law says such things cannot happen. It is thus more restrictive than the first law.

Heat Engines

But what does all this have to do with the question as to whether heat can be withdrawn from a heat reservoir and all of it be transformed into work? Chemical or nuclear combustion can provide a rough equivalent of a constant temperature heat reservoir; as heat is withdrawn more can be provided by burning more fuel. Can we arrange it so this continuous output of heat is cyclically or continuously converted into mechanical work? We shall see that the second law of thermodynamics answers this question in the negative, and that it provides a method of computing the maximum fraction of the heat output which *can* be changed into work in various circumstances.

At first sight this appears to contradict a result obtained in Chapter 4. There it was pointed out that a system, with internal energy that is a function of temperature only (such as a perfect gas or a perfect paramagnetic material), when placed in contact with a constant-temperature heat source, can isothermally transform all the heat it withdraws from the reservoir into useful work, either mechanical or electromagnetic. The trouble with such a process is that it cannot continue to do this *indefinitely*. Sooner or later the pressure gets too low or the tension gets greater than the elastic limit or the magnetic material becomes saturated, and the transformer's efficiency drops to zero. What is needed is a *heat engine*, a thermodynamic system that can operate cyclically, renewing its properties periodically, so it can continue to transform heat into work indefinitely.

Such an engine cannot be built to run entirely at one temperature, that of the heat source. If it did so the process would be entirely isothermal, and if we try to make an isothermal process cyclic by reversing its motion (compressing the gas again, for example) we find we are taking back all the work that has been done and reconverting it into heat; returning to the start leaves us with no net work done and all the heat given back to the reservoir. Our cycle, to result in net work done and thus net heat withdrawn, must have some part of it operating at a lower temperature than that of the source. And thus we are led to the class of cyclical operations called Carnot cycles.

Carnot Cycles

A Carnot cycle operates between two temperatures, a hotter, T_h, that of the heat source, and a colder, T_c, that of the heat sink. Any sort of material can be used, not just one having U a function of T only. And any pair of mechanical variables can be involved, P and V or J and L or $\mathfrak{H}$ and $\mathfrak{M}$ (we shall use P and V just to make the discussion specific). The cycle consists of four quasistatic operations: an isothermal expansion from 1 to 2 (see Figure 5-1) at temperature T_h, withdrawing heat ΔQ_{12} from the source and doing work ΔW_{12} (not necessarily equal to ΔQ_{12}); an adiabatic expansion from 2 to 3, doing further work ΔW_{23} but with no change in heat, and ending up at temperature T_c; an isothermal compression at

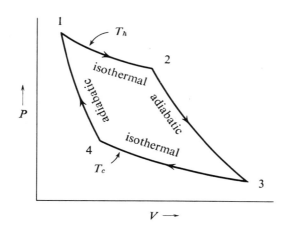

FIGURE 5-1
Example of a Carnot cycle, plotted in the P, V plane.

T_c from 3 to 4 requiring work $-\Delta W_{34} = \Delta W_{43}$ to be done on the system and contributing heat $-\Delta Q_{34} = \Delta Q_{43}$ to the heat sink at temperature T_c, ending at state 4, so placed that process 4 to 1 can be an adiabatic compression, requiring work

$$-\Delta W_{41} = \Delta W_{14}\ (\Delta Q_{41} = 0)$$

to be done on the system to bring it back to state 1, ready for another cycle (Figure 5-1). This is a specialized sort of cycle but it is a natural one to study and one that in principle should be fairly efficient. Since the assumed heat source is at constant temperature, part of the cycle had better be isothermal, and if we must "dump" heat at a lower temperature, we might as well give it all to the lowest temperature reservoir we can find. The changes in temperature should thus be done adiabatically.

This cycle, of course, does not convert all the heat withdrawn from the reservoir at T_h into work; some of it is dumped as unused heat into the sink at T_c. The net work done by the engine per cycle is the area inside the figure 1234 in Figure 5-1, which is equal to $\Delta W_{12} + \Delta W_{23} + \Delta W_{34} + \Delta W_{41} = \Delta W_{12} + \Delta W_{23} - \Delta W_{43} - \Delta W_{14}$ and which, according to the first law, is equal to $\Delta Q_{12} + \Delta Q_{34} = \Delta Q_{12} - \Delta Q_{43}$. The efficiency η with which the heat withdrawn from the source at T_h is converted into work is equal to the ratio between the work produced and the heat withdrawn.

$$\eta = \frac{\Delta W_{12} + \Delta W_{23} - \Delta W_{43} - \Delta W_{14}}{\Delta Q_{12}}$$

$$= \frac{\Delta Q_{12} - \Delta Q_{43}}{\Delta Q_{12}} = 1 - \frac{\Delta Q_{43}}{\Delta Q_{12}} \tag{5-1}$$

We note that, since all the operations are quasistatic, the cycle is *reversible*; it can be run backward, withdrawing heat ΔQ_{43} from the reservoir at temperature T_c and depositing heat ΔQ_{12} in the reservoir at T_h, requiring work $\Delta Q_{12} - \Delta Q_{43}$ to make it go.

There are a large number of Carnot cycles, all operating between T_h and T_c; ones using P and V to generate work, involving different substances with different equations of state; ones using $\mathfrak{H}$ and $\mathfrak{M}$ to produce magnetic energy, using different paramagnetic substances; and so on. One way of stating the second law is to say that all Carnot cycles operating between the temperatures T_h and T_c *have the same efficiency*. Another way is to say that *no engine*, or combination of engines, operating between a maximum temperature T_h and a minimum temperature T_c *can be more efficient than any Carnot cycle operating between these temperatures.*

Alternative Phrasings of the Second Law

By now we have mentioned four different paraphrases of the second law; it's about time we showed their equivalence. The equivalence of the last two, stated in the previous paragraph, can be demonstrated by an elementary application of logic, which the reader can surely supply. The linkage between the last two and the earlier statements may be demonstrated by working out what happens when we connect two Carnot-cycle engines together mechanically, as shown in Figure 5-2, one of them taking in heat from T_h and generating work, the other running backward, using all or part of this work to pump heat from T_c to T_h. Suppose, for a moment, that we could find two Carnot cycles, working between T_h and T_c, which differed in efficiency. Call the lower-efficiency one the "standard" engine and run it backward, using up work ΔW to pull heat ΔQ_{43} from T_c and donating heat $\Delta Q_{12} = \Delta Q_{43} + \Delta W$ to the reservoir at T_h. Let the more-efficient one do work, using heat $\Delta Q'$ from T_h, dumping heat $\Delta Q''$ to T_c and producing work $\Delta Q' - \Delta Q''$.

Now adjust $\Delta Q''$ so it equals the amount ΔQ_{43} picked up by the standard engine, so that the combination delivers as much

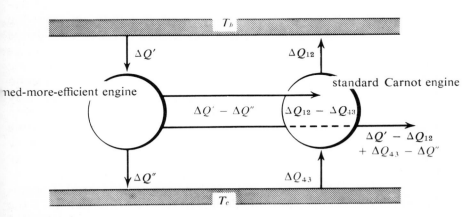

FIGURE **5-2**

Carnot engine (reversed) driven by an engine assumed more efficient; the combination would make a perfect engine, which is impossible.

heat to the reservoir at T_c as it takes away. Since we have assumed that the efficiency $1 - (\Delta Q''/\Delta Q') = 1 - (\Delta Q_{43}/\Delta Q')$ is greater than that of the standard engine, $1 - (\Delta Q_{43}/\Delta W_{12})$, this means that we are assuming that $\Delta Q'$ is *greater* than ΔQ_{12}, that the more efficient engine converts a greater proportion of heat into work. Therefore the combination of two engines will withdraw a net amount $\Delta Q' - \Delta Q_{12}$ of heat from the reservoir at T_h, will dump no net heat at T_c and will convert all the heat it withdraws into work, thus being an engine with 100 per cent efficiency. Figure 5-2 shows the assumed combination.

Thus if Carnot cycles between T_h and T_c can differ in efficiency, then we can devise a perfectly efficient engine, converting all the heat it receives from the upper reservoir into work, and not requiring a lower reservoir. Vice versa, if either of these statements is an impossibility, they both are.

Next we ask what happens if we run the assumed combination of Figure 5-2 with the presumed-more-efficient engine running just hard enough so that it runs the standard one (backward as before) with no mechanical energy to spare. In other words, let

$$\Delta Q' - \Delta Q'' = \varepsilon = \Delta Q_{12} - \Delta Q_{43}$$

(instead of $\Delta Q''$ equalling ΔQ_{43}). Since $(\Delta Q'/\Delta Q'') = \alpha$ is presumed to be larger than $(\Delta Q_{12}/\Delta Q_{43}) = \beta$, we see from

$$(\alpha - 1)\Delta Q'' = \varepsilon = (\beta - 1)\Delta Q_{43}$$

that $\Delta Q_{43} - \Delta Q''$, which equals $\Delta Q_{12} - \Delta Q'$, is positive. Therefore, if $\alpha > \beta$, a nonzero amount of heat is withdrawn from the lower reservoir and given to the upper one. What we now have is an engine which will withdraw $\Delta Q_{43} - \Delta Q''$ heat from the colder reservoir and deliver it to the hotter reservoir without requiring any work to be applied. Since our earlier phrasing of the second law (see page 48) states that this is impossible, it would thus deny the possibility of a combined engine of the sort we have been imagining and thus would require all Carnot engines, operating between equal temperatures, to have the same efficiency. We have but sketched the elements of the logical argument; the reader may add further details to make it as "airtight" as he may wish.

So, by various imagined combinations of idealized, Carnot-cycle engines, we are able to demonstrate the equivalence of the various ways in which the second law is phrased. One form, called *Clausius' principle*, is that it is impossible to transfer, in a continuous manner, heat from a lower temperature reservoir to one at a higher temperature without at the same time doing work to effect the transfer. This is equivalent to our earlier wording, on page 46. We have just shown that it is logically equivalent to another, apparently different statement, called *Kelvin's principle*, that it is impossible to convert, continuously, heat from a reservoir at a temperature T_h into work, without at the same time transferring additional heat from T_h to a colder temperature T_c. We note that the first law was used in working out the behavior of a Carnot cycle, and thus was necessary to demonstrate the equivalence of the two principles.

The Thermodynamic Temperature Scale

If all Carnot cycles operating between the same pair of temperatures, T_h and T_c, have the same efficiency, this efficiency must be independent of the material used in the Carnot cycle and must simply be a function of T_h and T_c;

$$\eta = 1 - \Psi(T_h, T_c); \qquad \Psi(T_h, T_c) = (\Delta Q_{43}/\Delta Q_{12}) \qquad (5\text{-}2)$$

The ratio of the heat dumped to that withdrawn must be the same for all these cycles. Determining the nature of the functional dependence will provide us with still another (this time the final) temperature scale.

To keep clear the different scales we go back to the mercury thermometer we talked about in Chapter 2, with its temperature reading Θ. The reading for the hot reservoir would then be Θ_h and that for the cold reservoir would be Θ_c. The efficiency debit Ψ is, as we have shown, a function only of those two readings, Θ_h and Θ_c; $\Psi = \Psi(\Theta_h, \Theta_c)$. To determine the functional dependence, we divide a Carnot cycle into two cycles, each using the same material, as shown in Figure 5-3. The upper one takes heat ΔQ_{12} from the upper reservoir at a temperature Θ_h (on the scale of the thermometer we are using) does work $\Delta Q_{12} - \Delta Q_{65}$, and delivers heat ΔQ_{65} to an intermediate reservoir at a measured temperature Θ_m. This reservoir immediately passes on this heat ΔQ_{65} to the second engine, which produces work $\Delta Q_{65} - \Delta Q_{43}$ and delivers heat ΔQ_{43} to a reservoir at temperature Θ_c as measured on our thermometer. The combination, which produces a total work of $\Delta Q_{12} - \Delta Q_{43}$, is thus completely equivalent to a single Carnot cycle, using the same material and operating between Θ_h and Θ_c on our scale, withdrawing ΔQ_{12} from the upper, exhausting ΔQ_{43}

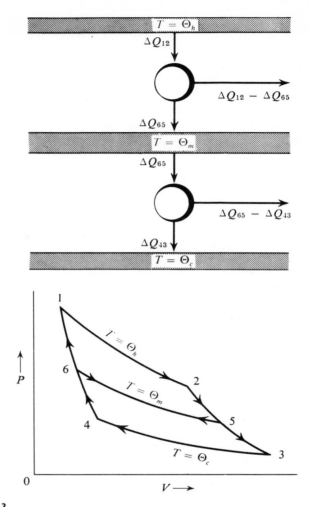

FIGURE **5-3**

Arrangement of two Carnot cycles so their combined effect is equivalent to one cycle between the temperature extremes.

at the lower, and doing work $\Delta Q_{12} - \Delta Q_{43}$. Therefore, according to Eq. (5-2), the efficiencies η_u and η_l of the two component cycles and the efficiency η_c of the combination, considered as a single engine, are related as follows:

$$\eta_u = 1 - \Psi(\Theta_h, \Theta_m); \qquad \Psi(\Theta_h, \Theta_m) = \frac{\Delta Q_{65}}{\Delta Q_{12}}$$

$$1 - \eta_l = \Psi(\Theta_m, \Theta_c) = \frac{\Delta Q_{43}}{\Delta Q_{65}}$$

$$1 - \eta_c = \Psi(\Theta_h, \Theta_c) = \frac{\Delta Q_{43}}{\Delta Q_{12}} = \Psi(\Theta_h, \Theta_m)\Psi(\Theta_m, \Theta_c) \qquad (5\text{-}3)$$

This last product must therefore be independent of Θ_m.

For the equation relating the three values of the function Ψ, for the three pairs of values of measured temperature Θ, to be valid. Ψ must have the functional form $\Psi(x, y) = [T(y)/T(x)]$, where $T(\Theta)$ is some single-valued, monotonically increasing function of Θ, the temperature reading on the thermometer used. For then $\Psi(\Theta_m, \Theta_c)\Psi(\Theta_h, \Theta_m)$ will equal

$$[T(\Theta_c)/T(\Theta_m)][T(\Theta_m)/T(\Theta_h)] = [T(\Theta_c)/T(\Theta_h)] = \Psi(\Theta_h, \Theta_c)$$

Therefore,

$$\Psi(\Theta_h, \Theta_c) = \frac{\Delta Q_{43}}{\Delta Q_{12}} = \frac{T(\Theta_c)}{T(\Theta_h)} \qquad \text{or} \qquad \frac{\Delta Q_{43}}{T(\Theta_c)} = \frac{\Delta Q_{12}}{T(\Theta_h)} \qquad (5\text{-}4)$$

The function $T(\Theta)$ is the basis of the scale of temperature called the *thermodynamic scale*. The ratio between the reading for the temperature of the cold reservoir, on this scale, to that of the hot reservoir is equal to the ratio between the heat exhaust ΔQ_{34} to the heat intake ΔQ_{12} for a Carnot cycle operating between the two reservoirs, as measured by the cycle's efficiency. For example, if $\eta = 90$ per cent, the thermodynamic temperature T_c of the cold reservoir must be one-tenth of the temperature T_h of the hot one, and so on. This procedure determines ratios of units at various parts of the scale, but does not settle the absolute size of the unit. This can be fixed, for convenience, to correspond to the two earlier scales, by requiring that the temperature difference between boiling water and melting ice be 100 degrees, as it was for the centigrade and the ideal gas scales.

We thus have at our disposal three temperature scales, that of the mercury thermometer, that of the ideal-gas thermometer, and that of the Carnot cycle, the thermodynamic scale. They all agree in having a difference of 100° between the temperature of boiling water and that of melting ice, but they may not be linearly related. The mercury thermometer scale has only the advantage of convenience; we know it is not linearly related to the ideal-gas scale (see Figure 3-2). But the other two scales both have the advantage of dependence on universal physical phenomena, so they can be reproduced at will. Which shall we choose?

Luckily the decision does not need to be made, for the two scales are identical. In other words, if (PV/nR) for an ideal gas in equilibrium with the cooler reservoir of a Carnot cycle is just half the value it would have when in equilibrium with the hotter reservoir, then it will turn out that the heat ΔQ_{43} dumped by the cycle is just half the heat taken in, ΔQ_{12}. This equality of the two scales will be proved in Chapter 6.

Therefore, from now on, we need make no differentiation as to whether a temperature T is measured by an ideal-gas thermometer or by means of a Carnot cycle or its equivalent. The scale is the same, the absolute or thermodynamic scale, in degrees Kelvin, in which the temperature of melting ice is 273.2° and that of boiling water is 373.2°. Of course both the ideal-gas thermometer and the Carnot cycle are idealized instrumentalities, and we must experimentally determine either scale by various limiting processes. For room temperatures and above, it is not difficult to find a gas which, when sufficiently rarified, will provide a good approximation to the absolute scale. Below 300° the problem of correcting for the difference between an actual gas and an ideal gas gets progressively more difficult as we approach absolute zero. Below about 4°K even helium is a liquid at ordinary pressures, and we must turn to Carnot cycles to extend the scale farther (the difficulties are logarithmic; it's just as hard to go from 0.4° to 0.04° as it is to calibrate between 4° and 0.4°).

The coupling between magnetization and heat in a paramagnetic substance, as illustrated in Eq. (4-13), provides the most easily manipulated process for the very low temperature range. Of course, the actual relationships are more complex than the approximate formulas of Eq. (4-13) but the appropriate corrections can be made more easily than the gas scale could be extrapolated. Therefore the temperature scale below 1°K has been calibrated by means of Carnot cycles of magnetization.

This now completes our series of definitions of temperature started in Chapter 1. From now on temperature T will always be measured in degrees Kelvin. In its terms, the efficiency of a Carnot cycle operating between T_h and T_c (both measured in degrees Kelvin) is

$$\eta = 1 - \Psi(T_h, T_c); \qquad \Psi(T_h, T_c) = \frac{\Delta Q_{43}}{\Delta Q_{12}} = \frac{T_c}{T_h} \qquad (5\text{-}5)$$

This is the maximum efficiency we can get from an engine that operates between T_h and T_c.

Thus the second law is a sort of relativistic principle. The minimal temperature at which we can exhaust heat is determined by the temperature of our surroundings, and this limits the efficiency of transfer of heat into work. Heat at temperatures high compared to our surroundings is "high-quality" heat; if we handle it properly a large portion of it can be changed into useful work. Heat at temperature twice that of our surroundings (on the Kelvin scale) is already half degraded; only half of it can be usefully employed. And heat at the temperature of our surroundings is useless to us for getting work done. Even heat at a million degrees Kelvin would be useless if the whole universe were at this same temperature.

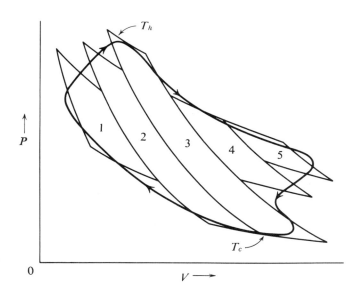

FIGURE 5-4

Reversible cycle (heavy line) simulated by a combination of several Carnot cycles.

Temperature *ratios* enable us to produce mechanical energy, not absolute magnitudes of average temperature.

In principle we can build up a combination of Carnot cycles to simulate any kind of reversible cycle, such as the one shown by the heavy line in Figure 5-4. In such a cycle, heat is taken on and given off at different temperatures, none of the elementary processes being isothermal or adiabatic. The maximum temperature reached is T_h, for the isothermal curve tangent to the top of the loop, and the minimum is T_c, for the lower tangent isothermal. The work produced is the area within the heavy line. This cycle is crudely approximated by the five Carnot cycles shown, with their isothermals and adiabatics as light lines; a better approximation could be obtained with a large number of Carnot cycles. The efficiency of subcycle 3 is greatest, because it operates between the greatest spread of temperatures; the others have less efficiency. Thus any cycle that takes in or gives off heat while the temperature is changing is not as efficient as a Carnot cycle operating between the same maximum and minimum temperatures, i.e., which takes on all its heat at T_h and which gives up all its heat at T_c.

CHAPTER

ENTROPY

We notice that, for a Carnot cycle, the relationship between each element dQ of heat taken on and the thermodynamic temperature T at which it is taken on (or given off) is such that the integral dQ/T taken completely around the cycle is zero. The heat taken on at T_h is ΔQ_{12} and the heat "taken on" at T_c is the negative quantity $\Delta Q_{34} = - \Delta Q_{43}$; Eq. (5-5) states that the sum $(\Delta Q_{12}/T_h)$ $- (\Delta Q_{43}/T_c) = 0$.

A Thermal-State Variable

Since any quasistatic, reversible cycle can be considered as a sum of Carnot cycles, as in Figure 5-4, we see that for any such cycle the integral of the quantity dQ/T around the whole cycle is zero. But for any thermodynamic state function $Z(x, y)$ (as in Figure 6-1) the integral of the perfect differential dZ around a closed path (such as ABA in Figure 6-1) is zero, as long as all parts of the path are reversible processes; alternatively any differential that integrates to zero around any closed path is a perfect differential and its integral is a state function of the variables x, y.

Therefore the quantity $dS = dQ/T$ is a perfect differential, where dQ is the heat given to the system in an elementary, reversible process and T is the thermodynamic temperature of the system during the process. The integral of this perfect differential, $S(x, y)$, is a state variable and is called the *entropy* of the system. It is an extensive variable, proportional to n.

This result, which is still another way of stating the second law, can be rephrased to answer the third of the questions posed

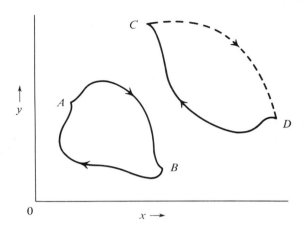

FIGURE 6-1

  *Paths in the xy plane for reversible processes (solid lines) and dashed line CD
  symbolizing a spontaneous (irreversible) process, which cannot be a path in
  the xy plane.*

in the first section of Chapter 5. There *is* an integrating factor for
dQ, *if* heat dQ is absorbed *in a reversible process*; it is the reciprocal
of the thermodynamic temperature, defined in Eq. (5-5). The resulting perfect differential dQ/T measures the change dS in the state
variable S, the entropy; and the difference $S_2 - S_1$ of entropy
between equilibrium states 1 and 2 is computed by integrating dQ/T
along any reversible path between 1 and 2. On the other hand,
there is no integrating factor for dQ for an irreversible process.

   The entropy S is the extensive variable that pairs with T as
V does with P and $\mathfrak{M}$ with $\mathfrak{H}$. The heat taken on by the system in a
reversible process is $dQ = T\,dS$, just as the work done by the system
is $dW = P\,dV - \mathfrak{H}\,d\mathfrak{M}$. And this is the answer to the first question
raised on page 47. The entropy is the extensive counterpart of the
intensive variable T. The product $T\,dS$ has the dimensions of energy
and equals the amount of heat given to the system, at temperature
T in the thermodynamic scale, during a quasistatic (reversible) process which produces a change dS in the system's entropy. If heat is
added quasistatically to a body, its entropy increases, not by much
if the temperature is high, by a lot if the temperature is low. Conversely if heat is subtracted quasistatically the body's entropy
decreases; if the body departs considerably from equilibrium while
the heat is given off, however, the body may gain entropy rather

than lose it, since the equation $dS = \dj Q/T$ only holds for reversible processes.

Reversible Processes

We are now in a position to be more specific about the adjective *reversible*, which we first used for a cycle (such as a Carnot cycle) and which we recently have been applying to processes. To see what it means let us first consider a few *irreversible* processes. Suppose a gas is confined at pressure P_0 within a volume V_0 of a thermally insulated enclosure, as shown in Figure 6-2. The gas is confined to V_0 by a diaphragm D; the rest of the volume, $V_1 - V_0$, is evacuated. We then break the diaphragm and let the gas undergo free expansion until it comes to a new equilibrium at volume V_1. This is a *spontaneous* process, going automatically in one direction only. It is obviously irreversible; the gas would never return by itself to volume V_0.

Next suppose we place an object, originally at temperature T_h, in thermal contact with a heat reservoir at temperature T_c, less than T_h. Here again the process is spontaneous; heat flows from the object until it comes to equilibrium at temperature T_c. This also is an irreversible process; it would take work (or heat from a reservoir at T_h) to warm the body up again.

We can thus define the adjective "reversible" in a negative way; a reversible process is one that has no irreversible portion. To expand the gas from V_0 to V_1 reversibly we could replace the diaphragm by a piston and move it slowly to the right. During the motion, as the volume is increased by dV, the gas is never far from equilibrium, and a reversal of motion of the piston (so the volume

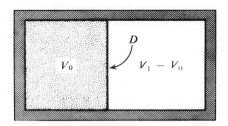

FIGURE **6-2**

The Joule experiment.

decreases again by dV) would bring the gas back to its earlier state. In such a case we could retrace every part of the process in detail. Every reversible process is quasistatic; not all quasistatic processes are reversible (see, for instance, the top of page 63).

In Eq. (6-1) we pointed out that the integral of dQ/T around a reversible cycle is zero. If the cycle is irreversible the integral differs from zero; the second law requires it to be *less than zero*. For example, suppose the irreversible cycle took in all its heat, an amount $\Delta Q'$ at T_h and exhausted an amount $\Delta Q''$ all at T_c. The efficiency of such a cycle would have to be less than the value $1 - (T_c/T_h) = 1 - (\Delta Q_{43}/\Delta Q_{12})$ for a Carnot cycle between the same temperatures. This means that $\Delta Q'$ would have to be smaller than ΔQ_{12} or $\Delta Q''$ would have to be larger than ΔQ_{43}, or both, so that the integral of dQ/T around the irreversible cycle would turn out less than that for the Carnot cycle, i.e., less than zero. The argument can be generalized for all closed cycles. Thus another way of stating the second law is that for all closed cycles the integral around the cycle,

$$\oint (dQ/T) \le 0 \qquad\qquad\qquad (6\text{-}1)$$

where the equality holds for reversible cycles and the inequality is for irreversible ones. Since dS is measured by the value of dQ/T *when* the process is reversible, we also have that

$$dS \ge dQ/T \qquad\qquad\qquad (6\text{-}2)$$

where again the equality holds for reversible processes, inequality for irreversible ones.

Irreversible Processes

For example, suppose the dotted line from C at D in Figure 6-1 represents a spontaneous process, during which no heat is absorbed or given up (as is the case with the free expansion of the gas in Figure 6-2), so that $\int (dQ/T) = 0$ for the dotted line. Since the process is irreversible, $\int dS = S_D - S_C$ must be larger than $\int (dQ/T) = 0$. In other words, *during a spontaneous process* taking place in a thermally isolated system, *the entropy always increases*.

The statement is not so simple when the spontaneous process involves transfer of heat, as with the irreversible cooling of a body from T_h to T_c mentioned above [see also the discussion four paragraphs below Eq. (4-1)]. In this case the body loses entropy and the reservoir gains it. If the heat capacity of the body is C_v, a constant, the heat reservoir gains $C_v(T_h - T_c)$ heat; since the reservoir is always at T_c, its gain in entropy is $C_v[(T_h - T_c)/T_c]$. The loss in entropy of the body is not much harder to compute. During its spontaneous discharge of heat to the reservoir, and before it comes to uniform temperature T_c, the body is not in equilibrium, so dS does not equal dQ/T. However we can devise a quasistatic process, placing a poor heat conductor between the body and the reservoir, so the heat flows into the reservoir slowly and the body, at any time, will have a nearly uniform temperature T, where T starts at T_h and gradually drops to T_c at the end. The loss of entropy from the body is thus the integral of $C_v(dT/T)$, which is $C_v \ln(T_h/T_c)$ if C_v is constant. This is always smaller than $C_v[(T_h - T_c)/T_c]$, the entropy gained by the reservoir, although the two approach each other in value as T_h approaches T_c. Thus, although the entropy of the body decreases during the spontaneous cooling of the body, the total entropy of body and reservoir (which we might call the entropy of the universe) increases by the amount

$$S = C_v x - C_v \ln(1 + x) = C_v(\tfrac{1}{2}x^2 - \tfrac{1}{3}x^3 + \tfrac{1}{4}x^4 - \cdots)$$

where $x = [(T_h - T_c)/T_c]$, which is positive for all values of $x > -1$.

Thus the statement at the end of the previous paragraph can be generalized by saying that in a spontaneous process of any kind, even if the entropy of the body decreases, the entropy of some other system increases even more, so that the *entropy of the universe always increases during an irreversible process*. This, finally, is the answer to the second question at the beginning of Chapter 5; the measure of the difference between a reversible and an irreversible process lies in the entropy change of the universe.

Entropy is a measure of the unavailability of heat energy. The entropy of a certain amount of heat at low temperature is greater than it is at high temperature, loosely speaking. Alternatively, entropy measures the degree of disorganization of the system. Irreversible processes increase disorder, increase the amount of low-temperature heat, and thus increase the entropy of the universe. Reversible processes, on the other hand, simply transfer entropy from one body to another, keeping the entropy of the universe constant. A few examples will familiarize us with these ideas.

The Internal Energy, Euler's Equation

But before we can work out these examples, we must finish the discussion we started in Chapter 4 regarding the change in the internal energy of a body during a reversible process. Substituting for dQ in Eq. (4-1) we obtain

$$dU = T\,dS - P\,dV + J\,dL + \mathfrak{H}\,d\mathfrak{M} + \sum_i \mu_i\,dn_i + \cdots \qquad (6\text{-}3)$$

which embodies both the first and second laws in a single, symmetric form. Every term in the right-hand side is now a product of an intensive variable and the differential of an extensive variable. No one of these terms is a perfect differential; yet the sum of all of them is the perfect differential dU.

As we shall demonstrate in several examples later, the entropy S is a function of T and of the various extensive variables. Alternatively T can be expressed as a function of S and the other extensive variables which enter into the situation under study. Therefore the set of extensive variables, including S, is a *complete set of independent variables*, which can be used to specify completely the state of the system. Because of the equations of state, all the related intensive variables can be expressed as functions of the independent variables, as can U itself. Thus all the intensive variables, as well as U, are dependent state variables, in this formulation.

The form of Eq. (6-3), plus the fact that dU is a perfect differential, indicates that if U is expressed as an explicit function of the appropriate extensive variables, $U(S, V, \mathfrak{M}, n_i, \cdots)$, then it has many of the properties of a potential function. Comparison between the expression of the perfect differential

$$dU = \left(\frac{\partial U}{\partial S}\right)_{V,\mathfrak{M},n_i} dS + \left(\frac{\partial U}{\partial V}\right)_{S,\mathfrak{M},n_i} dV + \left(\frac{\partial U}{\partial \mathfrak{M}}\right)_{S,V,n_i} d\mathfrak{M}$$
$$+ \sum_i \left(\frac{\partial U}{\partial n_i}\right)_{S,V,\mathfrak{M},n_j} dn_i + \cdots \qquad (j \neq i)$$

in terms of its partial derivatives, with Eq. (6-3) shows that the partial of the potential U, with respect to one of the extensive variables is equal to the corresponding intensive variable, which is thus analogous to a force component,

$$T = \left(\frac{\partial U}{\partial S}\right)_{V,\mathfrak{M},n_i,\cdots} \quad ; \qquad -P = \left(\frac{\partial U}{\partial V}\right)_{S,\mathfrak{M},n_i,\cdots}$$

$$\mathfrak{H} = \left(\frac{\partial U}{\partial \mathfrak{M}}\right)_{S,V,n_i,\cdots} \quad ; \qquad \mu_i = \left(\frac{\partial U}{\partial n_i}\right)_{S,V,\mathfrak{M},n_j,\cdots} \qquad \text{etc.} \qquad (6\text{-}4)$$

Thus, if we have determined the internal energy U as a function of the exclusively extensive, complete set of independent variables then the various partial derivatives of U will be the related intensive variables (including T), expressed as functions of the extensive variables. To use Eqs. (6-4) we must integrate Eq. (6-3). Since U also is an extensive variable, we can utilize a trick devised by Euler to perform the integration.

Suppose the amounts of all the component substances in the system were doubled, or halved, or in general all changed by a factor λ, without changing the values of any of the intensive variables. Then, by definition, the extensive variable U would be changed by a like factor λ and all the independent, extensive variables would also be changed by a like factor. For simplicity for the moment, we write the independent, extensive variables as X_1, X_2, $\cdots$ and the corresponding dependent, intensive variables as Y_1, Y_2, $\cdots$, $Y_j = (\partial U/\partial X_j)$, $\cdots$. Then we see that *if* U is expressed as a function of the X's,

$$\lambda U(X_1, X_2, \cdots) = U(\lambda X_1, \lambda X_2, \cdots)$$

Differentiating this equation with respect to λ we find

$$U(X_1, X_2, \cdots) = \frac{d}{d\lambda} U(\lambda X_1, \lambda X_2, \cdots) = \sum_j \left(\frac{\partial U}{\partial X_j}\right) X_j = \sum_j Y_j X_j$$

Translating this back to our familiar notation, we have

$$U(S, V, \mathfrak{M}, n_i, \cdots) = TS - PV + \mathfrak{H}\mathfrak{M} + \sum_i \mu_i n_i + \cdots \qquad (6\text{-}5)$$

where the intensive variables T, P, etc., are all expressed as functions of the extensive variable set. Not surprisingly, this equation is called *Euler's equation*. Once all the intensive variables are explicitly given in terms of the extensive variables, for a particular system, its thermal properties can all be calculated from the expression for U, by appropriate transformation and differentiation.

The fact that we have both a differential and an integrated equation for U implies that another differential relationship between the intensive and extensive variables can be· written. Taking the differential of Euler's equation, $dU = \Sigma Y_j\, dX_j + \Sigma X_j\, dY_j$, and subtracting Eq. (6-3), $dU = \Sigma Y_j\, dX_j$, we see that $\Sigma X_j\, dY_j = 0$, or

$$S\, dT - V\, dP + \mathfrak{M}\, d\mathfrak{H} + \sum_i n_i\, d\mu_i + \cdots = 0$$

In particular, for 1 mole of a single-component material ($n = 1$),

$$d\mu = -s\,dT + v\,dP - \mathfrak{m}\,d\mathfrak{H} - \cdots \qquad (6\text{-}6)$$

where s, v, $\mathfrak{m}$, etc., are the values of entropy, volume, magnetization, etc., per mole of material (for example, v is equal to M/ρ, where M is the molecular weight of the material and ρ is its density).

This differential equation for the chemical potential μ of a substance is called the *Gibbs–Duhem equation*. Once we know the equations of state and the expression for the entropy in terms of the intensive variables, we can integrate it to find μ, as will be seen later. In general, we can see that μ increases as the pressure increases, decreases as the temperature increases, and so on.

In practice it is not often convenient to use extensive variables exclusively as the independent variables. Temperature T is the variable which can be measured directly, rather than S. It is usually much easier to perform an experiment at constant pressure than at constant volume, and so on. Much of the algebraic manipulation in this and the next chapters arises from our desire to use a more appropriate set of independent variables than the exclusively extensive set. In any case the differential Eq. (6-3) for a reversible process is our starting point.

Entropy and Internal Energy

As a start let us work out the relationships between entropy and internal energy for a system having only P, V for mechanical variables. We start with the version of Eq. (6-3), $T\,dS = dU + P\,dV$ and use T and V for independent variables. First we express dS and dU in terms of their partials and the differentials of the independent variables T and V and equate coefficients of these differentials:

$$T\left(\frac{\partial S}{\partial T}\right)_V dT + T\left(\frac{\partial S}{\partial V}\right)_T dV = \left(\frac{\partial U}{\partial T}\right)_V dT + \left[\left(\frac{\partial U}{\partial V}\right)_T + P\right] dV$$

so

$$\left(\frac{\partial S}{\partial T}\right)_V = \frac{1}{T}\left(\frac{\partial U}{\partial T}\right)_V = \frac{C_v}{T}; \qquad \left(\frac{\partial S}{\partial V}\right)_T = \frac{1}{T}\left[\left(\frac{\partial U}{\partial V}\right)_T + P\right]$$

Now apply the highly useful Eq. (3-10), on these partials of S,

$$\frac{1}{T}\left[\frac{\partial}{\partial V}\left(\frac{\partial U}{\partial T}\right)_V\right]_T = \frac{1}{T}\left[\frac{\partial}{\partial T}\left(\frac{\partial U}{\partial V}\right)_T\right]_V$$

$$+ \frac{1}{T}\left(\frac{\partial P}{\partial T}\right)_V - \frac{1}{T^2}\left[\left(\frac{\partial U}{\partial V}\right)_T + P\right]$$

or

$$T\left(\frac{\partial P}{\partial T}\right)_V = \left(\frac{\partial U}{\partial V}\right)_T + P = T\left(\frac{\partial S}{\partial V}\right)_T \tag{6-7}$$

We also see that

$$(\partial C_v/\partial V)_T = T\left[\frac{\partial}{\partial T}\left(\frac{\partial S}{\partial V}\right)_T\right]_V = T(\partial^2 P/\partial T^2)_V \tag{6-8}$$

for any substance having only V and T as variables.

From these two equations a large number of interesting conclusions can be drawn. In the first place we can complete the discussion following Eq. (5-4), regarding the identity of the temperature as read by a perfect-gas thermometer and the thermodynamic temperature T, defined by a Carnot cycle and used in this chapter. A perfect gas, according to the discussion following Eq. (3-3), has an internal energy which depends only on T, not on V, and also obeys the ideal gas law $PV = nR\tau$, with τ being the temperature on the perfect-gas scale. What we must prove is that τ is proportional to the thermodynamic temperature T, not just a monotonically increasing function of T. If τ is a constant times T then the appropriate choice of units can make the constant equal unity and thus τ will be exactly equal to T, as we claimed.

To show this we take the first half of Eq. (6-7), which of course is expressed in terms of the thermodynamic temperature. For a perfect gas, as we just pointed out, $(\partial U/\partial V)_T = 0$, so that its pressure dependence on T must satisfy the equation $(\partial P/\partial T)_V = (P/T)$. Since $P = nR\tau(T)/V$, the equation for P corresponds to an equation for $\tau(T)$, $(d\tau/\tau) = (dT/T)$ or $\tau = \text{const.} \times T = T$ when we choose equal units. In other words a perfect gas, for which both U and PV are functions of T alone, can provide a thermometer which will read thermodynamic temperature T directly, since PV is proportional to T and the appropriate choice of R will make (PV/nR) exactly equal to T.

Reciprocally, we also see that an ideal gas (one for which PV is proportional to T) is also a perfect gas, one for which $(\partial U/\partial V)_T = 0$. Thus the adjectives *ideal* and *perfect*, as used here, are interchangeable.

Entropy as Function of T and X

The relationships Eqs. (6-7) and (6-8) may be generalized to any pair of mechanical variables, X being the extensive one and Y

the corresponding intensive one, forming one of the product terms in Eq. (6-5). Remembering that P is minus the corresponding Y, we see that the generalizations are

$$(\partial U/\partial X)_T = Y - T(\partial Y/\partial T)_X$$
$$(\partial S/\partial X)_T = -(\partial Y/\partial T)_X; \qquad (\partial S/\partial T)_X = (C_x/T) \qquad (6-9)$$
$$(\partial C_x/\partial X)_T = -T(\partial^2 Y/\partial T^2)_X$$

The first of these shows that if any intensive variable Y has an equation of state of the form $Y = f(X) \cdot T$, then the internal energy is independent of the extensive variable X, being a function of T alone (plus other variables, if they enter). This is the case, for example, for a perfect paramagnetic material (one which satisfies Curie's law exactly); since $\mathfrak{H} = (\mathfrak{M}/nD)T$, the internal energy U is independent of the magnetization $\mathfrak{M}$.

The second line of Eqs. (6-9) indicates that empirical determination of the heat capacity $C_x(T, X)$ and the equation of state $Y = F(T, X)$ for a system with T and X as independent variables, serve to determine the two partials of S with respect to T and X and thus, by integration, serve to determine the entropy, within an additive constant. Since a heat capacity C_x is always positive, S must always increase with temperature. If the equation of state requires the intensive variable to *decrease* as T increases (as it does for $-P$ for a gas) then S *increases* as the corresponding extensive variable (V for the gas) increases.

But perhaps it is better to put it the other way around. Since, as we have mentioned already and will show later, the entropy is a measure of the disorganization of a system, the second equation of (6-9) shows that if disorganization *increases* as a given extensive variable increases (as it does with V) then the corresponding intensive variable Y *decreases* as T is increased, holding X constant (as is the case with $-P$). On the other hand, if disorganization *decreases* as X increases, holding T constant (as is the case with magnetization $\mathfrak{M}$ of a paramagnetic substance; for as $\mathfrak{M}$ increases the atomic magnets tend more and more to line up) then the corresponding intensive variable Y will *increase* as T is increased, holding X constant (as does $\mathfrak{H} = T\mathfrak{M}/nD$ for a perfect paramagnetic substance). This interesting and not particularly obvious relationship between entropy change and equation of state for a pair of variables X, Y is a direct consequence of Eq. (6-3).

The last equation of (6-9) indicates that if the intensive variable Y, as a function of T and X, is linearly dependent on T

(as it is with $-P$ and $\mathfrak{H}$) then the corresponding heat capacity C_x is independent of X. If the dependence of Y is not linear then minus the curvature of the graph of Y against T, holding X constant, is proportional to the rate of change of C_x with X, holding T constant. Furthermore, unless the second derivative $(\partial^2 Y/\partial T^2)_X$ goes to infinity as $T \rightarrow 0$ (which it never does), the specific heat C_x is independent of X at absolute zero, no matter what the equation of state may be.

Entropy of a Monatomic Gas

The entropy of n moles of a perfect gas of point atoms, for which $U = (\frac{3}{2})nRT$ and $PV = nRT$, may be determined by integration of Eq. (6-1), which is for this case

$$T\,dS = dU + P\,dV \qquad \text{or} \qquad dS = \tfrac{3}{2}(nR/T)\,dT + (nR/V)\,dV$$

so

$$S = nR\ln[(T/T_0)^{3/2}(V/V_0)] + S_0 \tag{6-10}$$

where $S = S_0$ when $T = T_0$ and $V = V_0$. Increase in either T or V increases the entropy of the gas. Instead of T and V (and n), S and V (and n) can be used as independent variables, in which case

$$T = T_0\left(\frac{V_0}{V}\right)^{2/3} e^{2(S-S_0)/3nR}\,; \qquad P = \frac{nRT_0}{V_0}\left(\frac{V_0}{V}\right)^{5/3} e^{2(S-S_0)/3nR}$$

These formulas immediately provide us with the dependence of T and P on V for an adiabatic process. For a *reversible* adiabatic process, $dS = dQ/T = 0$, so S is constant during such a process. Setting $S = $ const., we obtain Eqs. (4-12). This also explains why the partial for an adiabatic process, as in Eq. (4-11), has a subscript S.

We can use Euler's Eq. (6-5) to calculate the chemical potential per mole of point atoms. For this system, $U = TS - PV + \mu n$, the atoms being all of one kind. Inserting the expressions for U, PV, and TS in terms of T, V, and n and dividing by n, we find that

$$\mu = -\,Ts_0 + RT\ln[e^{5/2}(V_0/V)(T_0/T)^{3/2}] \tag{6-11}$$

where e is the base of the natural logarithms ($\ln e = 1$) and $s_0 = (S_0/n_0)$ is the entropy per mole at T_0, V_0, and n_0. Therefore we can use μ for an independent variable and obtain, for example,

$$V = V_0(T_0/T)^{3/2} \exp\left(-\frac{5}{2} - \frac{s_0}{R} - \frac{\mu}{RT}\right)$$

We could also have obtained these equations by integrating the Gibbs–Duhem Eq. (6-6).

Finally we can write the internal energy as a function of the extensive variables, as in Euler's Eq. (6-5),

$$U(S, V, n) = TS - PV + \mu n$$
$$= n(b_0/V^{2/3})e^{2S/3nR} \tag{6-12}$$

where constant b_0 equals $\frac{3}{2}RT_0V_0^{2/3}e^{-2s_0/3R}$, independent of S, V, and n. If the gas is a perfect gas but its internal energy $U = \beta nRT$ is greater than $\frac{3}{2}nRT$, we can substitute β for $\frac{3}{2}$ in these formulas.

If the gas is not a perfect gas, we can compute its entropy by utilizing Eqs. (6-7) and the set preceding it, using the empirical formulas for the heat capacity C_v and the equation of state, in the form of P as a function of T and V. If it is a good enough approximation to use van der Waals Eq. (3-4), then

$$(\partial S/\partial V)_T = [nR/(V - nb)]$$

If this is not good enough, we can use the virial Eq. (3-5), and

$$\left(\frac{\partial S}{\partial V}\right)_T = \frac{nR}{V}\left\{1 + \frac{n}{V}\frac{d}{dT}[TB(T)] + \left(\frac{n}{V}\right)^2\frac{d}{dT}[TC(T)] + \cdots\right\}$$

To obtain the dependence of S on T we need an empirical formula for C_v, the capacity of the gas to store energy in the form of heat motion. For a gas a considerable fraction of this energy is in the kinetic energy of translation of the molecules, responsible for the $\frac{3}{2}nR$ of Eq. (4-7) for a perfect gas of point particles. An actual gas has other ways of storing energy, in molecular rotation and vibration, for example, and even in excitation of the electrons in the molecules. All these add a term which we can call $C_i(T)$, the internal heat capacity, which experiment shows is a monotonically increasing function of T. Therefore we have

$$(\partial S/\partial T)_v = (3nR/2T) + (1/T)C_i(T)$$

Integrating the two partials and combining them gives us an expression for the entropy of an actual gas in terms of T and V.

If the van der Waals formula is accurate enough, then

$$S = nR \ln\left(\frac{V - nb}{V_0 - nb}\right) + \tfrac{3}{2}nR \ln\left(\frac{T}{T_0}\right) + \int_{T_0}^{T} [C_i(T)/T] \, dT + S_0$$

(6-13)

which should be compared with Eq. (6-10) for a perfect gas. The logarithmic increase with V and T is still the major dependence, but the internal heat capacity adds a term which can be large at high temperatures.

In many cases the experimental measurements provide values of the heat capacity at constant pressure, C_P, rather than C_v. In this case we can use Eqs. (4-6) and (6-9) to express the difference between C_P and C_v in terms of partials which can be determined from the equation of state,

$$C_P - C_v = T\left(\frac{\partial V}{\partial T}\right)_P \left(\frac{\partial P}{\partial T}\right)_V = -T\frac{(\partial P/\partial T)_V^2}{(\partial P/\partial V)_T}$$

$$= -T\frac{(\partial V/\partial T)_P^2}{(\partial V/\partial P)_T}$$

(6-14)

where Eq. (3-9) has been used to obtain the second and third forms of the right-hand side. From this we can obtain C_v, given C_P.

Entropy of Mixing

As a final example, illustrating the connection between entropy and disorder, we shall demonstrate that mixing two gases increases their combined entropy. As shown in Figure 6-3, we start with αn moles of a gas of type 1 (such as helium) in a volume αV

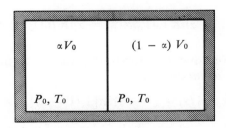

FIGURE 6-3

Arrangement to illustrate spontaneous mixing of two different gases.

at temperature T_0 and pressure $P_0 = \alpha n R T_0 / \alpha V$ on one side of a diaphragm and $(1 - \alpha)n$ moles of gas of type 2 (such as nitrogen) in volume $(1 - \alpha)V$ on the other side of the diaphragm, in equilibrium at the same temperature and pressure. We now destroy the diaphragm and let the gases spontaneously mix. According to our earlier statements the entropy should increase, since the mixing is an irreversible process of an isolated system. The total internal energy U of the combined system finally has the same value as the sum of the U's of the two parts initially. What has happened is that the type 1 gas has expanded its volume from αV to V and type 2 gas has expanded from $(1 - \alpha)V$ to V.

The statement that a gas is a perfect gas is equivalent to saying that each molecule of the gas behaves, on the average, as though no other molecule were present. There are, of course, collisions between molecules, but they are relatively rare and, between collisions, the only energy the molecule possesses is its kinetic energy of translation and its molecular energy of rotation, vibration, etc. If this is the case then in a mixture, each gas will behave as though the other gas were not present; each will have its own energy, its own partial pressure, and its own entropy, unchanged by the presence of the other gases. In actual gases, of course, there would be a small change in each, owing to the presence of the other gas, but for gases like helium and nitrogen at room temperatures this effect would be small.

For a perfect gas mixture, however, the entropy of the mixture would be the sum of the two entropies separately. The internal energies have not been changed by the mixing, so we integrate $(\partial S / \partial V)_U$ to find the change in individual entropies caused by the volume expansion. But, since $dU = T\,dS - P\,dV$, we have $(\partial S / \partial V)_U = (P/T)$, which equals (nR/V) for an ideal gas. Therefore the increase in entropy of the first gas is

$$\alpha n R \int_{\alpha V_0}^{V_0} dV/V = \alpha n R \ln(V_0/\alpha V_0) = \alpha n R \ln(1/\alpha)$$

and the total entropy increase for both gases upon mixing, which is called the *entropy of mixing*, is

$$\Delta S = nR\{\alpha \ln(1/\alpha) + (1 - \alpha) \ln[1/(1 - \alpha)]\} \qquad (6\text{-}15)$$

which is positive for $0 < \alpha < 1$. It is largest for $\alpha = \frac{1}{2}$, when equal mole quantities of the two gases are mixed. We note that for perfect gases, P and T of the final state are the same as those of the initial state; for nonperfect gases P and T change somewhat during the mixing (why?).

Entropy increase is to be expected when two *different* gases are mixed. But what if the two gases are the same? Does the removal of a diaphragm separating two parts of a volume V, filled with one sort of gas, change its entropy or not? When the diaphragm is in place a molecule on one side of it is more restricted in its travel than when the diaphragm is removed; but this difference is unnoticeable macroscopically. Does reinsertion of the diaphragm reduce the entropy again? We must postpone the resolution of this paradox (called Gibbs' paradox) until we treat statistical mechanics (page 312).

The Third Law of Thermodynamics

A plot on the TS plane is a useful way of visualizing the processes occurring in a Carnot cycle. If the cycle involves the

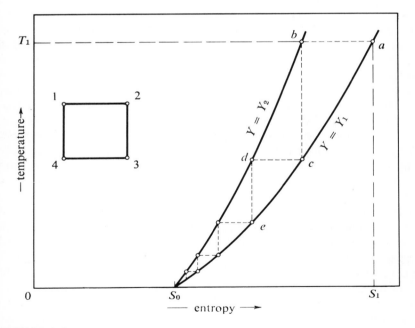

FIGURE **6-4**

Isothermal and isentropic processes plotted on the TS plane.

extensive variable X and the intensive variable Y, as well as T and S, we can use T and S as independent variables and let X and Y be dependent variables. A Carnot cycle is a rectangle on the TS plane, with the horizontal lines, such as 1-2 and 3-4 in Figure 6-4, representing the isothermal processes and the vertical lines 2-3 and 4-1 representing the reversible adiabatic processes. The heat transformed into work during the cycle is then the area of the rectangle. Incidentally, a process in which dQ is zero is adiabatic, as defined in Chapter 4; a process in which dS is zero is, strictly speaking, *isentropic*. Since $dS = (dQ/T)$ only for a reversible process, adiabatic processes are isentropic only when they are reversible.

The TS plane is also a useful one for visualizing what happens when we try to lower the temperature of a body to absolute zero, by manipulating one of the mechanical variables (such as Y) to produce alternate adiabatic and isothermal processes. Suppose we start at a, of Figure 6-4, with the intensive variable having the value Y_1 and the temperature at T_1. We place the body in contact with a reservoir of heat at temperature T_1 and change Y quasistatically from Y_1 to Y_2, in a direction to drive heat into the reservoir and thus reduce the body's entropy, reaching point b on the diagram. We next insulate the body thermally and change Y reversibly from Y_2 to Y_1 again. This adiabatic process, represented by the dotted line $b–c$, reduces the temperature, if the isothermal process $a–b$ reduced the entropy. To prove this we use Eq. (3-9) to show that the positive quantity $(\partial S/\partial T)_Y = (C_Y/T) = - [(\partial Y/\partial T)_S/(\partial Y/\partial S)_T]$ or $(\partial T/\partial Y)_S = - (T/C_Y)(\partial S/\partial Y)_T$ so if decrease of Y decreased S, increase of Y will decrease T (or vice versa) (see also page 68).

Next we go from c to d by another isothermal process, changing Y from Y_1 to Y_2 again and still further reducing the body's entropy. Then we use another adiabatic process, $d–e$, and so on, changing Y from Y_1 to Y_2 and back again, reducing entropy and temperature at each double step. The question arises as to whether we can reach absolute zero in a finite number of steps of this sort. The answer is another general negative, of the sort which cannot be verified by a simple, crucial experiment. It has been found in practice, in all cases so far tried, that the closer we get to absolute zero the harder it is to get further. As was said earlier, it seems to be as hard to go from 1°K to 0.1°K as it is to go from 10°K to 1°K, or from 100°K to 10°K. Heat capacities tend to go to zero as T goes to zero. This might sound as though it would be easier to lower temperatures at low temperatures; it takes less heat withdrawal to reduce the temperature a given amount. But the rapidly increasing difficulties in withdrawing heat, as the temperature is lowered, more than cancels the advantage of the lower heat capacity. The physical variables we have at our control at these low temperatures, such as pressure and the magnetic field, all seem to lose their effectiveness as T approaches zero.

Consequently Nernst, and later Fowler, were led to enunciate a third law of thermodynamics, one form of which is the statement that the temperature cannot be reduced to zero Kelvin in a finite number of operations, reversible or not. A glance at Figure 6-4 indicates that this means that the curves of T against S for $Y = Y_1$ and for $Y = Y_2$, between which the temperature-reducing steps were taken, get closer and closer together as they approach the S axis, until they meet at $T = 0$. In this way the successive steps get smaller and smaller as they approach $T = 0$ and, in the limit, become infinitesimal in size, so that an infinite number of steps are required to reach $T = 0$. Compare the general curves of Figure 6-3 with those of Figure 20-3 for the entropy of magnetization.

We thus see that another way of stating the third law is that the entropy change produced by a reversible, isothermal process approaches zero as the thermodynamic temperature approaches zero. At absolute zero entropy is independent of most of the thermodynamic variables. Therefore the heat capacity, $(\partial S/\partial T)$ times T, goes to zero at $T = 0$. When we come to study statistical mechanics we shall find reasons, rooted in atomic structure and quantum theory, which explain this behavior. Entropy is a measure of disorganization. Disorganization is a minimum at absolute zero; in fact for perfect crystals there is complete order at absolute zero; entropy would be zero there. The third law is not difficult to understand from this point of view.

CHAPTER

7

SIMPLE
THERMODYNAMIC
SYSTEMS

We have now assembled the theoretical machinery of thermodynamics; we next must get acquainted with its procedures and then
must survey some of its achievements in predicting and correlating
the thermal properties of matter. In this chapter we shall work out
the properties of a few simple systems, to gain facility with the
ideas and equations when used in less complex situations. In the
following chapter we shall return to the general theory, to put it
in its more general form, applicable to more complex systems.

The Joule Experiment

At first we return to the familiar behavior of a homogeneous
gas. In the last chapter we showed that, for an ideal gas, the internal
energy is independent of V. If the energy is entirely kinetic energy
of motion of the molecules, then it does not matter whether they
are close together or far apart, their energy will be the same; only
if there are forces of appreciable size between the molecules will
the internal energy depend on the mean distances between molecules,
i.e., on V.

How nearly do actual gases come to this limit of independence
of U on V? Joule first suggested an experiment to answer this
question regarding the value of $(\partial U/\partial V)_T$ for an actual gas. Since,
by Eqs. (3-9),

$$\left(\frac{\partial T}{\partial V}\right)_U = -\frac{(\partial U/\partial V)_T}{(\partial U/\partial T)_V}, \quad \text{then} \quad \left(\frac{\partial U}{\partial V}\right)_T = -C_v\left(\frac{\partial T}{\partial V}\right)_U$$

$$(7\text{-}1)$$

Therefore a measurement of the change of temperature of a gas with volume, at constant U, can measure the change in U with volume at constant temperature, which is our measure of a perfect gas.

To keep U constant during a process we can isolate it completely, can keep both dQ and dW zero during the process. Joule's way of doing this is shown in Figure 6-2. A gas, confined at temperature T_0 in a volume V_0, is allowed to expand freely into an evacuated volume $V_1 - V_0$ so that, when equilibrium is again established, the gas occupies volume V_1. If the system is insulated thermally, no heat leaves or enters, nor is work done on or by the system. Therefore no energy is lost or gained and, when the system again settles down, it still must have its original value U of internal energy.

The spontaneous expansion of the gas is, of course, an irreversible process, which we cannot follow in detail by thermodynamic calculations. Nevertheless the temperature T is a state variable and its change between the initial and the final equilibrium states can be computed by integrating the rate of change of T with V at constant U. As Eq. (7-1) shows, if U is independent of V, as it is for perfect gases, then the free expansion will produce no change of temperature in the gas. Careful measurement shows that the temperature of actual gases does drop by a small amount during a free expansion. For monatomic gases the quantity $n(\partial T/\partial V)_U$ (which is called the *Joule coefficient*) is less than 0.001 °K moles per m^3.

Joule's coefficient may be computed, for gases for which we know the heat capacity and equation of state, by using Eq. (6-7) to calculate $(\partial U/\partial V)_T$ and then using Eq. (7-1) to compute $(\partial T/\partial V)_U$.

For a gas satisfying the van der Waals Eq. (3-4), having a heat capacity C_v that is independent of V,

$$P - \left(T\frac{\partial P}{\partial T}\right)_v = -\frac{an^2}{V^2}$$

so

$$\left(\frac{\partial T}{\partial V}\right)_U = \frac{-an^2}{V^2 C_v(T)}$$

which is small because a is small for most gases. The temperature change will thus be small, so that we can consider C_v constant over

the limited range of temperature involved; in which case the total
temperature change caused by free expansion is

$$T_1 - T_0 \simeq - \frac{an^2}{C_v}\left(\frac{1}{V_0} - \frac{1}{V_1}\right)$$

This gives the small change in T during the expansion at constant
U for a van der Waals gas. Since $V_1 > V_0$ the temperature drops
during the process, although the drop is small. During the expansion a small amount of the molecular kinetic energy must be lost
in doing work against the small attractive forces between the
molecules. A better approximation for $(\partial U/\partial V)_T$ could be obtained
by using the virial Eq. (3-5).

The Joule–Thomson Experiment

It is difficult to obtain an accurate value for $(\partial U/\partial V)_T$ for a
gas by means of Joule's experiment. The temperature drop is very
small and it is hard to maintain the degree of thermal isolation,
during free expansion, which is necessary. The heat capacity of the
vessel walls is usually larger than that of the gas. A continuous
process, less susceptible to small heat loss and more amenable to
accurate measurement, was then devised by Joule and Thomson.
The basic plan of the experiment is shown in Figure 7-1. Gas is
forced through a nozzle N by moving pistons A and B to maintain
a constant pressure difference across the nozzle (or we can use
pumps working at the proper rates). The gas on the high-pressure
side is at pressure P_0 and temperature T_0; after going through the
nozzle it settles down to a pressure P_1 and temperature T_1. Suppose
we follow n moles of the gas as it goes through the nozzle, starting
from a state of equilibrium 0 at P_0, T_0 and ending at the state of
equilibrium 1 at P_1, T_1. We cannot follow the change in detail,
for the process is irreversible, but we can devise a reversible path
between 0 and 1 [or, rather, we can let Eq. (6-3) find a reversible
path for us] which will allow us to compute the difference between
states 0 and 1. In particular, we can compute the temperature
difference $T_1 - T_0$ in order to compare it with the measured
difference.

This process differs from free expansion because the energy
U does not stay constant; work P_0V_0 is done by piston A in pushing
the gas through the nozzle, and work P_1V_1 is done on piston B by
the time the n moles have all gotten through. Thus the net difference
in internal energy of the n moles between state 0 and state 1 is

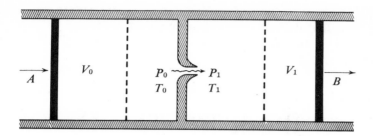

FIGURE 7-1

The Joule–Thomson experiment.

$U_1 - U_0 = P_0 V_0 - P_1 V_1$. Instead of U remaining constant during the process the quantity

$$H = U + PV \qquad\qquad (7\text{-}2)$$

is the same in states 1 and 0. This quantity, called the *enthalpy* of the gas, is an extensive state variable.

The experiment of measuring the difference $T_1 - T_0$, when all parts of the system shown in Figure 7-1 are thermally insulated, is called the Joule–Thomson experiment, and the relevant change in temperature with pressure when H is kept constant, $(\partial T/\partial P)_H$, is the *Joule–Thomson coefficient* of the gas. Experimentally, we find that for actual gases, the temperature increases slightly at high temperatures; at low temperatures the temperature T_1 is a little less than T_0. The temperature at which $(\partial T/\partial P)_H = 0$ is called the Joule–Thomson *inversion point*. Continuous processes, in which a gas, at a temperature below its inversion point, is run through a nozzle to lower its temperature, are used commercially to attain low temperature. We use work $P_0 V_0 - P_1 V_1$ to cool the n moles of gas, so Clausius' principle is not contradicted.

To compute the Joule–Thomson coefficient we manipulate the equation for H, or rather its differential,

$$dH = dU + P\, dV + V\, dP = T\, dS + V\, dP \qquad\qquad (7\text{-}3)$$

as we did with Eq. (6-3) to obtain Eq. (6-7). First we note that the change in heat $T\, dS$ in a system at constant pressure equals dH, so that C_P, the heat capacity of the system at constant pressure, is $(\partial H/\partial T)_P$, in contrast to Eq. (4-4). Just as internal energy U can be called the heat content of a system at constant volume, so enthalpy H can be called its heat content at constant pressure.

Next we manipulate Eq. (7-3) as we did $dU = T\, dS - P\, dV$, to obtain

$$\left(\frac{\partial S}{\partial T}\right)_P = \frac{1}{T}\left(\frac{\partial H}{\partial T}\right)_P = \frac{C_P}{T};$$

$$\left(\frac{\partial S}{\partial P}\right)_T = \frac{1}{T}\left[\left(\frac{\partial H}{\partial P}\right)_T - V\right] = -\left(\frac{\partial V}{\partial T}\right)_P \tag{7-4}$$

which are analogous to Eqs. (6-6), and

$$(\partial C_P/\partial P)_T = -T(\partial^2 V/\partial T^2)_P$$

From this we can compute the Joule–Thomson coefficient and the change in entropy during the process,

$$\left(\frac{\partial T}{\partial P}\right)_H = -\frac{(\partial H/\partial P)_T}{(\partial H/\partial T)_P} = \frac{-1}{C_P}\left[V - T\left(\frac{\partial V}{\partial T}\right)_P\right]$$

$$\text{and} \qquad \left(\frac{\partial S}{\partial P}\right)_H = -\frac{V}{T} \tag{7-5}$$

For a perfect gas $V = T(\partial V/\partial T)_P$, so that $(\partial H/\partial P)_T = 0$ and the Joule–Thomson coefficient $(\partial T/\partial P)_H$ is also zero; no change in temperature is produced by pushing it through a nozzle. The change in entropy of a perfect gas during the process is the integral of $-V/T = -nR/P$ with respect to P,

$$\Delta S = nR\ln(P_0/P_1) \tag{7-6}$$

Since $P_0 > P_1$ this represents an increase in entropy, as it must.

For a gas obeying van der Waals equation we can find $(\partial V/\partial T)_P$ by differentiating the equation of state and manipulating,

$$(V - nb)\,dP + \left[\frac{nRT}{V - nb} - \frac{2an^2}{V^3}(V - nb)\right]dV = nR\,dT$$

$$T\left(\frac{\partial V}{\partial T}\right)_P = (V - nb)\left[1 - \frac{2an}{V^3}\frac{(V - nb)^2}{RT}\right]^{-1}$$

$$\simeq (V - nb)\left(1 + \frac{2an}{RTV}\right)$$

so that

$$\left(\frac{\partial T}{\partial P}\right)_H \simeq \frac{1}{C_P}\left(\frac{2an}{RT} - nb\right)$$

since a and b are small quantities. For this same reason T and thus C_P do not change much during the process, and we can write

$$T_1 - T_0 \simeq \frac{nb}{C_P}\left(1 - \frac{2a}{RT_0 b}\right)(P_0 - P_1)$$

Since $P_0 > P_1$ this predicts an increase in temperature during the Joule–Thomson process if $T_0 > 2a/Rb$, a decrease if $T_0 < 2a/Rb$, the inversion temperature being approximately equal to $2a/Rb$. The approximation is not very good, since the van der Waals formula is not a very accurate equation of state for actual gases.

The corresponding increase in entropy during the process can be computed, using the same approximations as before,

$$\Delta S \simeq nR \ln\left(\frac{P_0}{P_1}\right) - \frac{nb}{T_0}(P_0 - P_1)\left(1 - \frac{a}{RT_0 b}\right) \qquad (7\text{-}7)$$

which is to be compared with the result of Eq. (7-6) for a perfect gas.

Black-Body Radiation

Let us now turn to a quite different sort of system, that called black-body radiation, electromagnetic radiation in equilibrium with the walls of an enclosure kept at temperature T. This is the radiation one would find inside a furnace with constant-temperature walls. It consists of radiation of all frequencies and going in all directions, some of it continually being absorbed by the furnace walls but an equal amount continually being generated by the vibrations of the atoms in the walls. This is a special kind of system, with special properties.

In the first place the energy *density* e of the radiation, the mean value of $\frac{1}{2}\mathfrak{E} \cdot \mathfrak{D} + \frac{1}{2}\mathfrak{H} \cdot \mathfrak{B}$, depends on the temperature but is independent of the volume of the enclosure. If the volume inside the furnace is enlarged, more radiation is generated by the walls, so that the energy density e remains the same. Therefore the total electromagnetic energy within the enclosure, $Ve(T)$, is proportional to the volume. This is in contrast to a perfect gas, for which U at a given temperature is constant, independent of volume. If the gas enclosure is increased in volume the density of atoms diminishes and so does the energy density; for radiation, if the volume is increased more radiation is produced to keep the density constant. We can, of course, consider the radiation to be a gas of photons, each with its own energy, but the contrast with atoms remains.

Extra photons can be created to fill up any added space; atoms are harder to create.

In the second place the radiation pressure, the force exerted per unit area on the container walls, is proportional to the energy density. In this respect it is similar to a perfect gas of point atoms, but the proportionality constant is different. For the gas [see Eq. (2-4)] $P = \frac{2}{3}[N\langle \text{K.E.}\rangle_{\text{tran}}/V] = \frac{2}{3}e$, where e is the energy contained per unit volume; for radiation it is $\frac{1}{3}e$. The difference comes from the fact that the kinetic energy of an atom, $\frac{1}{2}mv^2$, is *one-half* its momentum times its velocity. For a photon, if its energy is $\hbar\omega$, its momentum is $\hbar\omega/c$, where c is its velocity, $\omega/2\pi$ its frequency, and $h = 2\pi\hbar$ is Planck's constant; therefore the energy of a photon is the product of its velocity and its momentum, not half this product. Since pressure is proportional to momentum, $e = 3P$ for the photon, $= \frac{3}{2}P$ for the atom gas.

To compute U, S, and P as functions of T and V we start with the basic equation again, $dU = T\,dS - P\,dV$, inserting the appropriate expressions, $U = Ve(T)$ and $P = \frac{1}{3}e(T)$,

$$T\,dS = T\left(\frac{\partial S}{\partial T}\right)_V dT + T\left(\frac{\partial S}{\partial V}\right)_T dV = dU + P\,dV$$

$$= V\left(\frac{de}{dT}\right)dT + e\,dV + \tfrac{1}{3}e\,dV$$

or

$$\left(\frac{\partial S}{\partial T}\right)_V = \frac{V}{T}\left(\frac{de}{dT}\right); \qquad \left(\frac{\partial S}{\partial V}\right)_T = \frac{4}{3}\frac{e}{T}$$

Applying Eq. (3-10) we obtain a differential equation for $e(T)$.

$$\frac{1}{T}\left(\frac{de}{dT}\right) = \frac{4}{3}\frac{1}{T}\left(\frac{de}{dT}\right) - \frac{4}{3}\frac{e}{T^2} \qquad \text{or} \qquad \frac{de}{dT} = 4\frac{e}{T}$$

which has a solution $e(T) = aT^4$, so

$$U = aVT^4; \qquad P = \tfrac{1}{3}aT^4; \qquad S = \tfrac{4}{3}aVT^3 \qquad (7\text{-}8)$$

The equation for the energy density of black-body radiation is called *Stefan's law* and the constant a is Stefan's constant. In statistical mechanics [see Eq. (25-9)] we shall evaluate it in terms of atomic constants.

We see that the energy of black-body radiation goes up very rapidly with increase in temperature. Room temperature (70°F) is about 300°K. At the temperature of boiling water (373°K) the energy density of radiation is already $2\frac{1}{2}$ times greater; at dull red heat (920°K) it is 100 times that at room temperature. At the temperatures encountered on earth the pressure of radiation is minute compared to usual gas pressures. At temperatures of the center of the sun (10^7°K) the radiation pressure supports more than half the mass above it.

Reference to Eq. (6-3), $U = ST - PV + \mu n$ shows that the chemical potential of black-body radiation is

$$\mu = (1/n)(U - ST + PV)$$
$$= (1/n)(aVT^4 - \tfrac{4}{3}aVT^4 + \tfrac{1}{3}aVT^4) = 0$$

which is related to the freedom with which the number of moles of photons adjusts its value to keep the energy density constant (see page 81). In this matter, also, photons are a special kind of gas.

Paramagnetic Gas

To illustrate the additional complications which arise when systems of three independent variables are analyzed, let us work out the behavior of a paramagnetic, perfect gas. Here the two equations of state are $P = nRT/V$ and $\mathfrak{H} = T\mathfrak{M}/nD$. The heat capacity at constant V and $\mathfrak{M}$ is a constant. $C_{v\mathfrak{M}} = \tfrac{3}{2}nR$ for a monatomic gas. The basic equation is

$$dU = T\,dS - P\,dV + \mathfrak{H}\,d\mathfrak{M}$$

Three independent variables must be used; we first use T, V, and $\mathfrak{M}$. By methods that should be familiar by now we find

$$T\left(\frac{\partial S}{\partial T}\right)_{V\mathfrak{M}} = \tfrac{3}{2}nR = \left(\frac{\partial U}{\partial T}\right)_{V\mathfrak{M}}$$
$$T\left(\frac{\partial S}{\partial V}\right)_{T\mathfrak{M}} = P + \left(\frac{\partial U}{\partial V}\right)_{T\mathfrak{M}} = T\left(\frac{\partial P}{\partial T}\right)_{V} \qquad (7\text{-}9)$$
$$T\left(\frac{\partial S}{\partial \mathfrak{M}}\right)_{TV} = \left(\frac{\partial U}{\partial \mathfrak{M}}\right)_{TV} - \mathfrak{H} = -T\left(\frac{\partial \mathfrak{H}}{\partial T}\right)_{\mathfrak{M}}$$

Other manipulations result in

$$(\partial U/\partial V)_{T\mathfrak{M}} = (\partial U/\partial \mathfrak{M})_{TV} = 0$$
$$C_{P\mathfrak{H}} = C_{V\mathfrak{M}} + P\left(\frac{\partial V}{\partial T}\right)_{P} - \mathfrak{H}\left(\frac{\partial \mathfrak{M}}{\partial T}\right)_{\mathfrak{H}} = \tfrac{5}{2}nR + (\mathfrak{M}^2/nD)$$

Integrating the partials for U and S, we obtain

$$U = \tfrac{3}{2}nRT; \qquad S = nR \ln\left[\frac{V}{V_0}\left(\frac{T}{T_0}\right)^{3/2}\right] - \frac{\mathfrak{M}^2}{2nD} + S_0 \qquad (7\text{-}10)$$

The "natural variables" in terms of which to express U are S, V, $\mathfrak{M}$ rather than T, V, $\mathfrak{M}$. To do this we express T in terms of S, V, $\mathfrak{M}$ and obtain

$$U = nRT_0\left(\frac{V_0}{V}\right)^{2/3} \exp\left(\frac{2}{3}\frac{S - S_0}{nR} + \frac{\mathfrak{M}^2}{3n^2RD}\right) \qquad (7\text{-}11)$$

[see Eq. (6-12)].

The extensive variables S, V, $\mathfrak{M}$ are less easy to measure than are the intensive variables.

$$P = -(\partial U/\partial V)_{S\mathfrak{M}}; \qquad T = (\partial U/\partial S)_{V\mathfrak{M}}; \qquad \mathfrak{H} = (\partial U/\partial \mathfrak{M})_{SV}$$

which might be called the "experimental variables." Expressing the basic equation in terms of these (using $P\,dV = -(nRT/P)\,dP + nR\,dT$, for example) we have

$$T\,dS = \left(\tfrac{5}{2}nR + nD\frac{\mathfrak{H}^2}{T^2}\right)dT - \frac{nRT}{P}\,dP - \frac{nD\,\mathfrak{H}}{T}\,d\mathfrak{H} \qquad (7\text{-}12)$$

For isothermal operation, dT is zero. The heat contributed to the gas by the reservoir at temperature T, when the gas pressure is changed from P_0 to P_1 and the magnetic intensity is changed from $\mathfrak{H}_0$ to $\mathfrak{H}_1$, is

$$\Delta Q_{01} = nRT \ln(P_0/P_1) + (nD/2T)(\mathfrak{H}_0^2 - \mathfrak{H}_1^2) \qquad (7\text{-}13)$$

Increase in pressure squeezes out heat ($\Delta Q < 0$) as does an increase in magnetization. At low temperatures a change in magnetic field produces more heat than does a change of pressure.

The behavior of the system during an adiabatic process can be computed by setting $T\,dS = 0$. The integrating factor is $1/T$ and the integral of

$$n\left(\frac{5}{2}\frac{R}{T} + D\frac{\mathfrak{H}^2}{T^3}\right)dT - \frac{nD\mathfrak{H}}{T^2}\,d\mathfrak{H} - \frac{nR}{P}\,dP = 0$$

is

$$\left(\frac{P_0}{P_1}\right)\left(\frac{T_1}{T_0}\right)^{5/2} \exp\left[\frac{D}{2R}\left(\frac{\mathfrak{H}_0^2}{T_0^2} - \frac{\mathfrak{H}_1^2}{T_1^2}\right)\right] = 1 \qquad (7\text{-}14)$$

If the magnetic field is kept constant ($\mathfrak{H}_1 = \mathfrak{H}_0$), the gas undergoes ordinary adiabatic compression and T_1 is proportional to the two-fifths power of the pressure P_1. Or, if the pressure is kept constant, the temperature is related to the magnetic field by the formula

$$\mathfrak{H}_1^2 = T_1^2[(\mathfrak{H}_0^2/T_0^2) + (5R/D)\ln(T_1/T_0)] \tag{7-15}$$

At low-enough initial temperatures or high-enough initial fields, so that $(\mathfrak{H}_0/T_0)^2 \gg 5R/D$, the final temperature T_1 is approximately proportional to the final magnetic intensity $\mathfrak{H}_1$; an adiabatic reduction of $\mathfrak{H}$ proportionally lowers the temperature.

Finally, we can have a process which is *both* adiabatic and isothermal if we adjust the pressure continually so that, as the magnetic field is changed adiabatically, the temperature is kept at a constant value T_0 (the volume then changes inversely proportional to P). The relation between $\mathfrak{H}_1$ and P_1 that will keep T constant is

$$P_1 = P_0 \exp\left[\frac{D}{2RT_0^2}(\mathfrak{H}_0^2 - \mathfrak{H}_1^2)\right] \tag{7-16}$$

As $\mathfrak{H}_1$ decreases, P_1 will have to be increased exponentially to keep T constant. In this process, mechanical work is used to demagnetize the material.

Actual paramagnetic gases are not perfect gases, nor does their magnetic equation of state have exactly the form of the simple Curie law. In these cases one tries to find equations of state that do fit the data and then tries to integrate Eq. (7-12). Failing this, numerical integration must be used to predict the thermal properties of the material. A more accurate set of formulas, for low temperatures, is given in Eqs. (8-26) to (8-29).

Properties of Helium II

Finally, we can demonstrate how thermodynamics can correlate the peculiar properties of a very nonclassical substance, once the basic thermal model for the substance has been assumed. For the laws of thermodynamics apply to peculiar substances like liquid helium as well as to normal liquids, once we figure out how to apply them. Helium is the only substance which does not solidify at atmospheric pressure. Keeping the pressure at 1 atmosphere, the gas liquifies at about $4°$ absolute but, as the temperature is further reduced, instead of solidifying it turns, below about $2°K$, into a different fluid, called helium II, which has most peculiar properties. It can be solidified by compressing to about 20 atmospheres, but at 1 atmosphere it seems to stay fluid clear down to absolute zero. The behavior is plotted in Figure 9-7.

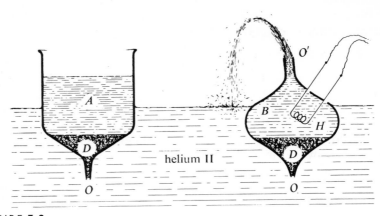

FIGURE 7-2

Apparatus for demonstrating mechanocaloric and thermomechanical, or fountain effect.

Some of the experimentally observed properties of helium II are the following. It conducts heat to an extraordinary degree, so well that the local "hot-spots" which accompany boiling in an ordinary liquid cannot occur; helium II never bubbles even when it is evaporating rapidly. In fact heat moves so rapidly that it seems to display properties of inertia, as though it really were a fluid; wave motion of heat can be set up in helium II. Also the liquid has a very low value of viscosity; part of the fluid seems to have no viscosity at all. Some of the fluid can glide through fine capillary tubing with no measurable frictional loss and, in the form of a thin film, this nonviscous portion can creep up the inside surface of any open vessel which holds it, and then down the outside surface to drip away below. And when this very slippery portion leaves the rest of the fluid it leaves its heat behind, for the residue warms up as it diminishes in volume. Only a part of the fluid is this slippery; the liquid as a whole has a finite, though small, viscosity.

Figure 7-2 illustrates how some of the peculiar effects can be demonstrated. A vessel *A* is made with a very fine orifice at its bottom, and this is made still more difficult for liquid to penetrate by covering the bottom with a layer of very finely divided powder *D*. The vessel is lowered into a bath of helium II until the level of liquid inside is the same as that outside, and the whole is in thermal equilibrium. Then vessel *A* is raised as shown, so the liquid level inside is higher than that outside. If there were no hole in the bottom, the thin film would creep up over the top edge of A and down the side, so the levels would gradually equalize. But when the capillary orifice is open the level in A will quite rapidly drop back to the outside level and, while it is dropping, the temperature of the liquid in A will perceptibly rise. On the other hand, if the vessel is lowered, so the level inside is below the outside level, the "superfluid" portion will slide in from outside and, while this is going on, the liquid in A becomes colder. Flow of the superfluid away from A raises the temperature of the remainder, addition of more superfluid lowers the temperature of the whole. This coupling of relative proportions of normal and superfluid with the temperature of the mix is called the *mechanocaloric effect*.

Vessel B of Figure 7-2 has a capillary orifice below and a wider tube above, capable of passing both fluids. Again the system is allowed to come to equilibrium. This time we introduce heat inside B, either by a resistor coil H or by shining light on the powder D, heating it somewhat. The vessel rapidly fills with liquid, sucked in through O and, if the dimensions are right, so much comes in that it squirts out the top orifice O', producing a fountain of helium II. This is, of course, called the fountain effect; a more bookish title is the *thermomechanical effect*. It is the converse of the mechanocaloric effect.

The Two-Fluid Model for Helium II

These various effects, and others we shall discuss shortly, may be correlated quantitatively if we assume that helium II is a mixture of two fluids. One, called the normal fluid, is a usual sort of liquid with viscosity, heat capacity and other usual properties; the other, called the *superfluid*, has no viscosity and *no entropy*. Both are helium, and a sample of helium II in bulk always is a mixture of the two. In fact the proportions of normal to superfluid are determined by the temperature; as the temperature is raised the proportion of superfluid reduces from 100 per cent at absolute zero to zero at about 4°K, when the liquid becomes helium I, with no unusual properties. This means that as heat is added, superfluid is changed to normal fluid. If s_n is the entropy per unit mass of normal fluid, it requires Ts_n units of heat to convert a gram of superfluid into a gram of normal fluid at temperature T. Thus, if ρ_n is the density of the normal fluid present and ρ_s the density of the superfluid, so that $\rho_n + \rho_s = \rho$ is the density of the fluid as a whole, then an amount dQ of heat will convert a volume dV of superfluid into dV of normal fluid, where

$$dQ = \rho_n T \dot{s}_n \, dV \qquad\qquad (7\text{-}17)$$

Alternatively, if we introduce an extra amount dV of the super-fluid into the vessel, the temperature will *reduce* by an amount $dT = (\rho_n T s_n / C) \, dV$, where C is the heat capacity of the whole fluid.

Now suppose we equalize the levels inside and outside of vessel A, and then insert some heat, at a controlled rate, by running current through a coil placed in A. Figure 7-3 shows what happens. If the rate of addition of heat is less than R_0, the flow of super-fluid into A, to be converted into normal fluid by the added heat, is sufficient to keep the temperature from *rising at all*. The rate of flow just takes up the heat emitted and there is none left over to warm up the mixture; the ratio of normal to superfluid remains constant. But there evidently is a limit to the free flow of the super-fluid; above a certain rate of heating R_0 the flow through the orifice is so great that turbulence (or something) interposes a drag. The temperature begins to rise and the rate of flow is no longer proportional to the rate of heating.

Knowing the heat generated and the rate of inflow of the superfluid (measured by the rate of rise of level of the liquid in A)

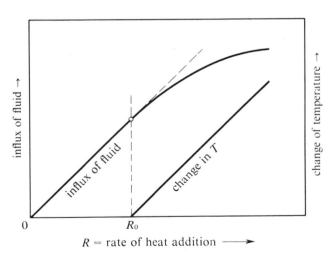

FIGURE 7-3

Rate of influx of superfluid and temperature change in response to addition of heat.

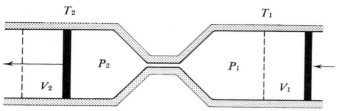

FIGURE 7-4

Reversible flow of superfluid from regions of differing temperatures and pressures.

we can calculate s_n, the entropy difference between the normal and the superfluid, by Eq. (7-17). To demonstrate that the entropy of the superfluid really is zero, the heat capacity C_v of helium II can be measured in the usual way (using thermometers calibrated by magnetic effects) and the entropy S of the whole fluid can be computed by integrating $(C_v/T) = (\partial S/\partial T)_v$. This value of S per unit volume turns out to equal $\rho_n s_n$, as measured using Eq. (7-17), to within the experimental error of a few per cent. Thus the superfluid contributes no measurable entropy.

Alternatively, as the fountain effect shows, the introduction of heat inside vessel B causes a flow of superfluid inward, raising the pressure there. The relationship between temperature difference and pressure difference on the two sides of the capillary can be worked out by a variant of the analysis we used to discuss the Joule–Thomson experiment, as shown in Figure 7-4. When equilibrium is reached on the two sides of the capillary, with pressure and temperature P_1, T_1 on one side and P_2, T_2 on the other, we move both pistons, quasistatically, so some superfluid goes from one side to the other. Note that this process would not be reversible if the superfluid did not have zero entropy.

The difference $U_2 - U_1$ in internal energy of this fluid, before and after passage, must equal the difference $P_1V_1 - P_2V_2$ of work done by and on the two pistons, plus the difference $T_2S_2 - T_1S_1$ in the amount of heat absorbed from and given off to the walls in order to change the temperature of the fluid from T_1 to T_2. Thus $U_2 - U_1 = P_1V_1 - P_2V_2 + T_2S_2 - T_1S_1$ or, for this reversible motion of superfluid, the quantity

$$G = U + PV - TS \qquad\qquad (7\text{-}18)$$

remains constant. The state variable G is called the *Gibbs function*. We will discuss its properties and uses in the next two chapters.

If there is only a small difference of pressure and temperature between the two sides of the capillary, the requirement that the Gibbs function does not change during the transfer of fluid is equivalent to saying that

$$dG = dU + P\,dV + V\,dP - T\,dS - S\,dT$$

$$= V\,dP - S\,dT = 0$$

Writing $S = \rho_n s_n V$ in terms of the entropy s_n per unit mass of normal fluid and ρ_n the density of the normal fluid into which the superfluid has been changed by the increase of temperature, we finally arrive at an equation relating the difference in pressure with the difference in temperature on the two sides,

$$dP = \rho_n s_n\,dT \qquad \text{or} \qquad (dP/dT) = \rho_n s_n \qquad\qquad (7\text{-}19)$$

which is the equation governing the fountain effect. Thus measurement of pressure versus temperature change provides still another method of measuring the difference in entropy between normal and superfluid. The two sets of measurements check over the available range of temperatures, thus providing another verification of the two-fluid model.

Wave Propagation of Pressure and of Temperature

We shall have to postpone until Chapters 25 and 27 showing how quantum statistical mechanics is able to predict the possibility that helium might be a two-fluid mixture at low temperatures, as was first pointed out by Tisza. The experiments just discussed, and many others, have verified that helium II behaves as though it consists of a mixture of normal and superfluid of densities ρ_n and ρ_s, respectively, the total density $\rho = \rho_n + \rho_s$ being the sum of the two and the proportion (ρ_s/ρ_n) being determined by (or, if you like, determining) the temperature. The superfluid can flow through the normal fluid without friction, so that the two can have different velocities. We need to take this fact into account when we come to consider wave motion in helium II.

If we move a piston, in contact with helium II, back and forth the two fluids will move in phase and an ordinary sound wave will be produced, with its accompanying pressure fluctuations. On the other hand, if a thin, electrically conductive film has its temperature varied in an oscillatory manner by sending a sequence of current pulses through it, then the two fluids in the helium II, in contact with it, will be out of phase, the superfluid going toward the film when it is hotter, being repelled when it is colder, and the normal fluid going in the opposite direction. No net change in density results, only a change in relative proportions of the two fluids; the resulting wave motion being one of temperature, not pressure. In such a wave both fluids move, in contrast to the fountain effect, where the normal fluid cannot pass through the lower capillary tube.

Referring to Eq. (4-13) for the equation of motion of an ordinary fluid, we see that the velocities $\mathbf{u}_n$ and $\mathbf{u}_s$ of the two fluids are related to the pressure and temperature gradients in the liquid by the following equations:

$$\begin{aligned}
\rho_n(\partial \mathbf{u}_n/\partial t) &= -(\rho_n/\rho)\,\text{grad}\,P - (\rho_n \rho_s s_n/\rho)\,\text{grad}\,T \\
\rho_s(\partial \mathbf{u}_s/\partial t) &= -(\rho_s/\rho)\,\text{grad}\,P + (\rho_n \rho_s s_n/\rho)\,\text{grad}\,T
\end{aligned} \tag{7-20}$$

where the relationship between pressure and temperature gradient for the second equation comes from Eq. (7-19) and the opposite sign on the grad T term in the first equation represents the tendency of the normal fluid to move away from heat.

The equation of motion for ordinary sound is obtained by adding both equations, which cancels out the thermal term. In this case $\mathbf{u}_n = \mathbf{u}_s = \mathbf{u}$ and we obtain $\rho(\partial \mathbf{u}/\partial t) = -\,\text{grad}\,P$, which is the same as Eq. (4-13) for an ordinary sound wave. Combining this with the compressibility equation div $\mathbf{u} = -\kappa(\partial P/\partial t)$, we arrive at the usual wave Eq. (4-15), for pressure waves generated by the motion of a piston or diaphragm, for example.

For the thermal waves, we subtract $(1/\rho_s)$ times the second equation of (7-20) from $(1/\rho_n)$ times the first, obtaining

$$\frac{\partial}{\partial t}(\mathbf{u}_n - \mathbf{u}_s) = - s_n\,\text{grad}\,T$$

The divergence of $(\mathbf{u}_n - \mathbf{u}_s)$ is the combined rate of inflow of superfluid and outflow of normal fluid from a unit volume of the liquid. A change in proportional amounts of the two fluids will change the temperature, as indicated in the discussion following Eq. (7-17). Therefore we obtain, for the equivalent to the divergence-compressibility Eq. (4-14), a divergence-caloric equation

$$\text{div}(\mathbf{u}_n - \mathbf{u}_s) = - (\rho c/\rho_n s_n T)(\partial T/\partial t)$$

where c is the heat capacity per unit mass of helium II. The combination of these two equations results in a wave equation for temperature

$$\nabla^2 T = (\rho c / \rho_n s_n^2 T)(\partial^2 T / \partial t^2) + (1/s_n)(\partial T / \partial t)\frac{\partial}{\partial t}(\rho c / \rho_n s_n T) \quad (7\text{-}21)$$

The second term on the right-hand side, prescribing wave attenuation, is small and may be neglected to the first order. The first term indicates that the wave motion has a wave velocity equal to $[\rho_n s_n^2 T / \rho c]^{1/2}$.

These waves are often called *second sound*, although no sound is produced; no pressure fluctuation accompanies them and no density fluctuation. Temperature is doing the "waving," as though heat actually were an elastic fluid, not something which just diffuses. Of course the reason for the behavior is the curious coupling between temperature and relative concentration of superfluid, because of its lack of entropy.

It should be pointed out that these two-fluid effects *only* occur in helium, and then only for helium with He^4 nuclei, consisting of two protons and two neutrons. Substances other than helium solidify before the temperature gets low enough for the effects to occur. Tritium, with He^3 nuclei, consisting of two protons and one neutron, stays liquid clear down to absolute zero, it seems, but the liquid is normal in behavior, with no fountain effect or second sound, as far down in temperature as has been measured (well below 1 °K). The reasons for all these peculiarities will be discussed in Chapter 25; in Chapter 27 we shall indicate why the effects are peculiar to He^4. The correlation of the effects is a fairly simple task for thermodynamics, as we have just seen, once we justify the use of the two-fluid model. The justification of the model requires the full paraphernalia of quantum statistics, as we shall see later. In fact the full theoretical details of the model are not yet satisfactorily worked out.

CHAPTER

8

THE THERMODYNAMIC POTENTIALS

In the discussion of Eqs. (6-3) and (6-4) we pointed out that the internal energy U had many of the properties of a potential function, that if it were expressed as a function of the extensive variables S, V, $\mathfrak{M}$, n, etc., the various partial derivatives of U were equal to the intensive variables T, $-P$, $\mathfrak{H}$, μ, etc., which thus played a role analogous to components of forces. It is not usually convenient, however, to use all the extensive variables as the independent variables; many times the experimental results are determined in terms of T instead of S and P instead of V, for example. And if U is expressed in terms of these other variables, then its partials are no longer equal to the intensive variables.

It would thus be desirable to find other functions, related to U, which behave like potential functions when expressed in terms of other sets of independent variables. The procedure for obtaining them is quite simple. For a two-variable system, with $U(S, V)$ for example, we have $(\partial U/\partial S)_v = T$ and $(\partial U/\partial V)_s = -P$. Consider the function $U - TS = F$. Its differential is $dF = dU - T\,dS - S\,dT = -P\,dV - S\,dT$, so that if F is expressed as a function of T and V, its partials are $(\partial F/\partial V)_T = -P$ and $(\partial F/\partial T)_V = -S$, which are the related dependent variables. Thus the function F, when T and V are the independent variables, is a potential function.

There are a whole set of such potentials, one for each set of independent variables we might choose. Since we usually deal with one pair of mechanical variables (call them X and Y) at a time, together with the pairs S, T and n, μ, the most-used potentials separate into sets of five, each set for a different pair of mechanical variables. For example $U = TS + YX + \mu n$ [see Eq. (6-5)], the *internal energy*, is the potential for the variables S, X, n, since $dU = T\,dS + Y\,dX + \mu\,dn$. For the variables S, Y, n the function $H = U - YX = TS + \mu n$, called the *enthalpy*, is the potential; for the variables T, X, n the function $F = U - TS = YX + \mu n$, called the *Helmholtz function*, is the potential, since $dF = -S\,dT + Y\,dX + \mu\,dn$; and for T, Y, n the function $G = \mu n = U - TS - YX$, called the *Gibbs function*, is the potential. Finally, for T, X, μ to be the independent variables, we must use the function $\Omega = U - ST - \mu n = YX$, called the *grand potential*, since $d\Omega = -S\,dT + Y\,dX - n\,d\mu$. There are other combinations for the three variables, and still more if four variables are needed, but the others are not used enough to have earned special names.

When the mechanical pair X, Y is V, $-P$, as with a gas or liquid, the potentials have the names just given. If the pair is $\mathfrak{M}$ and $\mathfrak{H}$, instead of V and $-P$, the names are preceded by the adjective *magnetic*. For example, the magnetic Gibbs function is $G_m = U - TS - \mathfrak{H}\mathfrak{M}$, so that $dG_m = -S\,dT - \mathfrak{M}\,d\mathfrak{H} + \mu\,dn$, and so on. When more than one pair of mechanical variables are involved there is no uniform nomenclature.

All these potentials have a number of interesting properties and can be used in a number of different ways, which will be the subject of this chapter.

The Internal Energy

We start with the potential already familiar to us, the internal energy. Equations (4-1), (6-1), and (6-3) show that

$$dU = dQ - P\,dV + \mu\,dn + J\,dL + \mathfrak{H}\,d\mathfrak{M} + \cdots$$
$$\leq T\,dS - P\,dV + \mu\,dn + J\,dL + \mathfrak{H}\,d\mathfrak{M} + \cdots \qquad (8\text{-}1)$$

where the equality holds for reversible processes, the inequality for irreversible ones. Consequently if *all* the extensive variables are held constant ($dS = dV = dL = \cdots = 0$) while the system is allowed to come spontaneously to equilibrium, then every change in U will have to be a decrease in value; it will only stop changing spontaneously when no further decrease is possible; and thus at equilibrium U is minimal. In other words, when the extensive variables are fixed, the *equilibrium* state is the state of *minimal U*.

It is usually possible (although not always easy) to hold the mechanical extensive variables, V, L, n, $\mathfrak{M}$, etc., constant, but it is more difficult to hold entropy constant during a spontaneous process. However if we allow the process to take place adiabatically, with $dQ = 0$ (and hold $dV = dL = dn = \cdots = 0$), then U does not change during the process and $T\,dS \geq 0$, so that S will reach a *maximum* value at equilibrium. If we had tried to keep S constant during the spontaneous process we would have had to withdraw heat during the process, enough to cancel out the gain in S during the process; this would have decreased U, until at equilibrium U would be minimal, less than the original U by the amount of heat that had to be withdrawn to keep S constant.

When all the mechanical extensive variables are held constant, an addition of heat to the system produces a corresponding increase in U. We thus can call U the heat content of the system at constant V, L, n,..., and can write the heat capacity

$$C_{Vn\cdots} = (\partial U / \partial T)_{Vn\cdots} \tag{8-2}$$

the subscripts indicating that all the extensive mechanical variables, which apply to the system in question, are constant. Moreover we can apply Eq. (3-10) to the various intensive variables of Eq. (8-1), obtaining a very useful set of relationships,

$$\left(\frac{\partial T}{\partial V}\right)_{Sn\cdots} = -\left(\frac{\partial P}{\partial S}\right)_{Vn\cdots}; \quad \left(\frac{\partial T}{\partial n}\right)_{SV\cdots} = \left(\frac{\partial \mu}{\partial S}\right)_{Vn\cdots}$$

$$\left(\frac{\partial T}{\partial \mathfrak{M}}\right)_{SV\cdots} = \left(\frac{\partial \mathfrak{H}}{\partial S}\right)_{V\mathfrak{M}\cdots}; \quad -\left(\frac{\partial P}{\partial n}\right)_{SV\cdots} = \left(\frac{\partial \mu}{\partial V}\right)_{Sn\cdots}$$

$$-\left(\frac{\partial P}{\partial \mathfrak{M}}\right)_{SV\cdots} = \left(\frac{\partial \mathfrak{H}}{\partial V}\right)_{S\mathfrak{M}\cdots}; \quad \text{etc.} \tag{8-3}$$

which are called *Maxwell's relations*. The relations, of course, hold for reversible processes. They are used to compute the differences in value of the various thermodynamic variables between two equilibrium states.

Enthalpy

Although the extensive variables are the "natural" ones in which to express U, they are often not the most useful ones to use as independent variables. We do not usually measure entropy directly, we measure T; and often it is easier to keep pressure constant during a process than it is to keep V constant. We should look for functions that have some or all of the intensive variables as the "natural" ones. Formally, this can be done as follows. Suppose we wish to change from V to P for the independent variable. We add the product PV to U, to generate the function $H = U + PV$, called the enthalpy [see Eq. (7-1)]. The differential is

$$dH = P \, dV + V \, dP + dU$$
$$\leq T \, dS + V \, dP + \mu \, dn + \mathfrak{H} \, d\mathfrak{M} + \cdots \qquad (8\text{-}4)$$

where again the equality holds for reversible processes, the inequality for irreversible ones. If a system is held at constant S, P, n,..., the enthalpy will be minimal at equilibrium (note that P, instead of V, is held constant). If H is expressed as a function of its "natural" coordinates S, P, n,..., then the partials of H are the quantities

$$\left(\frac{\partial H}{\partial S}\right)_{Pn\cdots} = T; \quad \left(\frac{\partial H}{\partial P}\right)_{Sn\cdots} = V; \quad \left(\frac{\partial H}{\partial n}\right)_{SP\cdots} = \mu; \quad \text{etc.}$$
$$(8\text{-}5)$$

and the corresponding Maxwell's relations are

$$\left(\frac{\partial T}{\partial P}\right)_{Sn\cdots} = \left(\frac{\partial V}{\partial S}\right)_{Pn\cdots}; \quad \left(\frac{\partial V}{\partial \mathfrak{M}}\right)_{SP\cdots} = \left(\frac{\partial \mathfrak{H}}{\partial P}\right)_{S\mathfrak{M}\cdots}; \quad \text{etc.}$$
$$(8\text{-}6)$$

Furthermore, since $dQ = dH - V \, dP - \mu \, dn - \mathfrak{H} \, d\mathfrak{M} - \cdots$ we see that if P, n, etc., are held constant, the change in heat is equal to dH, or

$$C_P = (\partial H/\partial T)_{P,n,\mathfrak{M},\cdots} \qquad (8\text{-}7)$$

so that enthalpy can be called the *heat capacity* of the *system* at *constant pressure*.

   Geometrically, the transformation from U to H is an example of a *Legendre transformation* involving the pair of variables P, V. Function $U(V)$, for a specific value of V (i.e., at point Q in Figure 8-1) has a slope $dU/dV = -P(V)$, which defines a tangent, HQ of Figure 8-1, which has an intercept on the U axis of $H = U + PV$. (We are not considering any other variables except P, V, so we can use ordinary derivatives for the time being.) Solving for V as a function of P from the equation $dU/dV = -P$, we can then express H as a function of the slope $-P$ of the tangent line. Since $dU = -P \, dV$ and $dH = dU + P \, dV + V \, dP = V \, dP$ we see that $dH(P)/dP = V(P)$. Thus enthalpy H is the potential that has P as a basic variable instead of V.

The Helmholtz and Gibbs Functions

At least as important as the change from V to P is the change from S to T. Adiabatic (constant S) processes are encountered, of course, but many thermodynamic measurements are carried out at constant temperature. The Legendre transformation appropriate for this produces the *Helmholtz function* $F = U - TS$, for which

$$dF \leq -S\,dT - P\,dV + J\,dL + \mu\,dn + \mathfrak{H}\,d\mathfrak{M} + \cdots \qquad (8\text{-}8)$$

$$\left(\frac{\partial F}{\partial T}\right)_{Vn\cdots} = -S; \qquad \left(\frac{\partial F}{\partial V}\right)_{Tn\cdots} = -P;$$

$$\left(\frac{\partial F}{\partial \mathfrak{M}}\right)_{TV\cdots} = \mathfrak{H}; \qquad \text{etc.}$$

The related Maxwell relations include

$$\left(\frac{\partial S}{\partial V}\right)_{Tn\cdots} = \left(\frac{\partial P}{\partial T}\right)_{Vn\cdots};$$

$$\left(\frac{\partial S}{\partial \mathfrak{M}}\right)_{TV\cdots} = -\left(\frac{\partial \mathfrak{H}}{\partial T}\right)_{V\mathfrak{M}\cdots}; \qquad \text{etc.} \qquad (8\text{-}9)$$

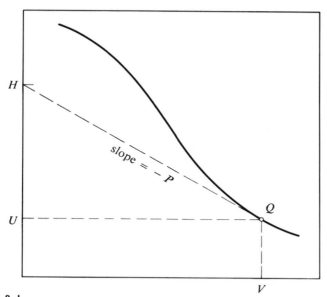

FIGURE 8-1

Legendre transformation from U as a function of V to H as a function of P.

Since the heat capacity $C_{V\mathfrak{M}\cdots}$ is equal to $T(\partial S/\partial T)_{V\mathfrak{M}\cdots}$ [see Eq. (6-8)], by differentiating these relations again with respect to T we obtain a set of equations

$$\left(\frac{\partial}{\partial V} C_{V\mathfrak{M}\cdots}\right)_{T\mathfrak{M}\cdots} = T\left(\frac{\partial^2 P}{\partial T^2}\right)_V$$

$$\left(\frac{\partial}{\partial \mathfrak{M}} C_{V\mathfrak{M}\cdots}\right)_{TV\cdots} = -T\left(\frac{\partial^2 \mathfrak{H}}{\partial T^2}\right)_{\mathfrak{M}}; \qquad \text{etc.} \qquad (8\text{-}10)$$

which indicate that the heat capacity is not completely independent of the equations of state. The dependence of $C_{V\mathfrak{M}\cdots}$ on T is something we must obtain directly by measurement; its dependence on the other variables, V, $\mathfrak{M}$, etc., can be obtained from the equations of state.

By the same arguments as before, the inequality of Eq. (8-8) shows that for a system in which T, V, n, $\mathfrak{M}$, etc., are held constant, the Helmholtz function F is minimal at equilibrium. If T is held constant, any change in F can be transformed completely into work, such as $-P\,dV$ or $\mathfrak{H}\,d\mathfrak{M}$, etc. Thus F is sometimes called the *free energy* of the system at constant temperature. The Maxwell relations (8-9) are particularly useful. For example, the second equation shows that, since an increase of magnetic polarization increases the orderliness in orientation of the atomic magnets and hence entropy (a measure of disorder) will decrease as $\mathfrak{M}$ increases, therefore the magnetic intensity $\mathfrak{H}$ required to produce a given magnetization $\mathfrak{M}$ will increase as T is increased. Similarly we can predict that since rubber tends to change from amorphous to crystal structure (i.e., becomes more ordered) as it is stretched, therefore the tension in a rubber band held at constant length will increase as T is increased.

Another potential of considerable importance is the *Gibbs function* $G = U + PV - TS = F + PV$, having T and P for its natural coordinates, instead of S and V, and for which

$$dG \leq -S\,dT + V\,dP + \mu\,dn + J\,dL + \mathfrak{H}\,d\mathfrak{M} + \cdots \qquad (8\text{-}11)$$

$$\left(\frac{\partial G}{\partial T}\right)_{Pn\cdots} = -S; \qquad \left(\frac{\partial G}{\partial P}\right)_{Tn\cdots} = V; \qquad \text{etc.}$$

The Gibbs function is a minimum at equilibrium for a system held at constant T, P, n, $\mathfrak{M}$,..., a property that we will utilize extensively in the next two chapters. The Maxwell relations include

$$\left(\frac{\partial S}{\partial P}\right)_{Tn\cdots} = -\left(\frac{\partial V}{\partial T}\right)_{Pn\cdots} ; \quad \left(\frac{\partial V}{\partial n}\right)_{TP\cdots} = \left(\frac{\partial \mu}{\partial P}\right)_{Tn\cdots} ; \quad \text{etc. (8-12)}$$

from which we can obtain the dependence of the heat capacity $C_{P\mathfrak{M}\cdots}$ on the mechanical variables

$$\left(\frac{\partial}{\partial P} C_{P\mathfrak{M}\cdots}\right)_{T\mathfrak{M}\cdots} = -T\left(\frac{\partial^2 V}{\partial T^2}\right)_P ; \qquad \text{etc.} \qquad (8\text{-}13)$$

The Gibbs function is sometimes called the free energy of the system at constant pressure and temperature, for the following reason. If the system is connected to a heat reservoir, to keep its temperature constant, and to a constant-pressure work reservoir (see page 40) to keep its pressure constant, then the work withdrawn from the combination must equal the sum of the changes of internal energy $-dU$ of the system, $-dU_h = -T\,dS_h$ of the heat reservoir and $-dU_p = P\,dV_p$ of the work reservoir. If the energy is withdrawn from the system quasistatically, the system will replenish, also quasistatically, some of its energy loss from the two reservoirs during the process. The entropy change $dS + dS_h$ of system plus heat reservoir will be zero, so $dS = -dS_h$. Also any change in volume of the work reservoir will produce an equal and opposite change in volume of the system, so $dV = -dV_p$. Therefore the work withdrawn, $-dU - dU_h - dU_p = -dU + T\,dS - P\,dV = -d(U - TS + PV) = -dG$, since T and P are constant, and it can be said that G represents the work which can be withdrawn, the *free energy*, at constant pressure and temperature.

This process can be continued for all the mechanical variables, obtaining a new potential each time, and a new set of Maxwell relations. For example, there is a *magnetic Gibbs function* $G_m = U - TS - \mathfrak{H}\mathfrak{M}$. From it we obtain

$$\left(\frac{\partial S}{\partial \mathfrak{H}}\right)_{TV\cdots} = -\left(\frac{\partial \mathfrak{M}}{\partial T}\right)_{\mathfrak{H}V\cdots} ; \quad \left(\frac{\partial P}{\partial \mathfrak{H}}\right)_{TV\cdots} = \left(\frac{\partial \mathfrak{M}}{\partial V}\right)_{T\mathfrak{H}\cdots} ; \quad \text{etc.} \quad (8\text{-}14)$$

also

$$\left(\frac{\partial}{\partial \mathfrak{H}} C_{\mathfrak{H}V\cdots}\right)_{TV\cdots} = -T\left(\frac{\partial^2 \mathfrak{M}}{\partial T^2}\right)_{\mathfrak{H}V\cdots}$$

Finally there is what is called the *grand potential* Ω, obtained by a Legendre transformation from n to μ as the independent variable, which is useful in the study of systems with variability of number of particles, such as some quantum systems exhibit:

$$\Omega = U - TS - \mu n;$$

$$d\Omega \leq - S\,dT - P\,dV - n\,d\mu + J\,dL + \cdots \qquad (8\text{-}15)$$

from which the following Maxwell relations come:

$$\left(\frac{\partial S}{\partial \mu}\right)_{TV\cdots} = \left(\frac{\partial n}{\partial T}\right)_{V\mu\cdots}; \qquad \left(\frac{\partial P}{\partial \mu}\right)_{TV\cdots} = \left(\frac{\partial n}{\partial V}\right)_{T\mu\cdots}; \qquad \text{etc.} \quad (8\text{-}16)$$

Procedures for Calculation

We have now displayed most of the techniques needed to work out relationships between one thermodynamic function and another, so that the thermal properties of the system can be explicitly given in terms of an experimentally determined heat capacity and various partials obtained from the equations of state. It may be useful to assemble these techniques in a sequence of recipes, which can be applied to any specific case. To keep the recipes simple we give them for a system in which (T, S) and one mechanical pair (Y, X) of variables are involved. The pair can be $(-P, V)$ or $(\mathfrak{H}, \mathfrak{M})$, etc., with Y the intensive and X the extensive variable. Cases where more variables are simultaneously involved can be worked out from the equations already given in this chapter.

A mnemonic device is useful to some people. One such for the system we are considering is given in Figure 8-2. The four variables T, S, Y, X are at the corners of a square and the four related potentials are at the sides. The variables adjacent to a potential are the natural ones for the potential and the arrows relate the other variables to the partials of the potential, with the arrow indicating the sign. For example $(\partial U/\partial S)_X = T$, $(\partial F/\partial T)_X = -S$, etc. It also indicates the nature of the various Maxwell relations, such as $(\partial X/\partial T)_Y = (\partial S/\partial Y)_T$ or $(\partial S/\partial X)_T = -(\partial Y/\partial T)_X$; the arrows this time connect the numerator of the partial derivative with the subscript on it and the direction of the arrow again indicates the sign. This diagram can be used for any pair of mechanical variables X, Y, such as $\mathfrak{M}$, $\mathfrak{H}$ or L, J or V, $-P$ (remember that, in contrast to most others, the pressure P is *minus* Y).

Using Euler's Eq. (6-5) for this case, the four related potentials are

$$U = ST + YX + \mu n; \qquad H = U - YX = TS + \mu n$$
$$F = U - ST = YX + \mu n; \qquad G = F - YX = \mu n \qquad (8\text{-}17)$$

Using the diagram we are now in a position to formulate a strategy for expressing any possible rate of change of a thermodynamic variable in terms of the immediately measurable quantities such as heat capacity and the partials $(dX/\partial T)_Y$ or $(\partial X/\partial Y)_T$, etc., coming from an equation of state relating X, Y, and T. Either C_X or C_Y can be considered basic ($C_P = C_Y$ is the one usually measured)

$$C_X = \left(\frac{\partial U}{\partial T}\right)_X = T\left(\frac{\partial S}{\partial T}\right)_X; \qquad C_Y = \left(\frac{\partial H}{\partial T}\right)_Y = T\left(\frac{\partial S}{\partial T}\right)_Y \quad (8\text{-}18)$$

The relation between them is given in terms of partials from the equation of state [see Eq. (7-3)]

$$C_Y = C_X - T\left(\frac{\partial Y}{\partial T}\right)_X\left(\frac{\partial X}{\partial T}\right)_Y = C_X + T\frac{(\partial X/\partial T)_Y^2}{(\partial X/\partial Y)_T}$$

$$= C_X + T[(\partial Y/\partial T)_X^2/(\partial Y/\partial X)_T] \quad (8\text{-}19)$$

The various tactics which can be used to express an unfamiliar partial in terms of an immediately measurable one are:

a. Replacing the partials of the potentials with respect to their adjoining variables in Figure 8-2 by the related variables, such as $(\partial F/\partial T)_X = -S$ or $(\partial U/\partial S)_X = T$, etc.

b. Replacing a partial of a potential with respect to a non-adjacent variable, obtainable from its basic equation, such as

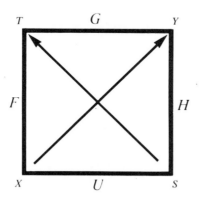

FIGURE 8-2

Diagram illustrating the thermodynamic potentials and their partials with respect to their natural variables and the Maxwell relations connecting the partials of these variables, for a two-variable system.

$dF = -S\,dT + Y\,dX$, from which we get $(\partial F/\partial X)_S = -S(\partial T/\partial X)_S$ $+\ Y$ and $(\partial F/\partial S)_Y = -S(\partial T/\partial S)_Y + Y(\partial X/\partial S)_Y$, etc.

c. Using one or more of the Maxwell relations, obtainable from Figure 8-2.

d. Using the basic properties of partial derivatives, as displayed in Eqs. (3-9) and (3-10).

In terms of these tactics, the appropriate strategies are:

1. If a potential is an independent variable in the given partial, make it the dependent variable by using (d) (this process is called *bringing the potential into the numerator*) and then use (a) or (b) to eliminate the potential. Examples:

$$(\partial T/\partial U)_X = (\partial U/\partial T)_X^{-1} = (1/C_X)$$

$$-\left(\frac{\partial T}{\partial X}\right)_U = \frac{(\partial U/\partial X)_T}{(dU/\partial T)_X} = \frac{T}{C_X}\left(\frac{\partial S}{\partial X}\right)_T + \frac{Y}{C_X}$$

2. Next, if the entropy is an independent variable in the given partial or in the result of step 1, bring S into the numerator and eliminate it by using (c) or Eq. (8-18). Examples:

$$\left(\frac{\partial Y}{\partial S}\right)_X = \frac{(\partial Y/\partial T)_X}{(\partial S/\partial T)_X} = \frac{T}{C_X}\left(\frac{\partial Y}{\partial T}\right)_X$$

$$\left(\frac{\partial T}{\partial X}\right)_S = -\frac{(\partial S/\partial X)_T}{(\partial S/\partial T)_X} = \frac{T}{C_X}\left(\frac{\partial Y}{\partial T}\right)_X = \left(\frac{\partial Y}{\partial S}\right)_X$$

3. If the measured equation-of-state partials have X in the numerator, bring X into the numerator of the result of steps 1 and 2, by using (d). If the equation of state is in the form of $Y = f(X, T)$, bring Y into the numerator.

The result of applying these successive steps will be an expression for the partial of interest in terms of measured, or measurable, quantities.

Examples and Useful Formulas

As examples of the way these procedures can be used, let us work out the various thermodynamic potentials for a perfect gas at high temperatures and for a perfect paramagnetic material at very low temperatures.

A perfect gas is, by definition, one which satisfies the equation of state for an ideal gas, $PV = nRT$, and for which C_V is a function of T alone. Of course no actual gas is "ideal" at low temperatures or high pressures; therefore we shall work out formulas which will be good approximations only for T larger than some value T_0 and for V larger than V_0. We assume that the values of S and U at T_0, V_0 can be computed by integration from $T = 0$ by the methods described in the next chapter.

The heat capacity of most gases is approximately independent of T in a range well above liquefaction and well below the temperature at which the molecules of the gas begin to dissociate or ionize. A monatomic gas, such as helium, which behaves as though its atoms were point particles, has a heat capacity at constant volume roughly equal to $\frac{3}{2}nR$ for T above about $100°K = T_0$ for He, as the discussion of Eq. (3-3) suggests. Polyatomic molecules can have energies of rotation as well as kinetic energy of translation; we shall find (see Chapter 22) that diatomic molecular gases have values of C_v more nearly equal to $\frac{5}{2}nR$, whereas gases with molecules having more than two atoms have values in the range $T_0 < T < T_d$ more nearly equal to $3nR$ (T_d is the temperature at which dissociation effects become apparent). Thus we assume that in the temperature range $T_0 < T < T_d$ the heat capacity of the gas at constant volume C_v is equal to $n\beta R$, where β is near $\frac{3}{2}$ for monatomic gases, near $\frac{5}{2}$ for diatomic, and near 3 for polyatomic gases [see Eq. (22-5)].

The basic partials, from which we build our thermodynamic functions, are

$$(\partial U/\partial T)_V = T(\partial S/\partial T)_V = C_v = n\beta R; \qquad (\partial U/\partial V)_T = 0$$
$$(\partial S/\partial V)_T = (\partial P/\partial T)_V = (nR/V); \qquad (\partial V/\partial T)_P = (V/T) \quad (8\text{-}20)$$

By use of Eq. (8-19) we see that $C_P = C_v + nR = n(1 + \beta)R$. The internal energy and entropy of the gas, in the temperature range of interest, are obtained by integration,

$$U = n\beta R(T - T_0) + nu_0 \tag{8-21}$$
$$S = n\beta R \ln(T/T_0) + nR \ln(V/V_0) + ns_0 \tag{8-22}$$

where u_0, s_0 are the internal energy and entropy of a mole of the gas at temperature T_0. For most gases s_0 is considerably larger than $(\beta + 1)R$. Both functions increase with temperature, the internal energy linearly and the entropy logarithmically. We should also invert these equations, expressing T and U as functions of the extensive variables S and V.

$$T = T_0(V_0/V)^{1/\beta} e^{(s-s_0)/\beta}$$
$$U = n\beta R T_0(V_0/V)^{1/\beta} e^{(s-s_0)/\beta} + n(u_0 - \beta R T_0) \tag{8-23}$$

where $s = (S/n)$ is the entropy per mole at temperature T.

We can now build up the other thermodynamic potentials by adding PV or subtracting ST and by changing to the independent variables appropriate to the potential,

$$F = U - ST = - n(s_0 - \beta R)(T - T_0)$$
$$+ n(u_0 - s_0 T_0) - nRT \ln[(T/T_0)^\beta (V/V_0)] \qquad (8\text{-}24)$$

$$G = F + PV = - n[s_0 - (\beta + 1)R](T - T_0)$$
$$+ n(u_0 - s_0 T_0 + nRT_0) - nRT \ln[(T/T_0)^{\beta + 1}(P_0/P)]$$

As we pointed out earlier (and shall explain in Chapter 9) s is greater than $(\beta + 1)R$ over the temperature range T_0 to T_d, for which these formulas are valid. Therefore both F and G *decrease* as T increases. They must, of course, since the entropy S, which is equal to $-(\partial F/\partial T)_V = -(\partial G/\partial T)_P$, is a large positive quantity for a gas.

Furthermore, since Euler's Eq. (6-5) shows that, for this system, $G = n\mu$, we can write down the expression for the chemical potential of the gas, as a function of T and P,

$$\mu = g_0 - [s_0 - (\beta + 1)R](T - T_0) - RT \ln[(T/T_0)^{\beta + 1}(P_0/P)]$$
$$(8\text{-}25)$$

where $g_0 = u_0 - s_0 T_0 + nRT_0$. This quantity also decreases as T increases.

Turning next to a perfect paramagnetic substance at low temperatures, we note that, by definition, the equation of state is Curie's law, $\mathfrak{M} = (nD\mathfrak{H}/T)$. This equation holds as long as $\mathfrak{M}$ is not so large as to approach saturation, where all the atomic magnets are lined up [see Eq. (13-18)]. We also assume that, in the low-temperature range V is sufficiently independent of P, so that we can neglect the effects of the pair of variables P, V. The heat capacity of a nonmagnetic crystal, as we shall demonstrate in Chapter 20 (see also Figure 3-1) goes to zero as T^3 for T small; thus the heat capacity of a paramagnetic crystal at low temperatures is primarily that of the atomic magnets, which are not yet "frozen in" (see the final sections of Chapters 18 and 20). Experimentally it is found that $C_{\mathfrak{M}P} \simeq C_{\mathfrak{M}v} \simeq n(B/T^2)$ for $T > T_0$. Of course, C must go to zero at $T = 0$ (see the end of Chapter 6), so this formula breaks down when T gets small enough. For some paramagnetic substances $T_0 < 0.05°K$, however. Therefore the basic equations are

$$C_{\mathfrak{M}} = T(\partial S/\partial T)_{\mathfrak{M}} = (\partial U/\partial T)_{\mathfrak{M}} = n(B/T^2); \qquad (\partial U/\partial \mathfrak{M})_T = 0$$
$$(\partial S/\partial \mathfrak{M})_T = - (\partial \mathfrak{H}/\partial T)_{\mathfrak{M}} = - (\mathfrak{M}/nD) \qquad (8\text{-}26)$$
$$(\partial \mathfrak{M}/\partial T)_{\mathfrak{H}} = - (nD\mathfrak{H}/T^2) = - (\mathfrak{M}/T)$$

so that

$$C_{\mathfrak{H}} = C_{\mathfrak{M}} + (\mathfrak{M}^2/nD) = (1/T^2)(B + nD\mathfrak{H}^2)$$

from Eq. (8-19). The entropy and internal energy of the substance (assuming no appreciable change in the variables P and V) are then

$$S = \frac{nB}{2}\left(\frac{1}{T_0^2} - \frac{1}{T^2}\right) - \frac{\mathfrak{M}^2}{2nD} + ns_0$$

$$= \frac{nB}{2}\left(\frac{1}{T_0^2} - \frac{1}{T^2}\right) - \tfrac{1}{2}nD\left(\frac{\mathfrak{H}}{T}\right)^2 + ns_0$$

$$U = nB[(1/T_0) - (1/T)] + nu_0 \tag{8-27}$$

$$= nu_0 + \frac{nB}{T_0}\left\{1 - \sqrt{1 - (1/B)[2(s - s_0) + D\mathfrak{M}^2]}\right\}$$

where $s = (S/n)$ as before and s_0 is the entropy per mole of the disordered magnets at $\mathfrak{M} = 0$ and $T = T_0$. As mentioned before (see also Chapter 20), this formula cannot hold clear down to $T = 0$, as S eventually goes to zero there. Both the Curie law and the equation for $C_{\mathfrak{M}}$ given in (8-26) are not valid for T less than T_0, even though T_0 is very small on the usual, linear Kelvin scale. Since the "true" scale of T is logarithmic, in terms of difficulty of achievement, we are just as far from $T = 0$ at $T = 0.01\,°\text{K}$ as we were at $100\,°\text{K}$.

The magnetic Helmholtz and Gibbs functions are

$$F_m = ns_0(T_0 - T) + \frac{nB}{2T_0}\left[1 - \left(\frac{T_0}{T}\right)\right]$$

$$+ \tfrac{1}{2}nD\frac{\mathfrak{H}^2}{T} + n(u_0 - s_0T_0)$$

$$G_m = ns_0(T_0 - T) + \frac{nB}{2T_0}\left[1 - \left(\frac{T_0}{T}\right)\right] \tag{8-28}$$

$$- \tfrac{1}{2}nD\frac{\mathfrak{H}^2}{T} + n(u_0 - s_0T_0)$$

The heat produced in a reversible, isothermal process is $T(S - S_0)$ so that the heat produced in changing the applied field from zero to $\mathfrak{H}$ isothermally is $\tfrac{1}{2}nD(\mathfrak{H}/T)^2$ for $T > T_0$. For reversible adiabatic demagnetization S is constant, so if the initial state of the substance is T_i, $\mathfrak{H}_i$ and the applied field is reduced to $\mathfrak{H}$ adiabatically, the temperature will be reduced to T, where

$$\left(\frac{T}{T_i}\right)^2 = \frac{B + D\mathfrak{H}^2}{B + D\mathfrak{H}_i^2} \tag{8-29}$$

as long as $T > T_0$. As the applied field is reduced the temperature decreases. If $\mathfrak{H}_i$ is large enough so that $D\mathfrak{H}_i^2 \gg B$ and if the final field is $\mathfrak{H} = 0$, then the final temperature is $T \simeq (B/D)^{1/2}(T_i/\mathfrak{H}_i)$ $= T_i\sqrt{(B/D\mathfrak{H}_i^2)}$, as long as $T > T_0$. Note the difference between these results and those of Eq. (4-17). The difference lies in the assumption regarding $C_{\mathfrak{M}}$; there for a gas we assumed it constant; here for the crystal we take it to be inversely proportional to T^2.

Entropy and Spontaneous Flow

The entropy also can be considered as a potential. Instead of letting U be a function of S and all the other extensive variables X_n, we can let S be a function of $U = X_0$ and all the other extensive variables $X_1,..., X_N$. Then, because

$$dS = (1/T)\, dU - \sum_{n=1}^{N} (Y_n/T)\, dX_n$$

we can relate the partials of S with respect to the extensive variables,

$$(\partial S/dU)_{X_i} = (1/T) = F_0; \qquad (\partial S/\partial X_n)_{UX_j} = -(Y_n/T) = F_n \tag{8-30}$$

where $j \neq n$. These quantities F_n, sometimes called *entropy parameters*, are state variables, each corresponding to an extensive variable of the system. For the internal energy, for example, it is the reciprocal temperature, for the volume it is (P/T), for magnetization it is $-(\mathfrak{H}/T)$, for the number of moles of material it is $-(\mu/T)$, and so on. These parameters are intensive variables.

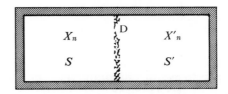

FIGURE 8-3

Spontaneous flow of extensive quantity X_n through the semipermeable partition D.

They play an important role in the analysis of spontaneous, irreversible processes, involving the flow of energy or matter or electricity. This can be demonstrated by considering the idealized situation shown in Figure 8-3. A system, completely isolated, consists of two parts, separated by a semipermeable partition D, which allows the extensive variable X_n to flow through it. The partition is made poorly conducting, so the flow is slow enough for the material on each side of D to be in equilibrium, though the flow through D itself is an irreversible process.

We now ask, "What function of the state variables will measure whether the system is in equilibrium as a whole and, if it is not, will tell us which direction the flow through D must go for equilibrium to be reached?"

From the discussion of this Chapter and of Chapter 6, we know that the entropy of an isolated system will increase spontaneously until it reaches its equilibrium value. Thus in this case if S is the entropy of the left-hand part of the system and S' that of the right-hand part, if X_n is flowing through D the entropy of the whole system will be increasing; if equilibrium has been reached $(S + S')$ will have reached its maximum value and will be constant. But a change in $(S + S')$ by a motion of dX_n through D will change $(S + S')$ by an amount

$$dS + dS' = (\partial S/\partial X_n)\, dX_n + (\partial S'/\partial X'_n)\, dX'_n$$
$$= F_n\, dX_n + F'_n\, dX'_n$$

where F_n is the entropy parameter, corresponding to X_n, for the left-hand part of the system and F'_n is that for the right-hand part. However, X_n is an extensive variable; the total amount $(X_n + X'_n)$ in the isolated system cannot change, so dX_n must equal $-dX'_n$. Therefore Eq. (6-2) for entropy increase takes the form

$$dS = (F_n - F'_n)\, dX_n \geq 0 \tag{8-31}$$

If the system is in equilibrium, $dX_n = 0$ and there is no flow; if the system is not in equilibrium, dX_n (the increase of X_n in the left-hand part) must *have the same sign* as $F_n - F'_n$. In other words if $F_n = (\partial S/\partial X_n)$ is larger than F'_n then X_n will tend to flow from right to left; if $F_n < F'_n$ then X_n will spontaneously flow from left to right through D. In a nonequilibrium situation the quantity X_n flows in the direction of *increasing* F_n. For example, if internal energy U is the extensive variable and the partition D is a heat conductor, heat will flow from right to left if $F_0 = (1/T)$ is larger than $F'_0 = (1/T')$, i.e., if the temperature T' on the right is larger than temperature T on the left. If the volume is the extensive variable and the partition is a piston, allowing volume changes on the two sides, the piston will move in a direction to *increase* the left-hand volume (i.e., will *move* to the right) if (P/T) is larger than (P'/T'). And, if matter itself, n, is the extensive variable and the partition is porous, then matter will flow from right to left if $(-\mu/T)$ is larger than $(-\mu'/T')$.

Thus $F_n - F'_n$ acts like a pressure gradient, driving X_n into spontaneous flow to reach equilibrium and producing an increase in entropy in the process. If the flow is small, to maintain equilibrium on the two sides, we would expect the rate of flow, the rate of increase of X_n, to be linearly proportional to the "pressure drop" $F_n - F'_n$. If the partition D allows multiple flow, so that several of the X_n's can change simultaneously, we would expect that the flows might interact, that thermal diffusion might affect material diffusion, for example, and vice versa. For small flows the various currents should be linearly dependent on the various "pressure drops," however, so that to the first approximation,

$$(dX_n/dt) = \sum_{m=0}^{N} L_{nm}(F_m - F'_m) = -(dX'_n/dt)$$

where the coefficients L_{nm} are determined by the properties of the partition D. The rate of increase in entropy is then equal to the sum of the products of $(F_n - F'_n)$ times the rate of flow (dX_n/dt), as indicated by Eq. (8-31).

Nonequilibrium Thermodynamics

Now let us look inside the partition D itself. Here the extensive quantity X_n is flowing through it, with a flux density $\mathbf{J}_n$ units of X_n per unit area per second. If $\mathbf{J}_n \neq 0$ there is no equilibrium in the partition but, if the flow is slow enough, we can consider each layer of the partition, dx thick, as itself a partition dividing two sides of the system, through which X_n is flowing with flux density $\mathbf{J}_n$, to maximize entropy and achieve equilibrium.

We now have defined a set of flows, of entropy and of the X's, through the interior of D; each flow vector must satisfy the law of continuity of flow vectors. As one of the "fluids", S or U or n, flows through D, the density ($\rho = s$ or u or n) of the fluid may differ from point to point and it may change with time as the fluid moves, but these changes are related. For example the density

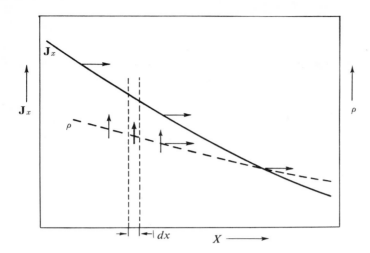

FIGURE 8-4

Flow of compressible fluid in the x direction. Both the density and the magnitude of the flow vector are plotted. Motion is to the right.

ρ, at a point fixed in space, will change with time if *either* the density of the fluid, as it moves along, actually changes with time, *or* if the flux **J** differs from point to point in D.

Figure 8-4 indicates this relationship. The flux **J** in the x direction decreases as x increases; thus, as the fluid moves to the right, more flows into the neighborhood of x than leaves. Also some fluid may be created as it flows along, as indicated by the vertical arrow within dx. This change in ρ, at a point moving with the fluid, is symbolized by the expression $(d\rho/dt)$, to distinguish it from the expression $(\partial\rho/\partial t)$, representing the net change of ρ at the point x, fixed in space. The figure indicates that

$$(\partial\rho/\partial t) = (d\rho/dt) - \text{div } \mathbf{J}$$

for any of the fluxes $\mathbf{J}_n$ and the related extensive-variable densities x_n engaged in the transport through D.

But entropy is the only one of the "fluids" which can increase as it goes along, it only can have a nonzero $(d\rho/dt)$ term. In fact, Eq. (8-31) indicates that the rate of creation of entropy, as the fluids move spontaneously toward equilibrium, is

$$(ds/dt) = \sum_n \mathbf{J}_n \cdot \text{grad } F_n \qquad (8\text{-}32)$$

But, from the definition of the F's, Eq. (8-30), we know that the rate of change of entropy density at x (fixed in space) is $(\partial s/\partial t) = \Sigma F_n(\partial x_n/\partial t)$. Therefore the divergence of the entropy flux through D is $(ds/dt) - (\partial s/\partial t) = \Sigma[\mathbf{J}_n \cdot \text{grad } F_n - F_n(\partial x_n/\partial t)] = \text{div } \mathbf{J}_s$.

However, all the other "fluids" participating in the flow are extensive variables, such as U or n, which can neither be created nor destroyed. Therefore $(dx_n/dt) = 0$ or $(\partial x_n/\partial t) = -\text{div } \mathbf{J}_n$, and the equation for div $\mathbf{J}_s$ becomes

$$\text{div } \mathbf{J}_s = \sum_n [\mathbf{J}_n \cdot \text{grad } F_n + F_n \text{ div } \mathbf{J}_n] = \sum_n \text{div}(\mathbf{J}_n F_n)$$

so that

$$\mathbf{J}_s = \sum_n F_n \mathbf{J}_n \qquad (8\text{-}33)$$

relates the flow of entropy through D to the fluxes of the other extensive variables.

As mentioned before, the entropy parameters, defined in Eq. (8-30), act like pressures, driving the various fluxes in a direction that will increase the entropy. Each of the fluxes $\mathbf{J}_n$ may be driven by the gradient of each of the pressures, grad F_n. If the system is sufficiently near equilibrium so that linear relationships are valid approximations, we can write

$$\mathbf{J}_m = \sum_n L_{mn} \text{ grad } F_n \qquad (8\text{-}34)$$

as the set of equations determining the spontaneous flow of the system toward equilibrium, in the direction of maximizing entropy. Thermodynamics does not evaluate the L's; they must be determined by experiment, or by kinetic theory, as we shall show in Chapter 14.

Equations (8-33) and (8-34) are the basis of nonequilibrium thermodynamics. They can become a powerful tool for flow analysis if we add to them a *principle of symmetry*, enunciated by Onsager,

$$L_{mn} = L_{nm} \qquad (8\text{-}35)$$

which has been verified experimentally in a number of cases. We should note that Eq. (8-35) holds *only* if the units of the fluxes and the entropy parameters are those defined in Eqs. (8-30), (8-32), and (8-34); otherwise L_{mn} is only proportional to L_{nm}, and we do not

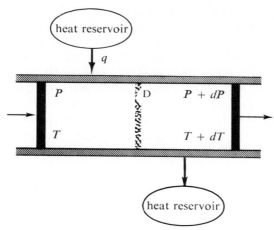

FIGURE 8-5

Percolation of fluid and heat through a porous partition D.

know what the proportionality constant is. Note also that, if magnetic fields $\mathfrak{H}$ or angular momenta ω are involved in any of the fluxes, the reversal of order of the subscripts of Eq. (8-35) is also accompanied by a change of sign of $\mathfrak{H}$ and ω. The procedure will be to choose the X's and $\mathbf{J}$'s appropriate for the problem, use Eqs. (8-30) and (8-32) to define the corresponding F's, and then relate Eqs. (8-34) to the appropriate experimental data, so as to determine the values of the L_{mn}'s.

Thermo-osmosis

As a first example of the use of these equations we analyze the tendency of a fluid to percolate through a porous partition under the combined influence of a pressure and a temperature gradient. Experimentally we find that if a temperature difference dT is maintained on the two sides of such a partition, fluid will flow through it until a pressure differential dP is built up, the equilibrium ratio (dP/dT) being a property of the percolating fluid. We use the flow Eq. (8-34) to relate (dP/dT) to other, measurable properties of the fluid.

In this case internal energy U and quantity of fluid n are flowing, so the equations for flux of U and flux of n are

$$\mathbf{J}_u = L_{uu}\,\mathrm{grad}(1/T) - L_{un}\,\mathrm{grad}(\mu/T)$$
$$\mathbf{J}_n = L_{nu}\,\mathrm{grad}(1/T) - L_{nn}\,\mathrm{grad}(\mu/T) \tag{8-36}$$

if we combine (8-34) and the appropriate formulas for F_u and F_n given following Eq. (8-30). Experimentally we find that if the temperature difference is maintained, by the two heat reservoirs of Figure 8-5, at T and $T + dT$, the fluid will flow through D until a steady-state condition is achieved, with a pressure difference dP across D, which is proportional to dT. At this steady-state condition the two pistons shown in Figure 8-5 are stationary, but equilibrium is not achieved, for heat is flowing through D; $\mathbf{J}_u$ is not zero, though $\mathbf{J}_n = 0$. The second of Eqs. (8-36) enables us to calculate the ratio between dP and dT, for this condition, in terms of the coefficients L_{nu} and L_{nn}. For $\mathbf{J}_n = 0$,

$$L_{nn} \operatorname{grad}(\mu/T) = L_{nu} \operatorname{grad}(1/T) = - L_{nu}(1/T^2) \operatorname{grad}(T)$$

or

$$\frac{L_{nu}}{L_{nn}} = - T^2 \frac{\operatorname{grad}(\mu/T)}{\operatorname{grad}(T)} = - T^2 \frac{d}{dT}\left(\frac{\mu}{T}\right)$$

$$= \mu - T\frac{d\mu}{dT} \tag{8-37}$$

To connect these quantities to other measurable quantities, we imagine another process, where we equalize the temperatures (make $dT = 0$) and let the pressure difference force fluid through D from left to right. During the passage of a mole of fluid from one side to the other we measure the heat q given to the left-hand side by the heat reservoir in contact there, in order to keep the temperature constant during the process. In this case both $\mathbf{J}_u$ and $\mathbf{J}_n$ differ from zero but grad $T = 0$, and Eqs. (8-36) show that the energy flow per mole of fluid transferred under these isothermal conditions is

$$(\mathbf{J}_u/\mathbf{J}_n) = (L_{un}/L_{nn}) = (L_{nu}/L_{nn}) = \mu - T(d\mu/dT)$$

from Eqs. (8-36) and from the principle of symmetry, Eq. (8-35).

However, the energy transferred across D per mole of fluid is the internal energy u per mole at P, T, plus the work $Pv = (MP/\rho)$ done by the left-hand piston in pushing the fluid through D ($v = M/\rho$ is the volume of a mole of fluid, M is its molecular weight, and ρ is its density) plus the heat q contributed by the reservoir during the process (q may be negative). Thus

$$(\mathbf{J}_u/\mathbf{J}_n) = u + Pv + q = \mu - T(d\mu/dT)$$

However, the Gibbs–Duhem Eq. (6-6) shows that, for a system of this sort $s\,dT - v\,dP + d\mu = 0$, or $s = v(dP/dT) - (d\mu/dT)$ and thus that

$$u = Tv(dP/dT) - T(d\mu/dT) - Pv + \mu$$

Inserting this in the equation for $(\mathbf{J}_u/\mathbf{J}_n)$ we finally have

$$u + Pv + q = Tv\left(\frac{dP}{dT}\right) - T\left(\frac{d\mu}{dT}\right) + \mu + q = \mu - T\left(\frac{d\mu}{dT}\right)$$

or

$$(dP/dT) = -(q/Tv) = -(q/M)(\rho/T) \qquad (8\text{-}38)$$

which gives the ratio between pressure and temperature differentials across the porous partition, for thermo-osmotic steady state, in terms of heat absorbed (q/M) per unit mass of fluid transferred through the partition isothermally, the fluid density ρ and the temperature T. This is, of course, the equation (7-19) that governs the fountain effect in helium II.

The ratio between the coefficients L_{un} and L_{nn} can also be determined, from Eqs. (8-37) and (8-38). The quantity $u + Pv = h = Ts + \mu$ is the enthalpy per mole of the fluid, a quantity which can be determined chemically, as will be demonstrated in Chapter 10. Therefore

$$(L_{un}/L_{nn}) = h + q$$

Also, for isothermal flow, one can measure the ratio between flow $\mathbf{J}_n$ through the partition in moles per second per unit area and pressure drop, $-(\mathbf{J}_n/\text{grad } P)$, which can be called the molar conductivity of the partition material, Q. But when grad $T = 0$ the Gibbs–Duhem equations $s\,dT - v\,dP + d\mu = 0$ becomes grad $\mu = v$ grad P and the second of Eqs. (8-36) becomes

$$L_{nn} = -T(\mathbf{J}_n/\text{grad } \mu) = (TQ/v) = (T\rho Q/M)$$

so that

$$L_{un} = (h + q)(T\rho Q/M) \qquad (8\text{-}39)$$

Finally we can measure the thermal conductivity of the partition material by establishing a temperature gradient across it and measuring the heat flux when no fluid flows. From Eq. (8-33), $\mathbf{J}_s = (1/T)\mathbf{J}_u - (\mu/T)\mathbf{J}_n$ and from the fact that T times the flux of entropy is equal to the heat flux $\mathbf{J}_q$, we have $\mathbf{J}_q = \mathbf{J}_u - \mu\mathbf{J}_n$. Therefore Eqs. (8-36) can be written

$$\mathbf{J}_q = -\frac{L_{uu} - 2\mu L_{un} + \mu^2 L_{nn}}{T^2} \text{ grad } T - \frac{L_{un} - \mu L_{nn}}{T} \text{ grad } \mu \tag{8-40}$$

$$\mathbf{J}_n = -\frac{L_{un} - \mu L_{nn}}{T^2} \text{ grad } T - \frac{L_{nn}}{T} \text{ grad } \mu$$

For the heat flow case, $\mathbf{J}_n = 0$, which enables us to substitute for grad μ in the first equation, obtaining

$$\mathbf{J}_q = -\left[\frac{L_{uu}L_{nn} - L_{un}^2}{T^2 L_{nn}}\right] \text{ grad } T$$

The ratio $-(\mathbf{J}_q/\text{grad } T)$ is the thermal conductivity κ of the material, so we can now calculate the value of the third coefficient L_{uu} in terms of measurable quantities,

$$L_{uu} = (h + q)^2(\rho T Q/M) + T^2\kappa \tag{8-41}$$

These values can then be substituted back in (8-40) or (8-36), to predict heat and fluid flow for any other combination of temperature and pressure gradients. We use the Gibbs–Duhem equation $s\, dT = v\, dP - d\mu$ in the form grad $\mu = v$ grad $P - s$ grad T, to obtain

$$\mathbf{J}_q = -\left[q(Ts + q)\left(\frac{\rho Q}{TM}\right) + \kappa\right] \text{grad } T - Q(Ts + q) \text{ grad } P$$

$$\mathbf{J}_n = -q\left(\frac{\rho Q}{TM}\right) \text{grad } T - Q \text{ grad } P \tag{8-42}$$

where s, of course, is the entropy content of a mole of fluid, related to the measurable enthalpy by the relation $h = \mu + Ts$. We can see that, although these equations are now expressed in terms of measurable quantities, they have not the symmetry of Eqs. (8-36), expressed in terms of entropy parameters.

One final comment. If we rearrange the last equations by eliminating grad P, we obtain

$$\mathbf{J}_q = -\kappa \text{ grad } T + (Ts + q)\mathbf{J}_n$$

or

$$\mathbf{J}_s = (\mathbf{J}_q/T) = \kappa T \text{ grad}(1/T) + [s + (q/T)]\mathbf{J}_n \tag{8-43}$$

Thus the entropy flux, $\mathbf{J}_s = (1/T)\mathbf{J}_u - (\mu/T)\mathbf{J}_n$, is equal to the entropy flow of heat conduction, proportional to grad$(1/T)$, plus the entropy $s + (q/T)$ carried per mole of fluid which passes through the partition.

Thermoelectric Effects

The flow of electrons in a metal can be analyzed in much the same way, at least to the extent that thermodynamics can deal with such things—without bringing in the atomic nature of the electric current. In Figure 8-6 wire A has ends at different temperatures. Wires B, of different metal, are attached to the two ends of A and brought to the two sides of an electric insulator which is a heat conductor, so the two ends are at the same temperature T', but have a difference $\Delta\mathfrak{B}$ of electric potential. In each wire we consider the flow of heat and electric current. The electric current is $e\mathbf{J}_n$, where $\mathbf{J}_n$ is the flow in moles, and e is the charge per mole of electrons. The chemical potential μ of the electrons is the sum of the usual thermodynamic terms, $\mu_c = u + Pv - Ts$, but in addition it includes the electric potential energy $e\mathfrak{B} = \mu_e$, where $\mathfrak{B}$ is the electric potential.

This time we shall set up the flow equations directly in terms of $\mathbf{J}_q$, the flow of heat along the wire, and $\mathbf{J}_e$ the electric current, which is $e\mathbf{J}_n$. As noted earlier, $\mathbf{J}_u = \mathbf{J}_q + \mu\mathbf{J}_n$, so that

$$(dS/dt) = \mathbf{J}_u \cdot \mathrm{grad}(1/T) - \mathbf{J}_n \cdot \mathrm{grad}(\mu/T)$$
$$= \mathbf{J}_q \cdot \mathrm{grad}(1/T) - (1/eT)\mathbf{J}_e \cdot \mathrm{grad}\,\mu$$

Comparison between Eqs. (8-32) and (8-34) shows that we can write the flow equations for each wire as follows:

$$-\mathbf{J}_e = (L_{ee}/eT)\,\mathrm{grad}\,\mu + L_{eq}\,\mathrm{grad}(1/T)$$
$$\mathbf{J}_q = (L_{eq}/eT)\,\mathrm{grad}\,\mu + L_{qq}\,\mathrm{grad}(1/T)$$

$$(8\text{-}44)$$

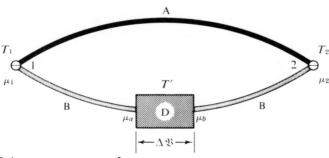

FIGURE 8-6

Circuit for thermoelectric effects. Junctions 1 and 2 are at different temperatures; wires A and B are of different metal; D is an electric insulator but a good heat conductor; voltage $\Delta\mathfrak{B}$ is read across D.

The electrical conductivity σ of either wire is defined as the electric current $\mathbf{J}_e$ per unit electric gradient at uniform temperature. Since for T constant the only variation of μ is for the electric part, $\mu_e = e\mathfrak{B}$, we see that $-\mathbf{J}_e = (\sigma/e)\text{grad } \mu$ so

$$L_{ee} = \sigma T \tag{8-45}$$

The thermal conductivity κ of the wire is the ratio between $\mathbf{J}_q$ and $-\text{grad } T$ when $\mathbf{J}_e = 0$. In this case we can eliminate $\text{grad } \mu$ in Eqs. (8-44) and obtain

$$\mathbf{J}_q = \frac{L_{eq}^2 - L_{qq}L_{ee}}{L_{ee}T^2}\text{ grad } T, \qquad \text{so that} \qquad \kappa = \frac{L_{qq}L_{ee} - L_{eq}^2}{L_{ee}T^2}$$

When no current flows, the change in μ along the wire is proportional to the temperature drop, for with $\mathbf{J}_e = 0$ Eqs. (8-44) have

$$\text{grad } \mu = (eL_{eq}/TL_{ee})\text{ grad } T$$

Integrating around the circuit from one side to the other of the insulator D, we find that the potential difference

$$\Delta\mathfrak{B} = \frac{1}{e}(\mu_b - \mu_a) = \int_1^2 (L_{qe}^A/TL_{ee}^A)\, dT$$

$$- \int_1^2 (L_{qe}^B/TL_{ee}^B)\, dT \tag{8-46}$$

where the superscripts indicate which wire the coefficient corresponds to.

From measurements of potential difference versus temperature difference for different metals we can find values of the fraction $(-L_{qe}/TL_{ee})$ (to within an arbitrary additive constant) which is called the *thermoelectric power* ε of the material. We have thus evaluated all the coefficients,

$$L_{ee} = \sigma T; \qquad L_{qe} = L_{eq} = -\varepsilon\sigma T^2$$
$$L_{qq} = \kappa T^2 + \varepsilon^2\sigma T^3 \tag{8-47}$$

in terms of measurable quantities. Then Eqs. (8-44) are

$$\mathbf{J}_e = -(\sigma/e)\text{ grad } \mu - \varepsilon\sigma\text{ grad } T$$
$$\mathbf{J}_q = -(\varepsilon\sigma T/e)\text{ grad } \mu - (\kappa + \varepsilon^2\sigma T)\text{ grad } T \tag{8-48}$$

and
$$\mathbf{J}_s = (\mathbf{J}_q/T) = \kappa T \operatorname{grad}(1/T) + \varepsilon \mathbf{J}_e$$

the last form of the equation shows that each portion of the electric current carries an entropy equal to ε times the magnitude of its charge.

The coupling between heat flow and electric current means that a current in the presence of a thermal gradient generates either more or less heat than when there is no such gradient. To show this we first produce a temperature gradient along a wire, but allow no current to flow. To keep the next part of the conceptual experiment simple, we imagine the wire imbedded in a sequence of heat reservoirs, which everywhere are at the same temperature as the wire, so that there is no heat transfer between wire and reservoirs as long as $\mathbf{J}_e = 0$. Now we pass a current $\mathbf{J}_e$ along the wire, in addition to the already present heat flow, but keep the reservoirs as they were, so the wire still has its no-current gradient of temperature. Now, however, heat flows from wire to reservoir (or vice versa) in order to keep this temperature distribution unchanged.

To find the heat flow out of the wire we compute minus the divergence of the energy current, $\mathbf{J}_u = \mathbf{J}_q + (\mu/e)\mathbf{J}_e$ from Eq. (8-33). Since div $\mathbf{J}_e = 0$ we have

$$-\operatorname{div} \mathbf{J}_u = -\operatorname{div} \mathbf{J}_q - (1/e)\mathbf{J}_e \cdot \operatorname{grad} \mu$$
$$= \operatorname{div}(\kappa \operatorname{grad} T) - \mathbf{J}_e \cdot \operatorname{grad}(\varepsilon T)$$
$$+ \mathbf{J}_e \cdot [(\mathbf{J}_e/\sigma) + \varepsilon \operatorname{grad} T]$$

where the quantity in brackets is $-(1/e)\operatorname{grad} \mu$, from Eq. (8-48). The first term is independent of $\mathbf{J}_e$; it must be zero since no heat is generated when $\mathbf{J}_e = 0$. Therefore the heat generated is

$$-T\mathbf{J}_e \cdot \operatorname{grad} \varepsilon + (\mathbf{J}_e^2/\sigma)$$
$$= -T(d\varepsilon/dT)\mathbf{J}_e \cdot \operatorname{grad} T + (\mathbf{J}_e^2/\sigma) \qquad (8\text{-}49)$$

The second term is the *Joule heat*, produced by the current flowing through the wire; it is independent of the temperature gradient. The first term is the heat generated (or absorbed) by the interaction between thermal gradient and electric current, called the *Thomson heat*. The *Thomson coefficient* $\mathfrak{X}$ is the Thomson heat absorbed, per unit current per unit temperature gradient,

$$\mathfrak{X} = \frac{-\text{heat}}{\mathbf{J}_e \cdot \operatorname{grad} T} = T(d\varepsilon/dT) \qquad (8\text{-}50)$$

As indicated by the last of Eqs. (8-48) the current carries with it entropy equal to ε per coulomb. If this current is pushed against a rising temperature, heat equal to $T\,dS = T(d\varepsilon/dT)\,dT$ is "created" by the rise in T. The Thomson coefficient is positive for some metals, negative for others.

There are many other combinations of flows which Eqs. (8-32) to (8-35) can analyze. Their use has extended the scope of thermodynamics from purely static problems to those of steady-state flow. We shall return to these questions in Chapter 14.

CHAPTER

9

CHANGES OF PHASE

Every substance we know about can exist in several different forms, called *phases*. Water can be a vapor, a liquid, or a solid. And, as a matter of fact, at high pressures solid water can exist in several different crystalline phases. Change of phase comes abruptly, as we know; just below 0°C, water is solid, just above 0°C it is liquid. It is our task, in this chapter, to work out the more obvious thermodynamic implications of these facts.

The Solid State

To see how the entropy and the thermodynamic potentials of a solid change as the temperature is increased, we shall utilize the simplified equation of state of Eq. (3-6).

$$V = V_0(1 + \beta T - \kappa P); \qquad P = (\beta/\kappa)T - [(V - V_0)/\kappa V_0]$$
$$(9\text{-}1)$$

Its heat capacities go to zero at zero temperature (see Figure 3-1) and C_v rises to $3nR$ at high temperatures. The capacity C_P is related to C_v according to Eq. (7-3). A simple formula that meets these requirements is

$$C_v = [3nRT^2/(\theta^2 + T^2)]; \qquad C_P = C_v + (\beta^2 V_0/\kappa)T \qquad (9\text{-}2)$$

The formulas do not fit well near $T = 0$, but they have the right general shape and can be integrated easily. Constant θ usually has a value less than 100°K so that by room temperature C_v is practically equal to $3nR$.

Using Eqs. (7-4) we compute the entropy as a function of T and P, remembering that $S = 0$ when $T = 0$ and $P = 0$,

$$S = \tfrac{3}{2}nR \ln[1 + (T^2/\theta^2)] + (\beta^2 V_0/\kappa)T - \beta V_0 P \qquad (9\text{-}3)$$

This formula makes it appear that S can become negative at very low temperatures if P is made large enough. Actually β becomes zero at $T = 0$, so that at very low temperatures S is independent of pressure. At moderate temperatures and pressures Eq. (9-3) is valid; curves of S as function of T for different values of P are shown in Figure 9-1.

Next we use Eqs. (8-11) to compute the Gibbs function G as function of T and P,

$$G = -\tfrac{3}{2}nRT \ln \left(1 + \frac{T^2}{\theta^2}\right) + 3nRT - 3nR\theta \tan^{-1}(T/\theta)$$

$$-(V_0\beta^2/2\kappa)T^2 + V_0 P(1 + \beta T - \tfrac{1}{2}\kappa P) + U_0 \qquad (9\text{-}4)$$

where U_0 is a constant of integration. Typical curves for this function are plotted in Figure 9-2 for different values of P. We note that G is nearly constant at low temperatures, dropping somewhat at higher temperatures, as indicated by Eq. (8-24).

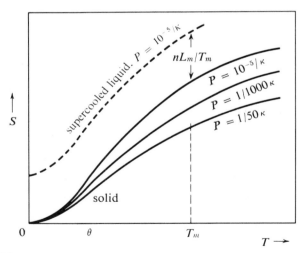

FIGURE 9-1

Solid lines plot entropy of a solid as a function of T and P.

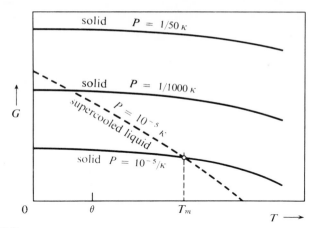

FIGURE **9-2**

Solid lines plot Gibbs function of a solid as a function of T and P.

Melting

If we add more and more heat (quasistatically) to the crystalline solid, holding the pressure constant at some moderate value, its temperature rises until finally it melts, turning into a liquid with none of the regularities of the crystal. During the melting, addition of more heat simply melts more crystal; the temperature does not rise again until all is melted. The temperature T_m at which the melting occurs depends on the pressure, and the amount of heat required to melt 1 mole of the crystal L_m is called the *latent heat of melting* (it also is a function of the pressure).

We wish to ask how thermodynamics explains these facts, next to find whether it can predict anything about the dependence of T_m, the temperature of melting, on P, and of the latent heat of melting L_m on any of the thermodynamic quantities.

The answer to the first question lies in the discussion of Eq. (8-11), defining the Gibbs function. At any instant during the quasistatic process of heat addition, the temperature and pressure are constant; thus the material takes up the configuration which has the lowest value of G for that T and P. Below T_m the G for the solid is less than the G for the liquid; above T_m the liquid phase has the lower value of G; if the process is carried out reversibly, all the material must melt at the temperature $T_m(P)$, at which the two G's are equal (for the pressure P).

If we could supercool the liquid to $T = 0$ we would find its entropy to be larger than zero, because a liquid is irregular in structure. Furthermore the entropy of the liquid increases more rapidly with temperature than does the S for the solid. Since $(\partial G/\partial T)_P = -S$, even though at $T = 0$ the G for the liquid is greater than the G for the solid, it will drop more rapidly as T increases (see the dotted line of Figure 9-2) until, at $T = T_m$ the two G's are equal; above T_m the liquid has the lower G and is thus the stable phase. Thus thermodynamics explains the sudden and complete change of phase at T_m.

The second question, raised earlier, is partly answered by pointing out that heat must be added to melt the material and that, at constant T an addition of heat dQ corresponds to an increase in entropy of the substance by an amount dQ/T. Thus the n moles of liquid, at the melting point T_m, has an entropy nL_m/T_m greater than the solid at the melting point, where nL_m is the heat required to melt n moles of the solid (at the specified P). Thus a measurement of the latent heat of melting L_m enables one to compute the entropy of the liquid at T_m, in terms of the entropy of the solid (which is computed by integration from $T = 0$), and further integration enables one to compute S for the liquid, for T greater than T_m, knowing the heat capacity and the equation of state of the liquid.

Clausius-Clapeyron Equation

To answer the rest of the second question we utilize the fact that at the melting point the Gibbs function $G_s(T_m, P)$ for the solid equals the Gibbs function $G_l(T_m, P)$ for the liquid, no matter what the pressure P. In other words, as we change the pressure, the temperature of melting $T_m(P)$, the Gibbs function G_s for the solid and G_l for the liquid all change, but the change in G_s must equal the change in G_l, in order that the two G's remain equal at the new pressure. Referring to Eq. (8-11) we see that this means that

$$dG_s = -S_s\,dT + V_s\,dP = dG_l = -S_l\,dT + V_l\,dP$$

or

$$(V_l - V_s)\,dP = (S_l - S_s)\,dT_m = (nL_m/T_m)\,dT_m$$

since the difference in entropy between a mole of liquid and a mole of solid is equal to the latent heat divided by the temperature of melting. Thus the equation relating to T_m to P is

$$\left(\frac{dT_m}{dP}\right) = \frac{T_m}{nL_m}(V_l - V_s) = \frac{MT_m}{L_m}\left(\frac{1}{\rho_l} - \frac{1}{\rho_s}\right) \qquad (9\text{-}5)$$

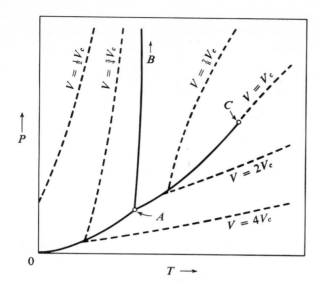

FIGURE 9-3

Phase diagram for a material that expands upon melting. Solid lines are the curves for phase change, dashed lines those for constant volume.

which is the *Clausius–Clapeyron equation*. We have used the relation $(V/n) = (M/\rho)$ from page 44.

 If the volume of the liquid is greater than the volume of the solid, then an increase of pressure will raise the temperature of melting so that, for example, if such a liquid is just above its melting point, an increase in pressure can cause it to solidify. Vice versa, if the solid is less dense than the liquid (as is the case with ice) an increase of pressure lowers the melting point and pressure can cause such a solid, just below its melting point, to melt. Thus the fact that ice floats is related to the fact that ice skating is possible; ice skates ride on a film of water which has been liquefied by the pressure of the skate. In general, since V_l differs but little from V_s, the pressure must be changed by several thousand atmospheres to change T_m by as much as 10 per cent.

 An extension of Eq. (9-5) allows us to calculate the change in the latent heat of melting with pressure, along the transition curve AB in the phase diagram of Figure 9-3. As we move along the curve representing T_m as a function of P, both T and P change, so that the rate of change of $L_m = T_m(S_l - S_s)/n$ with pressure as we move along the curve will include the related changes in both pressure and temperature,

$$\left(\frac{DL_m}{DP}\right) = \left(\frac{\partial L_m}{\partial P}\right)_T + \left(\frac{\partial L_m}{\partial T}\right)_P \left(\frac{dT_m}{dP}\right)$$

$$= \frac{T_m}{n}\left[\left(\frac{\partial S_l}{\partial P}\right)_T - \left(\frac{\partial S_s}{\partial P}\right)_T\right]$$

$$+ \left[(s_l - s_s) + (c_{pl} - c_{ps})\right]\frac{T_m(V_l - V_s)}{nL_m}$$

$$= \left[1 + \frac{T_m}{L_m}(c_{pl} - c_{ps})\right](v_l - v_s) - T_m(\beta_l v_l - \beta_s v_s)$$

$$\tag{9-6}$$

where we have used both Eqs. (9-5) and (8-12). The quantities $c_p = (T/n) \times (\partial S/\partial T)_P$ are the specific heats at constant pressure, $\beta = (1/V) \times (\partial V/\partial T)_P$ are the coefficients of thermal expansion, and $v = (V/n) = (M/\rho)$ are the volumes occupied per mole of the liquid and solid, respectively. If we know the dependence of L_m on P we can use this equation to calculate the difference in specific heats of liquid and solid.

Evaporation

If now heat is added to the liquid, its temperature will increase until another phase change occurs—the liquid *evaporates*. Here again the temperature remains constant at the temperature of vaporization T_V until all the liquid is converted into vapor. To be sure we understand what has taken place let us examine the process in more detail. We have tacitly assumed that the substance, first solid and then liquid, is confined in a container which adjusts its volume V so that it exerts a pressure P on *all parts* of the outer surface of the material. In other words we have assumed that the volume V is completely filled by the substance.

This may be difficult to do for the solid, but it is not hard to arrange it for the liquid. We provide the container with a piston, which exerts a constant force on the liquid and which can move to allow the liquid to expand at constant pressure P, and we make sure the liquid completely fills the container. In this case the liquid will stay liquid while we add heat, until its temperature reaches $T_v(P)$, when it must all be converted into gas (at a much greater volume but at the same pressure) before additional heat will raise the temperature beyond T_v. The temperature of vaporization $T_v(P)$ is related to the pressure by another Clausius–Clapeyron equation,

$$dT_v/dP = (T_v/nL_v)(V_g - V_l)$$

where here L_v is the latent heat of evaporation per mole of the material (at pressure P), V_l is the volume of the material as a liquid before evaporation, and V_g is its volume as a gas, after evaporation, at T_v and P. Since V_g is very much larger than V_l, T_v changes much more rapidly with P than does T_m.

However our usual experience is not with the behavior of liquids that entirely fill a container, but with evaporation from the free surface of a liquid. When a liquid (or solid) does not completely fill a container, some of the substance evaporates into the free space until there is enough vapor there so that equilibrium between evaporation and condensation is reached. This equilibrium is only reached when the temperature of the liquid and vapor is related to the pressure of the vapor in the free space above the liquid by the functional relationship we have been writing, $T_v(P)$. In the case we are now discussing it is better to reverse the functional relationship and write that the *vapor pressure* P_v is a function of the temperature T, and that the equation specifying this relationship is the reciprocal of the equation above,

$$dP_v/dT = [nL_v/T(V_g - V_l)] \tag{9-7}$$

This is the more-familiar form of the Clausius–Clapeyron equation.

The presence of another kind of gas in the space above the free surface of a liquid (or solid) only has an indirect effect on the amount of vapor present. The total pressure P on the liquid is now the sum of the partial pressures P_f of the foreign gas and P_v of the vapor. An addition of enough more of the foreign gas to increase this total pressure by dP (keeping T constant) will increase the Gibbs function of the liquid by $dG_l = V_l \, dP$ [see Eq. (8-11); $dT = 0$], but the Gibbs function of the same amount of material in gaseous form is not affected by the foreign gas, so $dG_g = V_g \, dP_v$. For liquid and vapor to remain in equilibrium dG_l must equal dG_g; consequently the relation between vapor pressure $P_v(P, T)$ in the presence of a foreign gas and the total pressure P is given by

$$dP_v/dP = V_l/V_g$$

which may be integrated from the initial state where no foreign gas is present [$P = P_v$ and P_v is the solution of Eq. (9-7)] to the final state where $P = P_f + P_v$. Since V_g is so much larger than V_l, P_v changes very little as the foreign gas is added, but what change there is, is positive. Addition of foreign gas squeezes a little more vapor out of the liquid, rather than pushing some vapor back into the liquid.

Water in a dish open to the air is not in equilibrium unless the partial pressure of water vapor in the air happens to be exactly equal to the vapor pressure $P_v(T)$ for the common temperature T of air and water (this is the condition of 100 per cent humidity). If the common temperature is above this, the water continues to evaporate until it is all gone, the evaporation proceeding more rapidly as the temperature is raised, until the boiling point is reached, when $P_v(T)$ is equal to the total atmospheric pressure; the gas immediately above the water is all water vapor, and the water boils rapidly away.

The latent heats of evaporation are usually from 10 to 50 times greater than the corresponding latent heats of melting, corresponding to the fact that it takes much more work to pull the material into a tenuous vapor than it does to change it into a liquid, which disrupts the crystal structure but doesn't pull the atoms much further apart.

It is also better to write down the change in latent heat L_v with T, instead of P, along the transition curve AC of Figure 9-3. Thus, instead of Eq. (9-6) we have

$$\left(\frac{DL_v}{DT}\right) = \frac{L_v}{T} + (c_{pg} - c_{pl}) - L_v\left(\frac{\beta_g v_g - \beta_l v_l}{v_g - v_l}\right)$$

This is known as *Clausius' equation*.

At very high pressures there are also phase changes in the solid state; the crystal structure of the solid changes, with accompanying latent heat, change of volume, and relationship between P and T for the change given by an equation such as (9-5).

Triple Point and Critical Point

We have just seen that the melting temperature is nearly independent of pressure, whereas the temperature of vaporization is strongly dependent on P. Therefore as P decreases, the two curves, one for T_v, the other for T_m, converge. This is shown in Figure 9-3, where the curve AB is the melting-point curve and AC that for vaporization. The two meet at the *triple point* A, which is the only point where solid, liquid, and vapor can coexist in equilibrium.

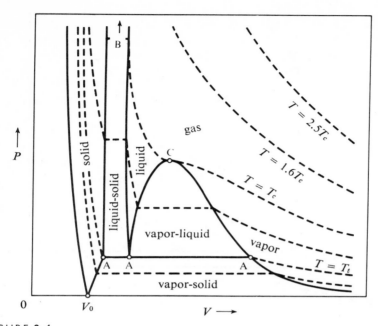

FIGURE **9-4**

PV curves (dashed lines) for the material of Figure 9-3. Solid lines are projections on the PV plane of the solid lines of Figure 9-3.

Below this pressure the liquid is not a stable phase, and along the curve 0A the solid transforms directly into the vapor (sublimation). The shape of curve 0A is governed by an equation similar to (9-7), with a latent heat of sublimation L_s (equal to $L_m + L_v$ at the triple point). The dashed lines of Figure 9-3 are lines of P against T for different values of V, intersections of the PVT surface by planes parallel to the PT plane.

As the pressure is increased, keeping $T = T_v(P)$ so that we follow the curve AC, the differences $(V_g - V_l)$ and $(S_g - S_l)$ $= (L_v/T_v)$ between gas and liquid diminish until at C, the *critical point*, there ceases to be any distinction between liquid and gas and the curve AC terminates. There seems to be no such termination of the curve AB for melting; the difference between the regularly structured solid and the irregular liquid remains for pressures up to the maximum so far attained; it may be that curve AB continues to infinity.

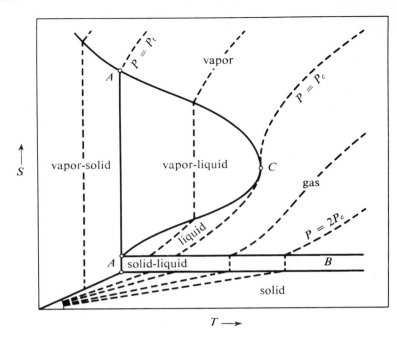

FIGURE **9-5**

> *Entropy as a function of T (dashed curves) for various values of P, for the material of Figure 9-3.*

The *PT* plane is only one way of viewing the *PVT* surface representing the equation of state. Another sometimes more useful projection is the one on the *PV* plane. In Figure 9-4 are plotted the dashed curves of *P* against *V* for different values of *T*, corresponding to intersections of the *PVT* surface with planes of constant *T*, parallel to the *PT* plane. The regions where the *PV* curves are horizontal are where there is a phase change. The boundaries of these regions, ABA and ACA, projected on the *PT* plane, are the curves AB and AC of Figure 9-3. It is clearer from Figure 9-4, why C is the critical point. The line AAA corresponds to the triple point A of Figure 9-3.

The *SPT* surface is also divided into the various phase regions. Figure 9-5 shows a part of this surface, projected on the *ST* plane, the surface being ruled with the dashed lines of constant pressure. We see that, as pressure is kept constant and *T* is increased,

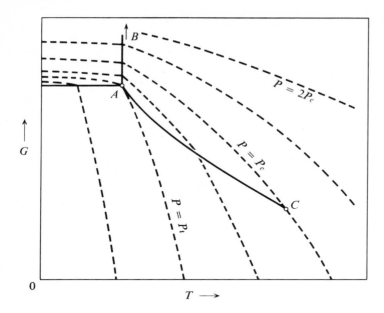

FIGURE **9-6**

 Gibbs function versus temperature (dashed lines) for various values of P,
 for the material of Figure 9-3.

S increases steadily until a phase change occurs, when S takes a
sudden jump, of amount L/T, and then continues its steady increase
with temperature in the new phase. Entropy change is largest
between solid and vapor, not because the latent heat is so much
larger for this phase change, but because it occurs at low temperature and $S_s - S_g = L_s/T_s$, where T_s is small.

The GPT surface for the Gibbs function, projected on the
GT plane, is shown in Figure 9-6. The dashed lines correspond to
the intersections of the surface with planes of constant P, parallel
to the GT plane. The solid lines correspond to the phase changes.
As noted earlier in this chapter, G does not change suddenly during
a phase change, as do V and S; only the slopes $(\partial G/\partial P)_T = V$ and
$(\partial G/\partial T)_V = -S$ (the slopes of the dashed lines of Figure 9-6)
change discontinuously across the phase-change boundaries. By
taking gradients the curves of Figures 9-4 and 9-5 can be obtained
from Figure 9-6 or, vice versa, the curves of Figure 9-6 can be
obtained by integration of the data on curves in Figures 9-4 and 9-5.

Phase Changes of the Second Kind

In a phase change of the sort we have just been discussing, the change involves a major rearrangement of structure of the substance, resulting in change of volume, viscosity or tensile strength, and so on. Since such changes involve energy input or output to produce, a finite amount of heat, the latent heat L, is required to raise the temperature of the substance from just below the transition temperature T_c to just above it. This latent heat represents a discontinuity in entropy, $L_{12} = T_c(S_2 - S_1)$, of the substance at the transition, as is shown by the dashed lines of Figure 9-5. Since $C_P = T(\partial S/\partial T)_P$ the heat capacity of the substance becomes infinite at the transition. These kinds of transitions are called *phase changes of the first kind*.

There are other changes of phase, involving the initiation of a different kind of ordering in a crystal lattice or the appearance of superfluid in helium II, which involve simply a change of *slope* of S against T at the transition point, not a change of value. In this case the heat capacity changes discontinuously but does not become infinite at T_c. Such changes are called *phase changes of the second kind*. For them the equation representing the constancy of the Gibbs function during the transition, $(V_2 - V_1)\, dP_c = (S_2 - S_1)\, dT$, which gave rise to the Clausius–Clapeyron Eq. (9-5) for transitions of the first kind, breaks down because there is no discontinuity in either volume $(V_2 = V_1)$ or entropy $(S_2 = S_1)$ during the transition.

However, in any such transition, not only is the value of the Gibbs function continuous, the rate of change of G along the transition curve must be the same on both sides of the curve. Thus we can take partials of the equation $V_2\, dP_c - S_2\, dT = V_1\, dP_c - S_1\, dT$ either with respect to P or to T,

$$\left[\left(\frac{\partial V}{\partial P}\right)_T dP_c - \left(\frac{\partial S}{\partial P}\right)_T dT\right]_2 = \left[\left(\frac{\partial V}{\partial P}\right)_T dP_c - \left(\frac{\partial S}{\partial P}\right)_T dT\right]_1$$

or

$$(dP_c/dT) = [(\beta_2 - \beta_1)/(\kappa_2 - \kappa_1)] \tag{9-8}$$

$$\left[\left(\frac{\partial V}{\partial T}\right)_P dP_c - \left(\frac{\partial S}{\partial T}\right)_P dT\right]_2 = \left[\left(\frac{\partial V}{\partial T}\right)_P dP_c - \left(\frac{\partial S}{\partial T}\right)_P dT\right]_1$$

or

$$(dP_c/dT) = [(C_{p2} - C_{p1})/VT(\beta_2 - \beta_1)]$$

where $\beta = (1/V)(\partial V/\partial T)_P = -(1/V)(\partial S/\partial P)_T$ is the coefficient of

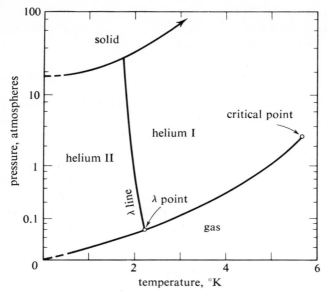

FIGURE 9-7

Phase diagram for He^4 at low temperatures.

thermal expansion, $\kappa = -(1/V)(\partial V/\partial P)_T$ the isothermal compressibility (see Problem 9-2) and $C_P = T(\partial S/\partial T)_P$ is the heat capacity at constant pressure of either phase 1 or phase 2. These are *Ehrenfest's equations* for the slope of the P_c, T curves for a transition of the second kind.

The phase diagram of helium at low temperatures, shown in Figure 9-7, illustrates some of these statements. The pressure scale is distorted in order to show both the liquid–gas and the solid–liquid transition curves. Both of these represent phase changes of the first kind, obeying equations of the form of (9-7), with a finite discontinuity in S and V. Of course, as T goes to zero, the difference in entropy between solid and liquid goes to zero, according to the third law, so the slope of the curve, $(dP_c/dT) = (\Delta S/\Delta V)$, goes to zero there, as shown by the extrapolation of the upper curve to $T = 0$. The liquid–gas curve extrapolates to the origin and ends at the critical point for helium, at 2.26 atm and 5.21 °K. The curve separating the two liquid phases separates the normal liquid, helium I, from the mixture of normal and superfluid, which is helium II. Since this transition is one of the second kind there is no discontinuity in S here, but there is a discontinuity in heat capacity. This curve is called a λ line and the point, analogous to the triple point, is called the λ point.

Superconductivity

Many metals and some alloys exhibit the phenomenon of superconductivity; the exceptions being the alkali and alkaline-earth metals and the magnetic group. The phenomenon manifests itself at low temperatures. As the temperature is reduced the electric resistivity diminishes, more or less linearly with the absolute temperature, until at some temperature T_0 it falls practically to zero, to remain there as T is reduced still further. In this region of superconductivity the resistance is so small that current loops, once started by magnetic induction, continue to flow for months. Here is a still different kind of phase change.

This change of phase turns out to be reflected in the magnetic behavior of the metal. In a substance of zero resistance no change of the magnetic field would be possible; for a change in $\mathfrak{B}$ would induce an electric field, according to Maxwell's electromagnetic equations, and this would create a current just sufficient to oppose the change. Experimenters at first expected that they could "freeze in" a magnetic field, by magnetizing the metal at temperatures above the transition temperature T_0 and then reducing T below T_0, after which the superconductivity would hold the field inside the metal even if it were reduced to zero outside.

To their surprise, however, they found that, as the metal became superconductive, it totally rejected whatever magnetic field it might have contained. In the superconductive metal the magnetic polarization $\mathfrak{P} = \chi \mathfrak{H}$ (see page 21) is exactly equal to $- \mathfrak{H}$ (in other words the magnetic susceptibility χ is exactly equal to -1) so that the magnetic induction $\mathfrak{B} = \mu_0(\mathfrak{H} + \mathfrak{P})$ inside the metal is zero and the magnetization is

$$\mathfrak{M} = \mu_0 V \mathfrak{P} = -\mu_0 V \mathfrak{H} \tag{9-9}$$

This exclusion of the magnetic field by a superconductive metal is known as the *Meissner effect*, after its discoverer. Thus during the transition the electric resistance drops to zero and at the same time the magnetization changes from nearly zero (for the normal metal) to $-\mu_0 V \mathfrak{H}$.

In addition, it was found that the presence of a magnetic field just outside the superconductor would tend to cancel out the

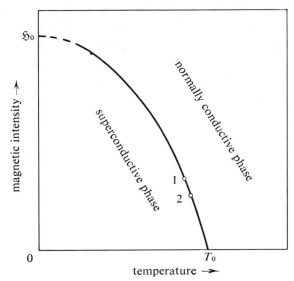

FIGURE **9-8**

Dependence of superconductivity on applied magnetic field.

superconductivity, if $\mathfrak{H}$ were large enough. But this could have been foreseen; for if the change of phase involved a discontinuous change in $\mathfrak{M}$, there would be a relationship between the temperature of transition and the magnetic field, governed by an equation analogous to the Clausius–Clapeyron equation. In other words the boundary between the superconductive phase and the normally conductive phase is a curved line in the $\mathfrak{H}T$ plane, as shown in Figure 9-8. At zero field the transition temperature $T_c(\mathfrak{H})$ has its maximum value T_0, and at $T = 0$ the magnetic field $\mathfrak{H}_c(T)$, above which the metal reverts to normal, has its maximum value $\mathfrak{H}_0$. A good approximation for the curve, for some metals, is

$$\mathfrak{H}_c(T) = \mathfrak{H}_0\left[1 - (1 - \alpha)(T/T_0)^2 - \alpha(T/T_0)^4\right] \tag{9-10}$$

We note that $(d\,\mathfrak{H}_c/dT) \to 0$ as $T \to 0$. We shall soon see that this is necessary, as is the requirement that ths lope $(d\mathfrak{H}_c/dT)$ is never positive.

The phase transition in this case involves the magnetic field and magnetization, instead of pressure and volume. The quantity which remains constant during the transition is the magnetic Gibbs function $G_m = U - TS - \mathfrak{H}\mathfrak{M} = -PV + \mu n$ (see page 99). Traversing the $\mathfrak{H}T$ plane just outside the transition curve, the change in G_m in going from point 1 to point 2, for example, is $dG_m = -S_n\,dT - \mathfrak{M}_n\,d\mathfrak{H}$, where the subscripts n indicate values for the normal metal, just outside the transition curve. This change in G_m must be the same as the change $dG_m = -S_s\,dT - \mathfrak{M}_s\,d\mathfrak{H}$ for the superconductive material just inside the curve, if G_m is to be the same on both sides of the curve, for every point on the curve.

Therefore the equivalent of the Clausius–Clapeyron equation for this transition, the differential equation for the phase diagram on the $\mathfrak{H}T$ plane, is

$$(S_n - S_s) = (\mathfrak{M}_s - \mathfrak{M}_n)(d\mathfrak{H}_c/dT)$$

or

$$T(S_n - S_s) = nL_c = -\mu_0 TV\mathfrak{H}_c(d\mathfrak{H}_c/dT) \tag{9-11}$$

using Eq. (9-9) and taking T times the difference in entropy between the phases to be the latent heat for the transition. Comparison between this equation and Figure 9-8 brings out a number of interesting points. In the first place, because of the third law (see page 99) the difference $(S_n - S_s)$ must go to zero at $T = 0$; therefore the slope $(d\mathfrak{H}_c/dT)$ of the transition curve must be zero at $T = 0$. Next, since $(d\mathfrak{H}_c/dT)$ is negative, for $T > 0$, the entropy S_n of the normal phase must be greater than S_s for the superconductive phase, which thus must be a more orderly phase (and this is born out by other properties of the superconductive phase). In the third place, we see that L_c goes to zero as $\mathfrak{H}_c$ goes to zero; thus the phase transition at zero magnetic field must be a transition of the second kind.

And, finally, the difference in heat capacity between normal and superconductive phases at the transition is

$$\begin{aligned}
C_s - C_n &= T(\partial S_s/\partial T) - T(\partial S_n/\partial T) \\
&= \mu_0 TV[(d\mathfrak{H}_c/dT)^2 + \mathfrak{H}_c(d^2\mathfrak{H}_c/dT^2)] \\
&\to \mu_0 T_0 V(d\mathfrak{H}_c/dT)^2 \qquad (\mathfrak{H}_c \to 0) \tag{9-12}
\end{aligned}$$

This formula, in giving the difference in heat capacities in terms of the properties of the transition curve, checks very well with directly measured values of the heat capacity difference. The heat capacity of the superconductive phase, just below the transition, is greater than that of the normal phase; but, of course, C_s decreases to zero as $T \to 0$. Equation (9-12) is the analogue, for magnetic fields, of the second Ehrenfest equation for $P - V$.

Careful readers may have noticed that we did not handle the differentiation with respect to T, as we followed along the transition curve, as carefully as we did the differentiation with respect to P to obtain Eq. (9-6). The reason is that $(\partial S/\partial \mathfrak{H})_T = (\partial \mathfrak{M}/\partial T)_{\mathfrak{H}} = 0$ both for the normal and the superconductive metal at these temperatures. Therefore we do not have to go through the procedure of setting

$$\frac{DS}{DT} = \left(\frac{\partial S}{\partial T}\right)_{\mathfrak{H}} + \left(\frac{\partial S}{\partial \mathfrak{H}}\right)_T \left(\frac{d \mathfrak{H}_e}{dT}\right)$$

since the second term is zero. Also, since $C_{\mathfrak{M}} = T(\partial S/\partial T)_{\mathfrak{M}}$ and $C_{\mathfrak{H}} = C_{\mathfrak{M}} - T(\partial \mathfrak{M}/\partial T)_{\mathfrak{H}} (\partial \mathfrak{H}/\partial T)_{\mathfrak{M}}$ are equal for both normal and superconductive metals, we need not bother to put a subscript $\mathfrak{M}$ or $\mathfrak{H}$ on the C's.

CHAPTER

CHEMICAL REACTIONS

In a chemical reaction one or more of the constituent substances in the system changes its molecular structure, with accompanying release or absorption of energy. Prior to the reaction the energy was present in potential form; during the reaction the energy is released as heat (or is absorbed, if the energy release is negative) which can be transformed in part into work. This change from potential energy to heat or work is represented in our equations by change in the number n_i of moles of substance i present, with a consequent change $\mu_i\, dn_i$ in the internal energy for each substance involved in the reaction. The chemical potential μ_i, as its name implies, is a measure of the chemical energy which substance i can generate per mole, in a reaction. If several constituents take part in the reaction, dn_1 and dn_2 moles of substances 1 and 2 disappearing and dn_3 and dn_4 moles of 3 and 4 appearing, for example, then dn_1 and dn_2 are negative and dn_3 and dn_4 are positive and the sum $\mu_1\, dn_1 + \mu_2\, dn_2 + \mu_3\, dn_3 + \mu_4\, dn_4$ is the amount of energy released during the reaction.

At constant pressure this energy (if positive) can be withdrawn as heat. In accord with the discussion of Eq. (8-7) the change in enthalpy; $dH = T\, dS - V\, dP + \Sigma \mu_i\, dn_i$ during the reaction equals the heat which can be withdrawn subsequently, reversibly, at constant pressure. The sum $\Sigma \mu_i\, dn_i$, the release of chemical energy during the reaction, is equal to the change in the Gibbs function, $dG = -S\, dT + V\, dP + \Sigma \mu_i\, dn_i$ if the reaction is carried out at constant pressure and temperature.

In principle it is possible to measure the chemical potential μ_i of substance i (within an arbitrary additive constant) by measuring its contribution to the heats evolved during a whole sequence of chemical reactions, carried out at standard temperature (25 °C) and pressure (1 atm). By subtraction one can obtain a μ_i for each chemical substance, which represents its contribution to the change in enthalpy produced during the reaction at constant pressure. The elements in their usual forms (H_2, O_2, S, Hg, etc.) have zero value at standard T and P; the *heats of formation* $-dH^0$ of compounds are then minus the heat produced per mole of their formation from their elements. These heats are additive, so that the heat evolved when two compounds combine to form a third is the heat of formation of the third minus the heat of formation of the first two.

But these heats of formation are not the chemical potentials μ of the compounds, because enthalpy $H = TS + \Sigma \mu_i n_i$ is not equal to the Gibbs function $G = \Sigma \mu_i n_i$. During the reaction several of the n's change in value, but also the entropy of the system changes. To find the values of the μ's we would have to subtract $T \Delta S$ from the ΔH obtained by heat measurement at constant P. This often is possible to do but is usually difficult experimentally. What is usually done is to find the value of μ_i for compound i by indirect means, using the thermodynamic equations we shall develop in this chapter. If we assume that compound i has a chemical potential μ_i, which satisfies the equations for μ_i [such as the Gibbs–Duhem Eq. (6-6)] and which contributes to the change $\Sigma \mu_i \, \Delta n_i = \Delta G$ during a reaction at constant T and P, we shall find more accurate means of measuring dG than by measuring the evolution of heat ΔH and change of entropy ΔS. Thus chemical thermodynamics is, in part, the development of a set of equations which enable one to measure values of the chemical potentials of compounds by methods that are more accurate than direct calorimetric measurements.

Chemical Equations

A chemical reaction is usually described by an equation, such as $2H_2 + O_2 \rightarrow 2H_2O$, which we can generalize as

$$- \sum_i v_i M_i \rightarrow \sum_j v_j M_j$$

stating that a certain number, $-v_i$, of molecules of the *initial reactants* M_i will combine to produce a certain number, v_j, of molecules of *final products* M_j. Of course the chemical reaction can run in either direction, depending on the circumstances; we have to pick a direction to call positive. This is arbitrarily chosen to be the direction in which the reaction generates heat. We then, also arbitrarily, place all the terms in the equation on the right, so that it reads

$$0 = \sum_{i=1}^{N} v_i M_i \qquad (10\text{-}1)$$

In this form the v's which are negative represent initial reactants and those which are positive represent final products (v for the H_2 in the example would be -2, v for the O_2 would be -1, and v for the H_2O would be $+2$). The number v_i is called the *stoichiometric coefficient* for M_i in the reaction.

For a chemical reaction to take place, more than one sort of material must be present either initially or finally, and the numbers n_i of moles of the reacting materials will change during the reaction. Referring to Eqs. (8-11) and (8-17) we see that the Gibbs function and its change during the reaction are

$$G = \sum_{i=1}^{N} n_i \mu_i; \qquad dG = -S\, dT + V\, dP + \sum_i \mu_i\, dn_i \qquad (10\text{-}2)$$

where μ_i is the chemical potential of the ith component of the reaction, which is the Gibbs function per mole of the material M_i.

A chemical reaction of the sort described by Eq. (10-1) produces a change in the n's in an interrelated way. If $dn_1 = v_1\, dx$ moles of material M_1 appear during a given interval of time while the reaction progresses, then $dn_i = v_i\, dx$ moles of material M_i will appear during the same interval of time (or will disappear, if v_i is negative). For example, if $2dx$ moles of H_2O appear, simultaneously $2dx$ moles of H_2 and dx moles of O_2 will disappear (i.e., $dn_H = -2dx$ and $dn_O = -dx$). In other words, during a chemical reaction at constant T and P, the change in the Gibbs function is

$$dG = \sum_{i=1}^{N} \mu_i v_i\, dx$$

Once started, most chemical reactions that generate heat proceed spontaneously until either some reactant runs out or else until the reaction reaches equilibrium. If the reaction proceeds at constant temperature and pressure this equilibrium will be reached when G reaches its minimal value. At equilibrium the reaction must be reversible; a small chemical change, represented by dx must represent a *zero change* in G at equilibrium. In other words, the chemical potentials of the reactants and products, at equilibrium, must be so related that

$$\sum_{i=1}^{N} \mu_i \nu_i = 0 \qquad \text{for equilibrium} \qquad (10\text{-}3)$$

Of course if one of the reactants runs out before equilibrium is reached, the reaction will perforce stop. For example, for the production of H_2O, we could start with 1 mole of H_2 and 1 mole of O_2. The number of moles of the constituents some time later would then be

$$n_{H_2} = 1 - 2x; \qquad n_{O_2} = 1 - x; \qquad \text{and} \qquad n_{H_2O} = 2x$$

The three μ's can be expressed as functions of P, T, and x and Eq. (10-3) can then be solved for x for equilibrium for the specified P and T. If this value of x is less than $\frac{1}{2}$ then equilibrium can be reached, and nonzero amounts of the three constituents will be present at equilibrium. But if the value of x to satisfy Eq. (10-3) is greater than $\frac{1}{2}$ the reaction will stop before equilibrium is reached because the hydrogen will have run out. The reaction will have run to completion but not to equilibrium.

This may be satisfactory if we are running the reaction to obtain the product. But if we are using the reaction to help us determine the chemical potentials of the constituents, we shall want the reaction to reach equilibrium, so we can use the properties of the equilibrium state as indirect measures of the thermodynamic properties of the constituents. Therefore, for the rest of this chapter we shall assume we have enough of each constituent present so the reaction can attain equilibrium.

Heat Evolved by the Reaction

During the progress of the reaction, if carried out at constant T and P, any evolution of heat would be measured as a change in enthalpy since, as we have already remarked, enthalpy is the heat content of the system at constant pressure. But since, from Eq. (8-17),

$$H = G + TS = G - T(\partial G/\partial T)_{Pn} \qquad (10\text{-}4)$$

the change in H as the parameter x changes at constant T and P is

$$dH = \left(\frac{dG}{dx}\right) dx - T\left[\frac{\partial}{\partial T}\left(\frac{dG}{dx}\right)\right]_{Pn} dx$$

$$= \left[\sum \mu_i v_i - T\left(\frac{\partial}{\partial T}\sum \mu_i v_i\right)_{Pn}\right] dx$$

At equilibrium $\sum \mu_i v_i = 0$ and the reaction ceases. If we measure the heat evolved per change dx when the system is close to equilibrium, the rate of evolution of heat becomes

$$dH/dx = -T\left(\frac{\partial}{\partial T}\sum \mu_i v_i\right)_{Pn} \tag{10-5}$$

Thus by measuring the rates of change with temperature, $(\partial \mu_i/\partial T)_{Pn}$, of the chemical potentials of the substances involved, we can predict the amount of heat evolved during the reaction. Or, vice versa, if we can measure (or can compute, by quantum mechanics) the heat evolved when v_i moles of substance M_i appear or disappear, by dissociating into their constituent atoms or by reassociating the atoms into the product molecules, we can predict the rate of change of the μ's with temperature.

Mixtures of Gases

We have already pointed out that if a number of perfect gases are mixed together their total entropy and internal energy are the sums of the individual entropies and energies of each constituent, as though each were unaware of the other's presence. If we refer each gas to the standard temperature T_0, volume V_0, and number of moles n_0 then, according to Eqs. (8-22), the energy and entropy of the ith gas, when it has temperature T, volume V, and is in amount n_i, are

$$U_i = (n_i/n_0)U_{0i} + n_i\beta_i R(T - T_0)$$

$$S_i = (n_i/n_0)S_{0i} + n_i\beta_i R \ln(T/T_0) + n_i R \ln(n_0 V/n_i V_0)$$

and the total energy and entropy of the mixture of perfect gases are

$$U = \sum_{i=1}^{N} U_i = \sum_i n_i(U_{0i}/n_0) + R(T - T_0)\sum_i n_i\beta_i$$

$$S = \sum_i n_i(S_{0i}/n_0) + R \ln(T/T_0)\sum_i n_i\beta_i + \sum_i n_i R \ln(n_0 V/n_i V_0) \tag{10-6}$$

where the independent variables are T, V, and the n_i's.

To check that this gives a self-consistent value for the temperature, we use Eqs. (6-4),

$$T = \left(\frac{\partial U}{\partial S}\right)_{V,n_i} = \frac{(\partial U/\partial T)_{V,n_i}}{(\partial S/\partial T)_{V,n_i}} = \frac{R\sum n_i\beta_i}{(R/T)\sum n_i\beta_i}$$

To obtain the pressure we use Eqs. (6-4) and (3-9)

$$P = -\left(\frac{\partial U}{\partial V}\right)_{S,n_i} = \frac{(\partial S/\partial V)_{U,n_i}}{(\partial S/\partial U)_{V,n_i}}$$

$$= \left(\frac{\partial U}{\partial T}\right)_{V,n_i} \cdot \left(\frac{\partial T}{\partial S}\right)_{V,n_i} \cdot \left(\frac{\partial S}{\partial V}\right)_{U,n_i}$$

$$= \left(\frac{\partial U}{\partial T}\right)_{V,n_i} \cdot \frac{(\partial S/\partial V)_{T,n_i}}{(\partial S/\partial T)_{V,n_i}} = \frac{RT}{V}\sum_{i=1}^{N} n_i = \frac{nRT}{V} \tag{10-7}$$

where we have used the fact that $U = $ const. is the same as $T = $ const. (for these gases) in going from the second to the third line and where we have defined n, the total molar quantity present, as being the sum Σn_i of the moles of all the constituents. Therefore the total pressure P is related to temperature and volume as though the mixture were a single-component, ideal gas. It is made up of the *partial pressures* $P_i = (n_i RT/V)$ of each constituent gas, as though each gas, in contributing its component to the total pressure, were unaware that the others were present. We can thus express the relative amount of the ith gas in the mixture by giving its *concentration* $\chi_i = (n_i/n)$ or its partial pressure $P_i = \chi_i P$.

We should note that although the entropy, as a function of T, V, and the n_i's is a sum of the entropies of the individual constituents, this is not the case if the entropy is given in terms of T, P, and the n_i's. In that case,

$$S = (S_{0i}/n_0)\sum n_i + R\ln(T/T_0)\sum n_i(\beta_i + 1)$$

$$+ R\ln(P_0/P)\sum n_i + R\sum n_i \ln(n/n_i)$$

where $P_0 = (n_0 RT_0/V_0)$. The first three terms are the sum of the entropies of each of the constituent gases, if each were at the temperature and pressure of the mixture. The last term is an added correction, which is the entropy of mixing, as per Eq. (6-15).

Reactions in Gases

If now the mixture of perfect gases undergoes a chemical reaction, it will come to equilibrium (if possible) at a combination of concentrations (or partial pressures) for which Eq. (10-3) is satisfied. To find the solution we must express the chemical potentials μ_i of the constituent gases as functions of P, T, and the n_i's. From the usual equation for internal energy, $dU = T\,dS - P\,dV + \Sigma\mu_i\,dn_i$ we obtain ($dV = 0$)

$$\mu_i = (\partial U/\partial n_i)_{T,V,n_j} - T(\partial S/\partial n_i)_{T,V,n_j} \qquad (j \neq i)$$

$$= (U_{0i}/n_0) + \beta_i R(T - T_0) - T(S_{0i}/n_0) - \beta_i RT \ln(T/T_0)$$

$$- RT \ln(n_0 V/n_i V_0) - RT$$

or

$$\mu_i = g_i(T) + RT \ln(P_i/P_0) = g_i(T) + RT[\ln(P/P_0) + \ln(\chi_i)]$$

$$(10\text{-}8)$$

where

$$g_i(T) = \frac{U_{0i} - n_0\beta_i R T_0}{n_0}$$

$$+ RT\left[\beta_i + 1 - \frac{S_{0i}}{n_0 R} - (\beta_i + 1)\ln(T/T_0)\right]$$

This expresses the chemical potential in terms of a function g_i of temperature only, plus a term dependent on the partial pressure P_i or the pressure P and the concentration χ_i. The value of g_i at the standard temperature T_0, $(U_{0i}/n_0) - (T_0 S_{0i}/n_0) + RT_0$, is the "free energy of formation" of the substance i which is used in chemical calculations.

The equation for chemical equilibrium (10-3) then takes on the form

$$RT \sum_{i=1}^{N} v_i[\ln(P_i/P_0) + (g_i/RT)] = 0$$

or

$$\left(\frac{P_1}{P_0}\right)^{v_1} \left(\frac{P_2}{P_0}\right)^{v_2} \cdots \left(\frac{P_N}{P_0}\right)^{v_N} = \exp\left\{-\sum_{i=1}^{N}\left[\frac{v_i g_i(T)}{RT}\right]\right\} \equiv K(T)$$

$$(10\text{-}9)$$

which gives an equation for the partial pressures of the constituents at equilibrium, in terms of a function $K(T)$ of the temperature only. The equation can also be expressed in terms of the concentrations $\chi_i = (P_i/P) = (n_i/n)$,

$$RT \sum_{i=1}^{N} v_i[\ln(P/P_0) + \ln \chi_i + (g_i/RT)] = 0$$

or

$$\chi_1^{v_1} \cdot \chi_2^{v_2} \cdots \chi_N^{v_N} = (P_0/P)^{\Sigma v_i} \cdot K(T) \tag{10-10}$$

where, as before

$$K(T) = \exp\left\{-\sum_{i=1}^{N} [v_i g_i(T)/RT]\right\}$$

The quantity $K(T)$ is called the *equilibrium constant* and the equation determining the χ's at equilibrium is called the *law of mass action*. To see how this goes, we return to the reaction between H_2 and O_2. Suppose initially n_1 moles of H_2 and n_2 moles of O_2 were present and suppose during the reaction x moles of O_2 and $2x$ moles of H_2 combined to form $2x$ moles of H_2O. The total number of moles then present would be $(n_1 - 2x) + (n_2 - x) + 2x = n_1 + n_2 - x$, so the concentrations at equilibrium and the stoichiometric coefficients for the reaction are

For H_2: $\chi_1 = [(n_1 - 2x)/(n_1 + n_2 - x)]$ $v_1 = -2$

For O_2: $\chi_2 = [(n_2 - x)/(n_1 + n_2 - x)]$ $v_2 = -1$

For H_2O: $\chi_3 = [2x/(n_1 + n_2 - x)]$ $v_3 = 2$

and the law of mass action becomes

$$\frac{4x^2(n_1 + n_2 - x)}{(n_1 - 2x)^2 (n_2 - x)} = \left(\frac{P}{P_0}\right) \exp\left[\frac{2g_1 + g_2 - 2g_3}{RT}\right]$$

$$= \left(\frac{P}{P_0}\right)K(T)$$

from which we can solve for x. Since this particular reaction is strongly exothermic, $2g_1 + g_2$ is considerably larger than $2g_3$ and the exponential $K(T)$ is a very large quantity unless T is large. Consequently, for $P \simeq P_0$ and for moderate temperatures, x will be close to $\frac{1}{2}n_1$ or n_2, whichever is smaller, i.e., the reaction will go almost to completion, using up nearly all the constituent in shorter supply. If there is a deficiency in hydrogen, for example, so that $n_1 < 2n_2$, then we set $x = \frac{1}{2}n_1 - \delta$, where δ is small, and

$$(2\delta)^2 \simeq \frac{n_1^2[n_2 + \frac{1}{2}n_1]}{[n_2 - \frac{1}{2}n_1]}\left(\frac{P_0}{P}\right)\exp\left(\frac{2g_3 - 2g_1 - g_2}{RT}\right)$$

$$\chi_1 \simeq \frac{2\delta}{n_2 + \frac{1}{2}n_1}; \qquad \chi_2 \simeq \frac{n_2 - \frac{1}{2}n_1}{n_2 + \frac{1}{2}n_1}; \qquad \chi_3 \simeq \frac{n_1}{n_2 + \frac{1}{2}n_1}$$

the system coming to equilibrium with a very small amount, 2δ moles, of H_2 left. This amount increases with decrease of pressure P and with increase of T.

To see how the equilibrium constant $K(T)$ changes with temperature, we can utilize Eqs. (10-3), (10-5), and (10-8). We have

$$\frac{d}{dT}\ln K = \frac{1}{RT^2}\left[\sum v_i g_i - T\sum v_i g'_i\right]$$

$$= \frac{1}{RT^2}\left[\sum v_i \mu_i - T\sum v_i\left(R\ln\frac{P_i}{P_0} + g'_i\right)\right]$$

$$= \frac{1}{RT^2}\left[-T\left(\frac{\partial}{\partial T}\sum v_i \mu_i\right)_{Pn}\right] = \frac{1}{RT^2}\left(\frac{\partial H}{\partial x}\right)_{Tp} \quad (10\text{-}11)$$

where $g'_i = (dg_i/dT)$. The rate of change of equilibrium constant $K(T)$ with temperature is thus (for gas reactions) proportional to the amount of heat evolved $(\partial H/\partial x)_{pT}$ per unit amount of reaction at equilibrium at constant T and P. This useful relationship is known as Van't Hoff's equation.

Electrochemical Processes

Some chemical reactions can occur spontaneously in solution, generating heat, or they can be arranged to produce electrical energy instead of heat. For example, a solution of copper sulfate contains free copper ions in solution. The addition of a small amount, Δn moles, of metallic zinc in powder form will cause Δn moles of copper to appear in metallic form, the zinc going into solution, replacing the Cu ions. At the same time an amount of $W\,\Delta n$ of heat is released. If the reaction takes place at constant pressure and temperature, Eq. (8-4) indicates that $W\,\Delta n$ must equal the difference in enthalpy between the initial state, with Cu in solution, and the final state, with Zn in solution,

$$H_1 - H_2 = W\,\Delta n \qquad (10\text{-}12)$$

However, the same reaction can take place in a battery, having one electrode of copper in a $CuSO_4$ solution and the other electrode (the negative pole of the battery) of zinc, surrounded by a $ZnSO_4$ solution, the two solutions being in contact electrically and thermally. If the battery now discharges an amount ΔC coulombs of charge through a resistor, or a motor to produce work, more Cu will be deposited on the Cu electrode and an equal number of moles, Δn, of Zn will leave the zinc electrode to go into solution. In this case the energy of the reaction goes into electromechanical work; for every charge ΔC discharged by the battery, $\mathfrak{B} \Delta C$ joules of work are produced, where $\mathfrak{B}$ is the equilibrium voltage difference between the battery electrodes.

If the battery is kept at constant temperature and pressure during the quasistatic production of electrical work, Eq. (8-11) (which in this case can be written $dG = -S\,dT + V\,dP + \mathfrak{B}\,dC$) shows that the work done equals the change in the Gibbs function caused by the reaction. In other words $\mathfrak{B}\,\Delta C = \Delta G$. But Eq. (10-4) shows the relationship between enthalpy and Gibbs function, which in this case can be written

$$\Delta G = H_1 - H_2 + T(\partial\, \Delta G/\partial T)_P \tag{10-13}$$

which, with Eq. (10-12), provides a relationship between the electrical properties of the battery and the thermal properties of the related chemical reaction.

If the ions have a valency z (for Zn and Cu, $z = 2$) then a mole of ions possesses a charge $z\mathfrak{F}$, where $\mathfrak{F}$ is the Faraday constant 9.65×10^7 coulombs per mole. Thus the charge ΔC is equal to $z\mathfrak{F}\,\Delta n$, where Δn is the number of moles of Zn that goes into solution (or the number of moles of Cu that is deposited). Combining Eqs. (10-12) and (10-13), we obtain the equation (using $\Delta G = z\mathfrak{F}\mathfrak{B}\,\Delta n$ and dividing by $z\mathfrak{F}\,\Delta n$)

$$\mathfrak{B} = (W/z\mathfrak{F}) + T(\partial\mathfrak{B}/\partial T)_P \tag{10-14}$$

relating the emf of the cell and the change in emf with temperature to the heat W evolved in the corresponding chemical reaction. Thus, by use of Eq. (10-5), we have derived an alternative means of obtaining empirical values of the chemical potentials μ_i. Electrical measurements, which can be made quite accurately, can be used instead of thermal measurements to measure heats of reaction. Equation (10-14) is called the Gibbs–Helmholtz equation.

2 KINETIC THEORY

CHAPTER

PROBABILITY AND DISTRIBUTION FUNCTIONS

We have now sketched out the main logical features of thermodynamics and have discussed a few of its applications. We could easily devote the rest of this text to other applications, covering all branches of science and engineering. But, as physicists, it is more appropriate for us to go on to investigate the connection between the thermal properties of matter in bulk and the detailed properties of the atoms that constitute this matter. The connection, as we saw in Chapter 2, must be a statistical one and thus will be expressed in terms of probabilities.

Probability

A basic concept is hard to define, except circularly. Probability is in part subjective, a quantization of our expectation of the outcome of some event (or trial) and only measurable if the event or trial can be repeated several times. Suppose one of the possible outcomes of a trial is A. We say that the probability that the trial results in A is $P(A)$ if we expect that, out of series of N similar trials, roughly $NP(A)$ of them will result in A. We expect that the fraction of the trials which do result in A will approach $P(A)$ as the number of trials increases. A standard example is the gambler's six-sided die; a 5 doesn't come up regularly every sixth time the die is thrown, but if a 5 comes up 23 times in 60 throws and also

85 times in the following 240 throws, we begin seriously to doubt the symmetry of the die and/or the honesty of the thrower.

If result A occurs for every trial, then $P(A) = 1$. If other events sometimes occur, such as event B, then the probability that A does *not* occur in a trial, $1 - P(A)$, is not zero. It may be possible that *both A and B* can occur in a single trial; the probability of this happening is written $P(AB)$. Simple logic will show that the probability of *either A or B or both* occurring in a trial is

$$P(A + B) = P(A) + P(B) - P(AB) \tag{11-1}$$

Relationships between probabilities are often expressed in terms of the *conditional probabilities* $P(A|B)$ that A occurs in a trial *if B* also occurs, and $P(B|A)$ the probability that B occurs *if A* occurs as well. We can see that

$$P(AB) = P(A|B)P(B) = P(B|A)P(A) = P(BA) \tag{11-2}$$

A simple example is in the dealing of well-shuffled cards. The chance of a heart being dealt is $P(H) = 1/4$, the chance that the card is a seven *if* it is a heart is $P(7|H) = 1/13$ and therefore the probability that the card dealt is the seven of hearts is $P(7H) = P(7|H)P(H) = (1/13)(1/4) = 1/52$.

If the probability of A occurring in a trial is not influenced by the simultaneous presence or absence of B, i.e., if $P(A|B) = P(A)$, then A and B are said to be *independent*. When this is the case,

$$P(A|B) = P(A); \qquad P(B|A) = P(B); \qquad P(AB) = P(A)P(B)$$
$$[1 - P(A + B)] = [1 - P(A)][1 - P(B)] \tag{11-3}$$

Saying it in words, if A and B are independent, then the chance of *both A and B* occurring in a trial is the *product* $P(A)P(B)$ of their separate probabilities of occurrence and the probability $1 - P(A + B)$ that *neither A nor B* occur is the product of their separate probabilities of nonoccurrence. In the example of the dealing of cards, since the chance $P(7)$ of a seven being dealt is the same as the conditional probability $P(7|H)$ that a seven is dealt *if* it is a heart, the occurrence of a seven is independent of the occurrence of a heart. Thus the probability that the card dealt is a seven of hearts is $P(7)P(H) = (1/13)(1/4) = 1/52$, and the chance that it is neither a seven nor a heart is $(12/13)(3/4) = 9/13$. If $A, B,..., M$ are all independent events, the probability that *all* of them occur is the product $P(A)P(B) \cdots P(M)$.

If the trial is such that, when A occurs B cannot occur and vice versa, then A and B are said to be *exclusive* and

$$P(AB) = P(A|B) = P(B|A) = 0$$

so that

$$P(A + B) = P(A) + P(B) \tag{11-4}$$

The chance of *either* A or B occurring, when A and B are exclusive, is thus the *sum* of their separate probabilities of occurrence. For example, the result that a thrown die comes up a 5 is exclusive of its coming up a 1; therefore the chance of either 1 or 5 coming up is $(1/6) + (1/6) = 1/3$. If $A, B,..., M$ are exclusive events, the probability that *one* of them occur is $P(A) + P(B) + \cdots + P(M)$.

Probabilities are useful in discussing *random* events. The definition of randomness is as roundabout as the definition of probability. The results of successive trials are randomly distributed if there is no pattern in the successive outcomes, if the only prediction we can make about the outcome of the next trial is to state the probabilities of the various outcomes.

Distribution Functions

The enumeration of the probabilities of all the possible outcomes of an event is called a *distribution function*, or a probability distribution. It is a means of specifying the statistical properties of a situation involving random events. A simple example is the number of passengers carried by a commercial airplane on a particular scheduled flight. This number might be any integer from zero up to the maximum number M which the plane can carry. A list of values of the probability P_n, that the number is equal to n, for all values of n from 0 to M, will describe the passenger-carrying characteristics of a given run as closely as the random nature of the situation will allow.

On some scheduled runs the plane is nearly always full; the probability P_M is larger than all the other P_n's. On another run the plane is nearly always half-full; on still another run there is great variability in the number carried, and so on. The distribution function, the set of values of all the $M + 1$ probabilities P_n, will display these characteristics in a quantitative manner.

Since the number of passengers carried has to be one of the integers from 0 to M, Eq. (11-4) indicates that the sum of all the probabilities,

$$\sum_{n=0}^{M} P_n = 1 \tag{11-5}$$

includes all the possibilities and hence must be certainty ($=1$).

Of course we can describe the situation in other ways. We can use the *expected* or average *number* $\langle n \rangle$, carried on a particular run, as a partial description. This is related to the distribution function by the equation

$$\langle n \rangle = \sum_{n=0}^{M} n P_n \tag{11-6}$$

But a notation of the expected number $\langle n \rangle$, carried on a particular run, may not be a complete enough description; we might need a measure of the variability of the number carried, whether the number of passengers is most likely to be nearly equal to $\langle n \rangle$ or whether almost any value is equally likely. A measure of variability is the *variance* σ_n^2 of the number of passengers n, the mean square of the deviation from the expected value of the number on any individual run,

$$\sigma_n^2 = \sum_{n=0}^{M} (n - \langle n \rangle)^2 P_n = \sum_{n=0}^{M} [n^2 - 2n\langle n \rangle + \langle n \rangle^2] P_n$$

$$= \sum_{n=0}^{M} n^2 P_n - \langle n \rangle^2 = \langle n^2 \rangle - \langle n \rangle^2 \tag{11-7}$$

The quantity σ_n, the square root of the variance, is called the *standard deviation* of the variable n. If variance σ_n^2 is small, it means that P_n has a peak for n near $\langle n \rangle$ and has small values for n very different from $\langle n \rangle$; if the variance is large then it is just as likely (or perhaps more likely) that $|n - \langle n \rangle|$ is large as that it is small.

Of course specifying the values of $\langle n \rangle$ and σ_n^2 does not completely determine the statistical behavior of the plane loading, as does the listing of values of all the P_n's, but it may be sufficient for our purposes. As we have seen, if we know the distribution function we can compute $\langle n \rangle$ and σ_n^2; the reverse is not possible.

Binomial and Multinomial Distributions

Several probability distributions of general usefulness deserve detailed study. One such distribution occurs when we combine several events, each of which involves an element of randomness. For example, suppose we carry out a sequence of N statistically equivalent "trials," each of which has a chance of "success" of p and thus a chance of "failure" $1 - p = q$. For example, the "trial" could be the tossing of a coin and the "success" could be that the coin landed with head showing. Or the trial could be the penetration of a target by a high-energy nuclear particle and the "success" could be the production of a particular nuclear reaction by collision of the particle with one of the target nuclei. We now ask for the probability that n of the N trials result in success.

This is a discrete probability distribution; n can have any integral value from 0 to N. The probability that all N trials are successes is, according to Eq. (11-3), p^N, if p is the probability of success in one trial. The probability that the first trial is a failure and the rest are successes is $p^{N-1}q$, where $q = 1 - p$, which also is equal to the probability that the second trial be a failure and the first and all the rest be successes, and so on. Usually we are not interested in the relative order of the successes and failures, we just wish to know the probability that n of the N trials are successes and $N - n$ are failures, without regard to order.

Now the number of different ways we can order N different things is $N!$ But the N trials are not all different, n of them are alike (successes) and the other $N - n$ are also alike (failures). We must thus divide out the ways in which the n and the $(N - n)$ things can be ordered and arrive at $[N!/n!(N - n)!]$ as the number of different ways we can have n successes and $N - n$ failures. Multiplying this by the probability $p^n q^{N-n}$ of occurrence of any one ordering, we finally obtain the distribution

$$P_n(N) = \frac{N!}{n!(N - n)!} p^n q^{N-n} \qquad (q = 1 - p) \qquad (11\text{-}8)$$

as the probability that N trials will result in n successes and $N - n$ failures. This is called the *binomial distribution*, because we can use the binomial expansion to prove Eq. (11-5) for it,

$$\sum_{n=0}^{N} P_n(N) = \sum_{n=0}^{N} \frac{N!}{n!(N - n)!} p^n q^{N-n} = (p + q)^N = 1^N = 1$$

In fact the coefficients in the binomial expansion are just the number of different ways in which n factors p and $N - n$ factors q can be ordered in the binomial expansion.

The expected value of n is not difficult to obtain

$$\langle n \rangle = \sum_{n=0}^{N} n P_n(N) = Np \sum_{n=1}^{N} \frac{(N-1)! p^{n-1} q^{N-1-n+1}}{(n-1)!(N-1-n+1)!} = Np$$

(11-9)

which is a quantitative paraphrase of the statement made in the second paragraph of this chapter. The expected number of successes in N trials equals the number of trials times the probability of success per trial. We can now go further, and find out how often the number of successes in N trials will differ from the expected result. We note that although $P_n(N)$ is largest for n near Np, the chance that n is quite different from Np is not zero.

A single measure of the tendency of n to stray from the expected value Np is the variance of n for this distribution,

$$\sigma_n^2 = \sum_{n=0}^{N} n^2 P_n(N) - \langle n \rangle^2 = \langle n(n-1) \rangle + \langle n \rangle - p^2 N^2$$

$$= p^2 N(N-1) + pN - p^2 N^2 = Np(1-p) = Npq$$

(11-10)

The square root of this is $\sigma_n = \sqrt{(pqN)}$, the standard deviation of n from its expected value in any single sequence of N trials. This is zero if $p = 1$ or $p = 0$; in either case the outcome is certain and there can be no variance in the results. The maximum variance is for $p = \frac{1}{2}$. The *fractional deviation*

$$(\sigma_n/N) = \sqrt{pq/N}$$

(11-11)

is a measure of the deviation of the fraction (n/N), of successes in any one sequence of N trials, from its expected value p. If it is large, a single sequence of N trials will not result in a very good experimental estimate of p; if it is small then the sequence of trials will likely result in a value of (n/N) close to p. We see that the fractional deviation decreases as N increases, but very slowly. To decrease our probable error of measurement of p by a factor of 10 we have to multiply the number of trials in the sequence by 100.

Now suppose the results of each trial are not just success or failure, but comprise v different outcomes, the single-trial probability that the ith outcome occurs being p_i

(where, of course, $\sum_{i=1}^{v} p_i = 1$)

Then the probability that, in N trials, outcome 1 occurs n_1 times, outcome 2 occurs n_2 times, outcome i occurs n_i times, and so on, up to $i = v$
(with the proviso that $\sum_{i=1}^{v} n_i = N$ in order to have all the trials accounted for) is the obvious generalization

$$P_{n_1 \cdots n_v}(N) = \frac{N!}{n_1! n_2! \cdots n_v!} \, p_1^{n_1} p_2^{n_2} \cdots p_v^{n_v} \tag{11-12}$$

where

$$\sum_{i=1}^{v} p_i = 1 \qquad \text{and} \qquad \sum_{i=1}^{v} n_i = N$$

We also have that

$$\sum P_{n_1 \cdots n_v}(N) = \left(\sum_i p_i\right)^N = 1$$

This is called the *multinomial distribution*. We shall find it useful when we come to talk about quantum statistics.

Random Walk

To make these generalities more specific let us consider the process called the *random walk*. We imagine a particle in irregular motion along a line; at the end of each period of time τ it either has moved a distance δ to the right or a distance δ to the left of where it was at the beginning of the period. Suppose the direction of each successive "step" is a random variable, independent of the direction of the previous one, that the probability that the step is in the positive direction is p and that the probability that it is in the negative direction is $1 - p$. Then the probability that during N periods the particle has made n positive steps, and thus $N - n$ negative steps, is the binomial probability $P_n(N)$ of Eq. (11-8).

The net displacement after these N periods, $x(n) = (2n - N)\delta$, might be called the "value" of the N random steps of the particle. The expected value of this displacement after N steps can be computed in terms of the expected value $\langle n \rangle$ of n, from Eq. (11-9), so

$$\langle x(n) \rangle = \langle (2n - N)\delta \rangle = (2p - 1)N\delta \tag{11-13}$$

When $p = 1$ (the particle always moves to the right) the expected displacement after N steps is $N\delta$; when $p = 0$ (the particle always moves to the left) it is $-N\delta$; when $p = \frac{1}{2}$ (the particle is equally likely to move right as left at each step) the expected displacement is zero. The variability of the actual result of any particular set of N steps can be obtained from $\langle n \rangle$ and $\langle n^2 \rangle$,

$$(\sigma_x)^2 = \langle [x(n) - N(2p - 1)\delta]^2 \rangle = \langle (2n - 2pN)^2 \delta^2 \rangle$$
$$= 4Np(1 - p)\delta^2 \qquad (11\text{-}14)$$

When $p = 1$ (all steps to right) or $p = 0$ (all steps to left) the variance $(\sigma_x)^2$ of the displacement is zero; when $p = \frac{1}{2}$ the variance is greatest, being $N\delta^2$ (the standard deviation is then $\sigma_x = \delta\sqrt{N}$). These results will be of interest when we come to talk about Brownian motion in Chapter 15.

The Exponential Distribution

Sometimes the variable of interest is continuous, not discrete. For example, we may be measuring the time interval between successive nuclear disintegrations in a sample of radioactive material, as measured by the successive clicks of a Geiger counter. Sometimes this interval will be quite short; two disintegrations coming practically simultaneously; occasionally there will be a long silence before the next click. Here the time interval t between clicks might have any value between 0 and ∞, and what we have to determine is the probability that the interval is of length between t and $t + dt$. This will be proportional to dt, so we can write it as $f(t)\,dt$, where $f(t)$ is called the *probability density* for the distribution. Thus the density $f(t)$ of the continuous variable t takes the place of the discrete distribution P_n. As before, if the range $0 \le t < \infty$ includes all possible values of the interval, we shall have

$$\int_0^\infty f(t)\,dt = 1 \qquad (11\text{-}15)$$

analogous to Eq. (11-5). The related equations for the expected length of interval and its variance are

$$\langle t \rangle = \int_0^\infty t f(t)\,dt; \qquad \sigma_t^2 = \int_0^\infty (t - \langle t \rangle)^2 f(t)\,dt = \langle t^2 \rangle - \langle t \rangle^2 \qquad (11\text{-}16)$$

The simplest example of a continuous distribution is that of the intervals between dots placed at random along a line. The line may be the distance traveled by a molecule of a gas and the dots may represent collisions with other molecules, or the line may be the time axis and the dots may be the instants of emission of a gamma ray from a piece of radioactive material; in any of these cases a dot is equally likely to be found in any element dx of length of the line. If the mean density of dots is $(1/\lambda)$ per unit length (λ being thus a mean distance between dots) then the chance that a randomly chosen element dx of the line will contain a dot is dx/λ, independent of the position of dx.

We can now compute the probability $P_0(x/\lambda)$ that *no* dot will be found in an interval of length x of the line, by setting up a differential equation for P_0. The probability $P_0[(x + dx)/\lambda]$ that no dots are in an interval of length $x + dx$ is equal to the product of probability $P_0(x/\lambda)$, that no dots are in length x, times the probability $1 - (dx/\lambda)$, that no dot is in the additional length dx [see Eq. (11-3)]. Using Taylor's theorem we have

$$P_0\left(\frac{x + dx}{\lambda}\right) = P_0\left(\frac{x}{\lambda}\right) + \frac{d}{dx}P_0\left(\frac{x}{\lambda}\right)dx = P_0\left(\frac{x}{\lambda}\right)\left[1 - \left(\frac{dx}{\lambda}\right)\right]$$

or

$$\frac{d}{dx}P_0\left(\frac{x}{\lambda}\right) = -\left(\frac{1}{\lambda}\right)P_0\left(\frac{x}{\lambda}\right) \quad \text{so that} \quad P_0\left(\frac{x}{\lambda}\right) = e^{-x/\lambda}$$

$$(11\text{-}17)$$

The probability that no dot is included in an interval of length x, placed anywhere along the line, thus decreases exponentially as x increases, being unity for $x = 0$ (it is certain that no dot is included in a zero-length interval), being 0.368 for $x = \lambda$, the mean distance between dots (i.e., a bit more than a third of the intervals between dots are larger than λ) and dropping off rapidly to zero for x larger than λ. Obviously this distribution is called the *exponential* distribution.

The probability density for this distribution is the derivative of P_0; it supplies the answer to the following question. We start at an arbitrary point and move along the line: What is the probability that the *first* dot encountered lies between x and $x + dx$ from the start? Equation (11-3) indicates that this probability, $f(x)\,dx$, is equal to the product of the probability $e^{-x/\lambda}$ that no dot is encountered in length x, times the probability dx/λ that a dot *is* encountered in the next interval dx. Therefore

$$f(x) = (1/\lambda)e^{-x/\lambda} \qquad (11\text{-}18)$$

is the probability density for encountering the first dot at x (i.e., $f\,dx$ is the probability that the first dot is between x and $x + dx$).

From Eq. (11-16) we find that the expected distance one goes, from an arbitrarily chosen point on the line, before a dot is encountered is

$$\langle x \rangle = \int_0^\infty x f(x)\,dx = \lambda \int_0^\infty u e^{-u}\,du = \lambda \qquad (11\text{-}19)$$

for the randomly distributed dots. The variance of this distance is

$$(\sigma_x)^2 = \int_0^\infty (x - \lambda)^2 f(x)\, dx = \lambda^2 \tag{11-20}$$

so the standard deviation of x, $\sigma_x = \lambda$, is as large as the mean value λ of x, an indication of the variability of the interval sizes between the randomly placed dots.

The Poisson Distribution

We can go on to ask what is the probability $P_1(x/\lambda)$ of finding just one dot in an interval of length x of the line. This is obtained by using Eq. (11-3) again to show that the probability of finding the first dot between y and $y + dy$ from the beginning and no other dot between this and the end of the interval x is equal to the product $f(y)\, dy\, P_0[(x - y)/\lambda]$. Then we can use Eq. (11-4) to show that the probability $P_1(x/\lambda)$ that the one dot is somewhere within the interval x is the integral

$$P_1(x/\lambda) = \int_0^x (1/\lambda)e^{-y/\lambda}\, dy\, e^{-(x-y)/\lambda} = (x/\lambda)e^{-x/\lambda}$$

An extension of this argument quickly shows that the probability that there are exactly n dots in the interval x is

$$P_n(x/\lambda) = \int_0^x f(y)\, dy P_{n-1}\left(\frac{x-y}{\lambda}\right) = \frac{(x/\lambda)^n}{n!}e^{-x/\lambda}$$

We have thus derived another discrete distribution function, called the *Poisson distribution*, the set of probabilities $P_n(x/\lambda)$, that n dots, randomly placed along a line, occur in an interval of length x of this line. More generally, a sequence of events has a Poisson distribution if the outcome of an event is a positive integer and if the probability that the outcome is the integer n is

$$P_n(E) = (E^n/n!)e^{-E}; \qquad \sum_{n=0}^\infty P_n(E) = 1 \tag{11-21}$$

If $E < 1$, $P_0(E)$ is larger than any other P_n; if $E > 1$, P_n is maximum for $n \simeq E$, tending toward zero for n larger or smaller than this. Quantity E is the expected value of n, for

$$\langle n \rangle = \sum_{n=0}^\infty n P_n(E) = E \qquad \text{and}$$

$$(\sigma_n)^2 = \sum_{n=0}^{\infty} (n - E)^2 P_n(E) = E \tag{11-22}$$

Arrivals of customers to a store during some interval of time T or the emission of particles from a radioactive source in a given interval of time—both of these have Poisson distributions. We shall encounter distributions such as these later on.

The Normal Distribution

In the limit of large values of N for the binomial distribution or of large values of E for the Poisson distribution, the variable n can be considered to be proportional to a continuous variable x, and the probability P_n approaches, in both cases, the same probability density $F(x)$. This function $F(x)$ has a single maximum at $x = \langle x \rangle$, the expected value of x (which we shall write as X, for the time being, to save space). It drops off symmetrically on both sides of this maximum, the width of the peak being proportional to the standard deviation σ, as shown in Figure 11-1. Function F has the simplest form compatible with these requirements, plus the additional requirement that the integrals representing expected values converge,

$$F(x - X) = \frac{1}{\sigma \sqrt{2\pi}} e^{-(x-X)^2/2\sigma^2}; \qquad \int_{-\infty}^{\infty} F(x - X)\, dx = 1$$

$$\langle x \rangle = \int_{-\infty}^{\infty} xF(x - X)\, dx = X; \qquad \langle (x - X)^2 \rangle = \sigma^2 \tag{11-23}$$

This is known as the *normal*, or Gaussian, *distribution*. It is typical of the behavior of a system subject to a large number of small, independent random effects. As an example, we might take the limiting case of a symmetric random walk ($p = \frac{1}{2}$, where the number N of steps is large but the size of each step is small, say $\delta = \sigma/\sqrt{N}$. Then, if a particular sequence of steps turns out to have had n positive steps and $N - n$ negative ones, the net displacement would be $x = \sigma(2n - N)/\sqrt{N}$ and the probability of such a displacement would be

$$P_n(N) = \frac{N!\,\frac{1}{2}^N}{[\frac{1}{2}N + (x/2\sigma)\sqrt{N}]!\,[\frac{1}{2}N - (x/2\sigma)\sqrt{N}]!}$$

since $n = \frac{1}{2}N + (x/2\sigma)\sqrt{N}$.

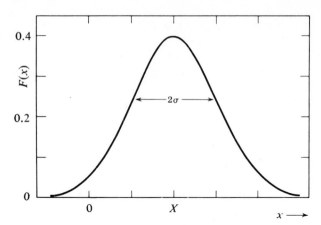

FIGURE 11-1

The normal distribution.

When N is very large, x tends to become practically a continuous variable and in the limit $P_n(N)\, dn \to F\, dx$, where F is the probability density

$$F = \lim_{N \to \infty}\left[\frac{\sqrt{N}}{2\sigma}\, P_n(N)\right]$$

$$= \lim_{N \to \infty}\left[\frac{(\sqrt{N}/2\sigma)N!\frac{1}{2}^N}{[\frac{1}{2}N + (x/2\sigma)\sqrt{N}]![\frac{1}{2}N - (x/2\sigma)\sqrt{N}]!}\right]$$

To evaluate this for large N we use the asymptotic formula for the factorial function,

$$n! \simeq \sqrt{2\pi n}\, n^n e^{-n}; \qquad n > 10 \tag{11-24}$$

which is called *Stirling's formula*. Using it for each of the three factorials and rearranging factors, we obtain

$$F = \lim\left[\frac{1}{\sigma\sqrt{2\pi}}\,\frac{1}{\sqrt{1 - (x^2/\sigma^2 N)}}\left(1 + \frac{x}{\sigma\sqrt{N}}\right)^{-(1/2)N - (x/2\sigma)\sqrt{N}}\right.$$

$$\left. \times \left(1 - \frac{x}{\sigma\sqrt{N}}\right)^{-(1/2)N + (x/2\sigma)\sqrt{N}}\right]$$

$$= \lim \left[\frac{1}{\sigma \sqrt{2\pi}} \left(1 - \frac{x^2}{\sigma^2 N} \right)^{-(1/2)N - 1/2} \left(1 - \frac{x}{\sigma \sqrt{N}} \right)^{x \sqrt{N}/2\sigma} \right.$$

$$\left. \times \left(1 - \frac{x}{\sigma \sqrt{N}} \right)^{-x \sqrt{N}/2\sigma} \right]$$

By using the limiting definition of the exponential,

$$e^z = \lim_{n \to \infty} \left(1 + \frac{z}{n} \right)^n \tag{11-25}$$

this expression reduces to that of Eq. (11-23), with $X = 0$. The terminal point of a random walk with a large number of small steps is distributed normally, as is any other effect which is the result of a large number of independent, random components. A proof that the limiting form of the Poisson distribution also is normal constitutes one of the problems. The distribution of random errors in a series of measurements is usually normal, as is the distribution of shots at a target. In fact the normal distribution is well-nigh synonymous with the idea of randomness.

CHAPTER

12

VELOCITY
DISTRIBUTIONS

Probability distributions are the connecting link between atomic characteristics and thermodynamic processes. We mentioned in Chapter 1 that each thermodynamic state of a system corresponded to any of a large number of microstates, macroscopically indistinguishable but microscopically different configurations of the system's atoms. If we had an assembly of thermodynamically equivalent systems, any one of the systems may be in any one of this large set of microstates; indeed each one of the systems will pass continuously from one microstate to another of the set. All we can specify is the *probability* f_i of finding the system (any one of them) in microstate i. In fact a specification of the distribution function f_i, specifying the value of f_i for each microstate possible to the system in the given macrostate, will serve to specify the thermodynamic state of the systems of the assembly.

Momentum Distribution for a Gas

In the case of a perfect gas of N point atoms, each atom is equally likely to be anywhere within the volume V occupied by the gas, but the distribution-in-velocity of the atoms is less uniformly spread out. What is needed is a probability density function, prescribing the probability that an atom, chosen at random from those present, should have a specified momentum, both in magnitude and direction. We can visualize this by imagining a three-dimensional *momentum space,* as shown in Figure 12-1, wherein the momentum **p** of any atom can be given either in terms of its rectangular components p_x, p_y, p_z or else in terms of its magnitude p and its spherical direction angles α and β. The probability density f is then a function of p_x, p_y, p_z or of p, α, β (or simply of the vector **p**) such that $f(\mathbf{p}) \, dV_p$ is the probability that an atom of gas will turn out to have a momentum vector **p** whose head lies within the volume element dV_p in momentum space (i.e., whose x component is between p_x and $p_x + dp_x$, y component is between p_y and $p_y + dp_y$, and z component is between p_z and $p_z + dp_z$, where $dV_p = dp_x \, dp_y \, dp_z = p^2 \, dp \sin \alpha \, d\alpha \, d\beta$).

We do not assume that any given atom keeps the same momentum forever; indeed its momentum changes suddenly, from time to time, as it collides with another atom or with the container walls. But we do assume, for the present, that these collisions are rather rare events and that if we should observe a particular atom the chances are preponderantly in favor of finding it between collisions, moving with a constant momentum, and that the probability that it has momentum **p** is given by the distribution function $f(\mathbf{p})$ which has just been defined.

If the state of the gas is an equilibrium state we would expect that f would be independent of time; if the state is not an equilibrium one, f may depend on time as well as on **p**. By the basic definition of a probability density, we must have that

$$\iiint f(\mathbf{p}) \, dV_p = \int_{-\infty}^{\infty} dp_x \int_{-\infty}^{\infty} dp_y \int_{-\infty}^{\infty} dp_z \, f(\mathbf{p})$$

$$= \int_0^{2\pi} d\beta \int_0^{\pi} \sin \alpha \, d\alpha \int_0^{\infty} f(\mathbf{p}) p^2 \, dp = 1 \qquad (12\text{-}1)$$

since it is certain that a given atom must have some value of momentum. The distribution function will enable us to calculate all the various average values characteristic of the particular thermodynamic state specified by our choice of $f(\mathbf{p})$. For example, the mean kinetic energy of the gas molecule (between collisions, of course) is

$$\langle \text{K.E.} \rangle_{\text{tran}} = \int_0^{2\pi} d\beta \int_0^{\pi} \sin \alpha \, d\alpha \int_0^{\infty} f(\mathbf{p})(p^2/2m) p^2 \, dp \qquad (12\text{-}2)$$

and the total energy of a gas of point atoms would be $N\langle \text{K.E.} \rangle$, where N is the total number of atoms in the system [see Eq. (2-1)].

If the gas is moving as a whole, there will be a *drift velocity* **V** superimposed on the random motion, so that $f(\mathbf{p})$ is larger in

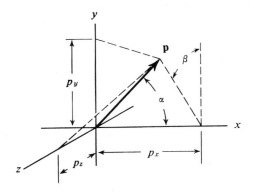

FIGURE 12-1
Coordinates in momentum space.

one direction of **p** than in the opposite direction, more atoms going in the positive x direction (for example) than in the negative x direction. In this case the components of the drift velocity are

$$V_x = \langle p_x/m \rangle = \int_{-\infty}^{\infty} (p_x/m)\, dp_x \int_{-\infty}^{\infty} dp_y \int_{-\infty}^{\infty} dp_z\, f(\mathbf{p})$$

$$= \int_0^{2\pi} d\beta \int_0^{\pi} \sin \alpha\, d\alpha \int_0^{\infty} (p/m) \cos \alpha\, f(\mathbf{p}) p^2\, dp \qquad (12\text{-}3)$$

and similarly for V_y and V_z.

If the gas is in equilibrium and its container is at rest the drift velocity must of course be zero. In fact at equilibrium it should be just as likely to find an atom moving in one direction as in another. In other words, for a gas at equilibrium in a container at rest we should expect to find the distribution function independent of the direction angles α and β and dependent only on the magnitude p of the momentum. In this case $\mathbf{V} = 0$ and

$$4\pi \int_0^{\infty} f(\mathbf{p}) p^2\, dp = 1; \qquad \langle \text{K.E.} \rangle = \frac{2\pi}{m} \int_0^{\infty} f(\mathbf{p}) p^4\, dp \qquad (12\text{-}4)$$

The Maxwell Distribution

We now proceed to obtain, by a rather heuristic argument, the momentum distribution of a gas of point atoms in equilibrium at temperature T. A more "fundamental" derivation will be given later; at present understandability is more important than rigor. We have already seen that at equilibrium the distribution function should be a function of the magnitude p of the momentum, indedependent of the angles α and β. One additional fact can be brought to bear: Eq. (11-3) states that, if the magnitudes of the three components of the momentum are distributed independently, then $f(\mathbf{p})$ should equal the product of the probability densities of each component separately, $f(\mathbf{p}) = F(p_x) \cdot F(p_y) \cdot F(p_z)$, each of the factors being similar functions of the three components.

Moreover, since the atomic motions are entirely at random, it would seem reasonable that the function F should have the form of the normal distribution, Eq. (11-23), which represents the effects of randomness. Thus we would expect that the equilibrium momentum distribution would be

$$f_0(\mathbf{p}) = F(p_x)F(p_y)F(p_z) = \frac{1}{(2\pi\sigma^2)^{3/2}} \exp\left[\frac{-p_x^2 - p_y^2 - p_z^2}{2\sigma^2}\right]$$

(12-5)

where σ is the standard deviation of either of the momentum components from its zero mean value. This result strengthens our impression that we are on the right track, for the sum $p_x^2 + p_y^2 + p_z^2 = p^2$ is independent of the angles α and β, and thus f is a function of the magnitude of $\mathbf{p}$ and independent of α and β. In order to have $f(\mathbf{p})$ be, at the same time, the product of functions F of the individual components and also a function of p alone, f must have the exponential form of Eq. (12-5) (or, at least, this is the simplest function which does so).

To find the value of the variance σ^2 in terms of the temperature of the gas, we have recourse to the results of Chapters 2 and 3, in particular of Eq. (3-2), relating the mean kinetic energy of translational motion of the atoms in a perfect gas with the temperature T. To compute mean values for the normal distribution we write down the following integral formulas:

$$\int_0^\infty e^{-u^2/a}\, du = \tfrac{1}{2}\sqrt{\pi a}$$

$$\int_0^\infty e^{-u^2/a}u^{2n}\, du = \tfrac{1}{2}\sqrt{\pi a}\,\tfrac{1}{2}\cdot\tfrac{3}{2}\cdot\tfrac{5}{2}\cdots\frac{2n-1}{2}\,a^n$$

(12-6)

$$\int_0^\infty e^{-u^2/a}u^{2n+1}\, du = \tfrac{1}{2}n!\,a^{n+1}$$

Therefore the mean value of the atomic kinetic energy is

$$\langle \text{K.E.} \rangle = 4\pi \int_0^\infty (p^2/2m) f(\mathbf{p}) p^2 \, dp = \frac{1}{m\sigma^3 \sqrt{2\pi}} \int_0^\infty p^4 e^{-p^2/2\sigma^2} \, dp$$

$$= \tfrac{3}{2}(\sigma^2/m)$$

which must equal $\tfrac{3}{2}kT$, according to Eq. (3-2). Therefore the variance σ^2 is equal to mkT, where m is the atomic mass, k is Boltzmann's constant, and T is the thermodynamic temperature of the gas.

Thus a rather heuristic argument has led us to the following momentum distribution for the translational motion of atoms in a perfect gas at temperature T,

$$f(\mathbf{p}) = (2\pi mkT)^{-3/2} e^{-p^2/2mkT} \tag{12-7}$$

which is called the *Maxwell distribution*. It is often expressed in terms of velocity $\mathbf{v} = \mathbf{p}/m$ instead of momentum. Experimentally we find that it corresponds closely to the velocity distribution of molecules in actual gases. The distribution is a simple one, being isotropic, with a maximum at $p = 0$ and going to zero at $p \to \infty$.

The expression (12-7) is the probability density of molecules in momentum space. In other words, $f(\mathbf{p}) \, dV_p$ is the probability that a molecule has momentum vector $\mathbf{p}$ with its head in the volume element $dV_p = dp_x \, dp_y \, dp_z = p^2 \sin \alpha \, d\beta \, d\alpha \, dp$, at the point $\mathbf{p} = \{p_x, p_y, p_z\} = \{p \cos \alpha, p \sin \alpha \cos \beta, p \sin \alpha \sin \beta\}$. We can translate this into other probability densities, for other coordinate geometries. For example, we may wish to compute the probability density for molecular speed, $v = p/m$, independent of direction. To obtain this we integrate over the direction angles α and β and change scale from p to v,

$$\int\int\int f \, dV_p = \int_0^{2\pi} d\beta \int_0^{\pi} \sin \alpha \, d\alpha [f(mv) \cdot m^3 v^2 \, dv]$$

$$= \frac{4\pi m^3}{(2\pi mkT)^{3/2}} [e^{-mv^2/2kT} v^2 \, dv] = f_v(v) \, dv$$

where

$$f_v(v) = \frac{4}{\sqrt{\pi}} \left(\frac{m}{2kT} \right)^{3/2} v^2 e^{-mv^2/2kT} \tag{12-8}$$

Use of Eqs. (12-6) will show that f_v satisfies Eq. (11-15).

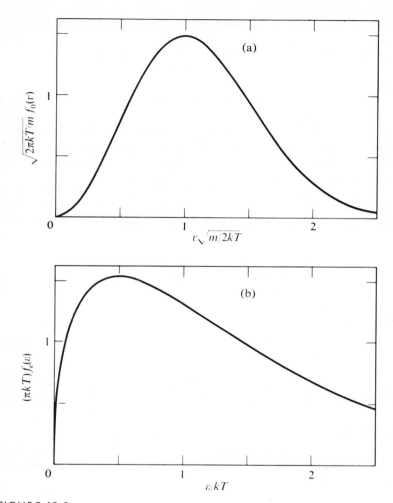

FIGURE 12-2

Maxwell distribution in speed (a) and energy (b).

A plot of $f_v(v)$ is given in Figure 12-2a. It starts at zero for $v = 0$, in distinction to $f(\mathbf{p})$, which is maximum at $p = 0$. The difference is of course because we have included all directions in $f_v(v)$ and the area of the spherical shell of radius $p = mv$ in momentum space is proportional to the square of v. The maximum value of $f_v(v)$ is at $v^2 = (2kT/m)$; for larger values it drops rapidly to zero. In fact the fraction of molecules having speed larger than v_0,

$$\int_{v_0}^{\infty} f_v(v)\, dv = \frac{2}{\sqrt{\pi}} e^{-mv_0^2/2kT} \int_0^{\infty} \left[\left(\frac{mv_0^2}{2kT}\right) + u\right]^{1/2} e^{-u}\, du$$

$$\rightarrow 1.13\sqrt{mv^2/2kT}\ e^{-mv_0^2/2kT} \qquad \text{when} \qquad mv_0^2 \gg kT$$

where we have set $v^2 = v_0^2 + (2kTu/m)$ in the integral. Thus about half of the molecules have speeds greater than $\sqrt{(2kT/m)}$, about $1/25$ of them have speeds greater than $2\sqrt{(2kT/m)}$ and only one molecule in about 2400 has a speed greater than $3\sqrt{(2kT/m)}$.

At times we wish to compute the probability density in terms of kinetic energy $\varepsilon = \frac{1}{2}mv^2 = p^2/2m$. This can be obtained from Eq. (12-8). The probability that a molecule has kinetic energy between ε and $\varepsilon + d\varepsilon$ is

$$f_v(\sqrt{2\varepsilon/m})\, d(\sqrt{2\varepsilon/m}) = \frac{4}{\sqrt{\pi}} \left(\frac{m}{2kT}\right)^{3/2} \frac{2\varepsilon}{m} e^{-\varepsilon/kT} \frac{d\varepsilon}{\sqrt{2m\varepsilon}}$$

$$= f_e(\varepsilon)\, d\varepsilon$$

where

$$f_e(\varepsilon) = \frac{2}{\sqrt{\pi}} (kT)^{-(3/2)} \sqrt{\varepsilon}\ e^{-\varepsilon/kT} \qquad\qquad (12\text{-}9)$$

This is plotted in Figure 12-2b, and it will be useful in Chapter 24, to compare with the distributions obtained from quantum statistics.

Mean Values

The mean velocity of the atoms is, of course, zero, since the momentum distribution is symmetric. The mean speed and the mean-squared speed are

$$\langle v \rangle = \langle p/m \rangle = \frac{4\pi}{m} \int_0^{\infty} p^3 e^{-p^2/2mkT} \frac{dp}{(2\pi mkT)^{3/2}} = \sqrt{\frac{8kT}{\pi m}}$$

$$\langle v^2 \rangle = \langle p^2/m^2 \rangle = 3(kT/m) \qquad\qquad (12\text{-}10)$$

We note that the mean of the square of the speed is not exactly equal to the square of the mean speed ($8/\pi$ is not exactly equal to 3, although the difference is not large). The mean molecular kinetic energy of translation is proportional to T; the mean molecular speed is proportional to $\sqrt{T}$.

If the gas is a mixture of two kinds of molecules, one with mass m_1, the other with mass m_2, then each species of molecule will have its own distribution in velocity, the one with m_1, instead of m in the expression of Eq. (12-7), the other with m_2 instead of m. This is equivalent to saying that the mean kinetic energy of translational motion of each species is $\frac{3}{2}kT$, no matter what the molecular weight of each is, as long as the two kinds are in equilibrium at temperature T. In fact if a dust particle of mass M is floating in the gas, being in equilibrium with the molecules of the gas, it will be in continuous, irregular motion (called *Brownian motion*), which is equivalent to the thermal motion of the molecules, so its mean kinetic energy of translation will also be $\frac{3}{2}kT$. Its mean-square speed, of course, will be less than the mean-square speed of a gas molecule, by a factor equal to the ratio of the mass of the molecule to the mass of the dust particle.

Finally, we should check to make sure that a gas with molecules having a Maxwell distribution of momentum will have a pressure corresponding to the perfect gas law of Eq. (3-1). Of those molecules which have an x component of momentum equal to p_x, $(N/V)\,dA(p_x/m)$ of them will strike per second on an area dA, perpendicular to the x axis, N/V being the number per unit volume and $(p_x/m)\,dA$ being the volume of the prism, in front of dA, which contains all the molecules that will strike dA in a second. Each such molecule, striking dA, would impart a momentum $2p_x$ if dA were a part of the container wall, so that the average momentum given to dA per second, which is equal to the pressure times dA, is (see Figure 2-1)

$$P\,dA = (N/V)\,dA \int_{-\infty}^{\infty} dp_z \int_{-\infty}^{\infty} dp_y \int_{0}^{\infty} (2p_x^2/m) f(\mathbf{p})\,dp_x$$

or

$$P = \frac{N/V}{(2\pi mkT)^{3/2}} \int_{0}^{\infty} \exp[-p_x^2/2mkT](2p_x^2/m)\,dp_x$$

$$\times \int\!\!\int_{-\infty}^{\infty} \exp[-(p_y^2 + p_z^2)/2mkT]\,dp_y\,dp_z$$

$$= \frac{N}{V}\sqrt{\frac{2}{\pi m^3 kT}} \int_{0}^{\infty} p_x^2\,\exp[-p_x^2/2mkT]\,dp_x$$

$$= NkT/V = nRT/V$$

where integration is only for positive values of p_x since only those with positive values of p_x are going to hit the area dA in the next second; the ones with negative values have already hit. Thus a gas with a Maxwellian distribution of momentum obeys the perfect gas law.

Experimental Verification

The most straightforward experiment to verify the Maxwell distribution is to let a beam of atoms, emergent from a furnace H at temperature T into a vacuum, pass through a slit S and eventually be caught, in plane P, by a device which measures the number of atoms falling per second per unit area of P, as shown in Figure 12-3. If the emergent beam is nearly horizontal, each atom will fall downward in the gravitational field, the faster ones falling less in the distance L, than the slower ones do. The reading of atomic flux $F(y_L)$ at M, as a function of y_L, will thus be related to the velocity distribution of the emergent beam.

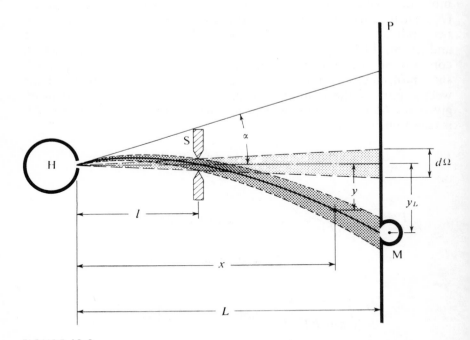

FIGURE 12-3

Molecular beam experiment for verification of the Maxwell distribution.

For the atoms which come out with a speed v, at an angle α with respect to the horizontal, the vertical displacement y below the horizontal is related to the horizontal distance of travel x by the equation

$$y = \frac{gx^2}{2v^2 \cos^2 \alpha} - \frac{\sin \alpha}{\cos \alpha} x$$

where g is the gravitational acceleration (9.8 m/sec). The slit S picks out those atoms that have a value of y near zero at $x = l$ and therefore relates the angle α to the speed v for those molecules that continue on to be measured at P. Setting $y = 0$ for $x = l$ we have $\sin 2\alpha = 2 \sin \alpha \cos \alpha = (gl/v^2)$. Now the peak of the Maxwell speed distribution f_v is for $v^2 = (2kT/m)$, so the angle α for the more prevalent molecular speed is such that $\sin 2\alpha = (mgl/2kT)$. For a furnace of 1600°K temperature and for cesium atoms (molecular weight 132) the fraction $(mgl/2kT)$ comes out to be approximately $10^{-4}l$, with l in meters. Therefore, unless l is extremely large, the angle α for most of the molecules which get through S to fall on P, is small enough so we can consider $\sin \alpha \simeq \alpha$ and thus $\alpha \simeq (gl/2v^2)$. Therefore the vertical displacement of an atom of speed v, by the time it strikes plane P, is

$$y_L = (L - l)(gL/2v^2)$$

The slit S allows through those atoms having velocity directions in a solid angle $d\Omega$ about the angle α, so that the number of molecules per unit volume in F, having velocity between v and $v + dv$, which are pointed in the solid angle $d\Omega$ is

$$N_0 \, d\Omega (m/2\pi kT)^{3/2} e^{-mv^2/2kT} v^2 \, dv$$

if the molecules in H have a Maxwell distribution and have a density N_0 per unit volume. The number of these falling per unit area *per second* on P is thus

$$K N_0 \, d\Omega (m/2\pi kT)^{3/2} e^{-mv^2/2kT} v^3 \, dv$$

where K is determined by the geometry of the slits.

Translating this into displacement y, we see that the flux falling on P between y_L and $y_L + dy_L$, as measured by M, is

$$A\left[\frac{(L-l)mgL}{4y_L kT}\right]^3 \exp\left[-\frac{(L-l)mgL}{4y_L kT}\right] dy_L = F(y_L) \, dy_L \quad (12\text{-}11)$$

where

$$A = K N_0 \left(\frac{kT}{2\pi m}\right)^{3/2} \frac{8}{(L-l)gL}$$

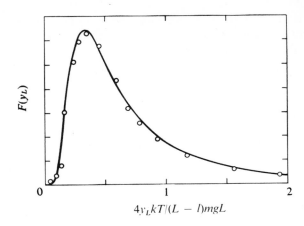

FIGURE 12-4

> *Molecular flux F and M for experiment of Figure 12-3. Solid line is for Maxwell distribution; circles are measurements of Estermann, Simpson, and Stern.*

The curve of flux density $F(y_L)$ is plotted in Figure 12-4. We see that it has a fairly sharp maximum at $y_L = [(L - l)mgL/12kT]$. The circles are experimental values, for cesium, as measured by Stern and his students. We see that the assumption of the Maxwell distribution fits the experimental results quite well, both in regard to the location of the maximum and also in regard to the general shape of F.

Collisions between Gas Molecules

The techniques of probability theory must also be used in describing the trajectory of a molecule in a gas, as it moves from one collision to the next. Most of the time, of course, the gas molecule is moving freely, unaffected by the presence of other molecules of the gas. Occasionally, two molecules collide, bouncing off with changed velocities. Roughly speaking, if two molecules come within a certain distance R of each other their relative motion is affected and we say they have collided; if their centers are farther apart than R they are not affected. To each molecule, all other molecules behave like targets, each of area $\sigma_c = \pi R^2$, perpendicular to the path of the molecule in question. If the path of this molecule's center of mass happens to intersect one of these targets, a collision has occurred and the path changes direction. Since there are N/V molecules in a unit volume, then in a disk-like region, unit area wide and dx thick, there are $(N/V)\,dx$ molecules. Therefore, the fraction of the disk obstructed by targets is $(N\sigma_c/V)\,dx$ and consequently the chance of the molecule in question having a collision while moving a distance dx is $(N\sigma_c/V)\,dx$. Target area σ_c is called the *collision cross section* of the molecules.

Thus a collision comes at random to a molecule as it moves around; the density $(1/\lambda)$ of their occurrence along the path of its motion is $N\sigma_c/V$. Reference back to the discussion before and after Eqs. (11-17) and (11-18) indicates that if the chance of encountering a "dot" (i.e., a collision) is dx/λ, λ is the mean distance between collisions (or dots) and the probability that the molecule travels a distance x without colliding and then has its next collision in the next dx of travel is

$$f(x)\,dx = (dx/\lambda)e^{-x/\lambda} \qquad \text{where} \qquad \lambda = (V/\sigma_c N) \qquad (12\text{-}12)$$

The mean distance between collisions λ is called the *mean free path* of the molecule. We see that it is inversely proportional to the density (N/V) of molecules and also inversely proportional to the molecular cross section σ_c. This mean free path is usually considerably longer than the mean distance between molecules in a gas. For example, σ_c for O_2 molecules is roughly 4×10^{-19} m^2 and N/V at standard pressure and temperature (0°C and 1 atm) is approximately 2.5×10^{25} molecules per m^3. The mean distance between molecules is then the reciprocal of the cube root of this, or approximately 3.5×10^{-9} m, whereas the mean free path is $\lambda = V/N\sigma_c \simeq 10^{-7}$ m, roughly 30 times larger. The reason, of course, is that the collision radius R is about 4×10^{-10} m, for O_2, about one-tenth of the mean distance between molecules; thus only about 1/1000 of the volume is "occupied" by molecules at standard conditions. The difference is even more marked at low pressures. At 10^{-7} atm, λ is 1 m in length, but there are still 2.5×10^{18} molecules per m^3 at this pressure, hence the mean distance between molecules is roughly 0.7×10^{-6} m. Neither λ nor σ_c is strongly dependent on the velocity distribution of the molecules.

We can also talk about a mean time τ between collisions, although this quantity does depend on the molecular velocity distribution. For a molecule having a speed p/m it would take, on the average, $m\lambda/p$ seconds for it to go from one collision to the next. Therefore the mean free time τ for a gas with a Maxwellian distribution of momentum is

$$\tau = \langle m\lambda/p \rangle = 4\pi m\lambda \int_0^\infty p e^{-p^2/2mkT} \frac{dp}{(2\pi mkT)^{3/2}} = \lambda\sqrt{\frac{2m}{\pi kT}}$$

$$(12\text{-}13)$$

which is not exactly equal to either $\lambda/\langle v \rangle$ or $\lambda/\sqrt{\langle v^2 \rangle}$ but it is not very different from either. The mean free time decreases with increasing temperature because an increase in temperature increases the mean molecular speed, whereas it does not change λ. Since the mean speed of an oxygen molecule at standard conditions is about 400 m/sec, the mean free time for these conditions is about 3×10^{-10} sec; at 0°C and 10^{-7} atm it is about (1/300) sec.

The Poisson distribution of Eq. (11-21) can be used to evaluate the probability that a molecule will have n collisions while it has traveled a distance x, measured along its trajectory. The expected number of collisions in this length is (x/λ), so the probability of having n collisions is $P_n(x/\lambda) = [(x/\lambda)^n/n!]e^{-x/\lambda}$. Likewise the probability that the molecule has n collisions in time t is $[(t/\tau)^n/n!]e^{-t/\tau}$, where τ is an average value of mean free time for the molecule during time t.

CHAPTER

13

DISTRIBUTION IN PHASE SPACE

In the previous chapter we discussed the probability density in momentum for point molecules in a gas subject to no forces, except for the infrequent ones incident to intermolecular collisions. An obvious question arises as to what must be the effect of the presence of a force field, such as that of gravity, which acts on every molecule. If the force is a conservative one, so that each molecule possesses a potential energy, $\phi(\mathbf{r})$, as well as a kinetic energy, between collisions, we would expect that the probability density would be a function of ϕ as well as of kinetic energy, that it would be less likely to find a molecule in regions of high potential energy than in regions of low.

Thus the probability density f should be a function, not only of the momentum $\mathbf{p} = \{p_x, p_y, p_z\}$ of translation of the point molecule but also of its position $\mathbf{r} = \{x, y, z\}$, so that $f(\mathbf{r}, \mathbf{p})\, dV_r\, dV_p$ is the probability that a molecule has momentum in the volume element $dV_p = dp_x\, dp_y\, dp_z$ at $\mathbf{p}$ and has position in the volume element $dV_r = dx\, dy\, dz$ at point $\mathbf{r}$ in position space. If the Maxwell distribution $N \exp\{-\text{kinetic energy}/kT\}$ is an appropriate probability density for momentum space when $\phi(\mathbf{r})$ is constant, we might guess that $N \exp\{-[(p^2/2m) + \phi(\mathbf{r})]/kT\}$ is the probability density $f(\mathbf{r}, \mathbf{p})$ in both momentum and position space when ϕ varies with $\mathbf{r}$. If this is the case, then f decreases whenever the molecular energy, either kinetic or potential, increases, and the decrease is the more abrupt the smaller the temperature T. Not only is it unlikely that a molecule, chosen at random, should have a kinetic energy larger than kT, it is also unlikely that a molecule be found in a region where the potential energy is larger than its minimal value by an amount equal to kT.

This distribution function, for molecules in a perfect gas, does turn out to be the correct form, but we shall have to wait before we can demonstrate the range of physical conditions over which it is valid. Meantime, in this chapter, we shall concern ourselves with preliminary matters, such as the investigation of the general properties of a probability density which is a function of both momentum and position.

Phase Space

The probability density f is thus a function of both the position $\mathbf{r}$ and the momentum $\mathbf{p}$ of the molecule (and, perhaps, also of time t, if the distribution changes with time). In principle, the specification of a molecule's momentum and position is sufficient to determine completely its future motion, if it were not for the presence of the other molecules. Although collisions are rare in the life of a gas molecule, they are frequent enough to make the detailed calculation of the trajectory of an individual molecule a quite impractical task. What we intend to do, instead, is to determine the distribution, in momentum and position, of the molecular motions between collisions, assuming that the effect of collisions is to make both position and momentum random, in the sense of the normal distribution of Eq. (11-23).

Thus we propose to define the status of an individual molecule at any instant in terms of six coordinates, x, y, z, p_x, p_y, p_z (and, perhaps, of a seventh, time). This six-dimensional space has been called *phase space*. Each molecule is pictured as following a trajectory in phase space, between collisions, as it moves under the influence of the general field of force represented by the potential energy $\phi(\mathbf{r})$. Because of the nature of the equations of motion these trajectories in phase space have certain properties which we shall find useful in our analysis. To demonstrate some of these properties let us take the simplest case, of a particle of mass m, constrained to move along a line, so its phase space is two-dimensional, the coordinates being position x and momentum $p_x = m\dot{x}$, as shown in Figure 13-1 (the dot over a letter stands for time differentiation).

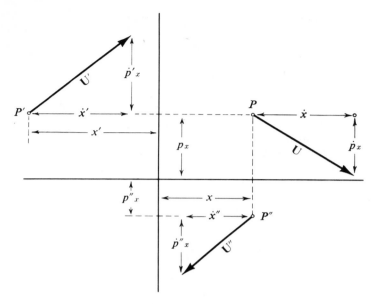

FIGURE 13-1

Representation of a one-dimensional molecule's position (x, p_x) and its velocity $\mathbf{U} = \{\dot{x}, \dot{p}_x\}$ at three points in phase space. Note that $\dot{x}' = \dot{x}$ and $\dot{p}''_x = \dot{p}_x$ but $\dot{p}'_x \neq \dot{p}_x$ and $\dot{x}'' \neq \dot{x}$.

If the particle is at P of Figure 13-1, it will not stay there; its velocity *in phase space* $\mathbf{U}$ will have two components, $\dot{x}$ and $\dot{p}_x$. But since $p_x = m\dot{x}$ there is a direct relationship between the x component of the molecule's *velocity* in phase space and its p_x component of *position* in phase space, $\dot{x} = (p_x/m)$. Several useful properties of molecular trajectories in phase space follow: (1) Velocities in phase space of all points above the x axis point to the right (i.e., have horizontal components $\dot{x}$ that are positive) and those below the x axis point to the left; (2) the horizontal components $\dot{x}$ of phase-space velocity $\mathbf{U}$ for every point along a horizontal line in phase space (such as P and P') are equal; and (3) the velocity vectors $\mathbf{U}$ for points along the x axis point vertically (i.e., all trajectories in phase space cross the x axis vertically).

But there is more to come. The p_x component of velocity in phase space, $\dot{p}_x$, is the rate of change of the molecule's momentum which, by Newton's equation, must equal the x component of the force $F_x(x)$ acting on the molecule. This force is usually conservative and can be obtained from a potential energy function $\phi(x)$, $F_x = -(\partial\phi/\partial x)$, which thus is a function of x, not of p_x,

$$\dot{p}_x = F_x = -(\partial\phi/\partial x) \tag{13-1}$$

We therefore have, in addition, that: (1) the vertical components $\dot{p}_x$ of phase-space velocity $\mathbf{U}$, of all points in a vertical line in phase space (such as P, P'') are equal; (2) the velocity vectors for the two points (x, p_x) and $(x, -p_x)$ have equal vertical components and horizontal components which are equal in magnitude but opposite in sign (which results in the trajectories in phase space being symmetric about the x axis); (3) if $\phi(x)$ has a minimum at $x = x_0$, then all $\mathbf{U}$ vectors for points on the line $x = x_0$ are horizontal (thus all trajectories cross this line horizontally).

Trajectories and Densities in Phase Space

As we have said before, if we know that a one-dimensional molecule is at point (x, p_x) in phase space, we shall be able to predict exactly where it will be any time t later. In other words, for a given one-dimensional molecule and force field, there passes through any point in phase space one *and only one* trajectory, representing the subsequent sequence of positions and momenta of the particle in future time (until the next collision). In fact we can write out the equation for the trajectory by using the equation for conservation of mechanical energy,

$$E = (p_x^2/2m) + \phi(x) = \text{constant} \tag{13-2}$$

Therefore, solving for p_x as a function of x,

$$p_x = \sqrt{2m[E - \phi(x)]}$$

which is the equation for the trajectory for total energy $= E$.

A given trajectory is thus specified by a particular value of E, its total energy. If a molecule starts at $t = 0$ from point P of Figure 13-2, it will inevitably pass through point P' at some later time t' and through point P'' at a still later time t''. If the molecule were at P at a time t, its motion would be similar but delayed; it would pass through P' at time $t + t'$ and through P'' at time $t + t''$. It is obvious that no two trajectories can cross.

The distance in phase space between two trajectories, as for instance the distance d between P and Q, is determined by computing the partials of E with respect to x and p_x. From Eq. (13-2),

$$\left(\frac{\partial E}{\partial x}\right)_{p_x} = \left(\frac{\partial \phi}{\partial x}\right) = -\dot{p}_x; \qquad \left(\frac{\partial E}{\partial p_x}\right)_x = \left(\frac{p_x}{m}\right) = \dot{x} \qquad (13\text{-}3)$$

These equations are called *Hamilton's equations*. They show that the length d, the distance in phase space between two trajectories,

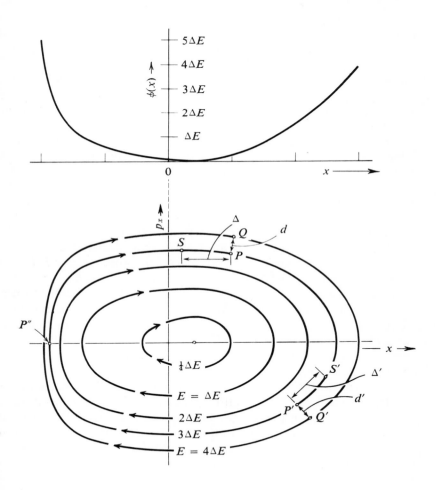

FIGURE 13-2

Potential function $\phi(x)$ for a one-dimensional molecule, and corresponding trajectories in phase space for differing total energies E.

differing by an amount ΔE in energy, at P, is

$$d = \frac{\Delta E}{\mathrm{grad}\ E} = \frac{\Delta E}{\sqrt{(\partial E/\partial x)^2 + (\partial E/\partial p_x)^2}} = \frac{\Delta E}{U} \qquad (\Delta E \to 0)$$

where $U = \sqrt{(\dot{p}_x^2 + \dot{x}^2)}$ is the magnitude of the velocity of the molecule's point in phase space. Thus if two molecules are at P and Q, a distance $(\Delta E/U)$ apart at time $t = 0$, at time t' they will be a distance $(\Delta E/U')$ apart, where U' is the phase-space speed of the molecule when at point P' (unless one or both molecules have suffered collision meanwhile). The faster the molecule is traversing phase space the closer together are the trajectories for differing values of E. Remember that speed in phase space, U, is not equal to the speed $\dot{x}$ of the molecule in ordinary space; U can be quite large even when $\dot{x}$ is zero, in regions where $\dot{p}_x$ is large (as at P'').

On the other hand, if two molecules are started along the same trajectory, but one delayed a time Δt behind the other (as with points P and S) then their distance Δ apart (until one or both suffer collision) is $\Delta = U\Delta t$ at P and $\Delta' = U'\Delta t$ at P', where U is again the phase-space speed given in the equation above.

We thus arrive at an interesting and important property of collective motions of molecules in phase space. A molecule at P at $t = 0$ is surrounded by other molecules, spaced distances $\Delta = U\,\Delta t$ apart along the trajectory E and spaced distances $d = \Delta E/U$ apart on parallel trajectories, thus having a density $(1/d\Delta) = (1/\Delta E\,\Delta t)$ of molecules per unit area of phase space. When this collection has moved on to point P' (assuming no collisions have occurred meanwhile) the spacing along the trajectory $U'\,\Delta t$ will be shortened but the cross-trajectory spacing $(\Delta E/U')$ will be correspondingly lengthened, so the density $(1/\Delta E\,\Delta t)$ of molecules per unit area of phase space *has remained the same*.

The trouble with a one-dimensional model is that collisions would inevitably and frequently occur if more than one molecule were to move along the x axis. During the collision both molecules would change their momentum and energy and thus they would disappear from one pair of points in phase space and reappear at two other points (all on the same vertical line, however). In strictly one-dimensional motion, conservation of energy and momentum for an elastic collision would require that the molecules would exchange places. This is illustrated in Figure 13-3. Molecule A starts in trajectory E and molecule B starts with energy E'. They collide at

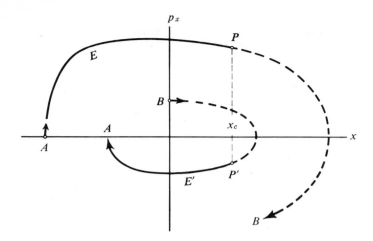

FIGURE 13-3

Phase-space representation of a one-dimensional elastic collision.

$x = x_c$, when A is at P and B at P'; they interchange momenta and therefore interchange phase-space trajectories.

The corresponding situation in the six-dimensional phase space for three-dimensional molecules is simpler in one sense and more complicated in others. There is much more room to dodge, so the number of collisions, for a given number of molecules, is much less frequent. As we pointed out in the discussion just above Eq. (12-13) the mean distance between collisions in a three-dimensional gas is usually much longer than the mean distance between molecules. The trajectories in phase space are each on hypersurfaces (five-dimensional "surfaces") of constant energy E. The trajectories covering a hypersurface have many of the properties of the two-dimensional ones of Figure 13-2: none of the trajectories cross; through each point one and only one trajectory passes, and so on. If the motion is confined inside a vessel, the hypersurfaces of constant E are closed surfaces, since they enclose a finite volume of the six-dimensional phase space. The trajectories on these surfaces may be closed on themselves (periodic motion) or they may not be (nonperiodic motion) depending on the nature of the potential energy $\phi(\mathbf{r})$.

The collisions, when they do occur, are not as rigidly pre-scribed as in the one-dimensional case. Energy and momentum are still conserved, for elastic collisions, but the two molecules, originally on hypersurfaces E_a and E_b, find themselves on hypersurfaces E'_a and E'_b, such that $E_a + E_b = E'_a + E'_b$. The exact values of E'_a and E'_b are determined by the requirements of conservation of momentum, as will be shown later. In general, during a collision, a pair of molecules will disappear from their original trajectories and reappear elsewhere, rather than replacing each other, as it was in the one-dimensional case.

Thus we have the picture of a swarm of N points in phase space, each corresponding to one of the N molecules of the gas, each point traversing a trajectory in phase space for the time between collisions, then disappearing, pairwise, to reappear at two other points, to take up two new trajectories. Each region of phase space, of volume $dV_r \, dV_p$ at point $\mathbf{r}$, $\mathbf{p}$, contains an average number of points equal to $Nf(\mathbf{r}, \mathbf{p}, t) \, dV_r \, dV_p$, where $f(\mathbf{r}, \mathbf{p}, t)$ is the probability density of molecules at the point $\mathbf{r}$, $\mathbf{p}$ in phase space at time t, and where $Nf = \rho(\mathbf{r}, \mathbf{p}, t)$ is the *density* of points in phase space. As usual, the integral of f over all accessible regions of phase space is unity, so that the integral of ρ is N.

Since the trajectories in a small region are all nearly parallel and since the points in a small region are all moving with nearly the same velocity $\mathbf{U} = \{\dot{x}, \dot{y}, \dot{z}, \dot{p}_x, \dot{p}_y, \dot{p}_z\}$, the neighboring points will tend to move along together, only changing slowly in relative position as the trajectories gradually move apart or as small differences in speed U become apparent; except that, from time to time, one of the points will disappear as one of the original set of molecules suffers a collision, or a new point will appear because a molecule, originally with different $\mathbf{p}$, has been knocked into this region by a collision. In situations of thermodynamic equilibrium the number of appearances will exactly equal the number of disappearances at each point of phase space, so it may seem that the same molecules are continuing along their trajectories forever. The effect of the collisions is there nonetheless, for it determines what form f must have in order that equilibrium can be achieved.

In the meantime we must investigate the equations governing the behavior of the probability density f, and its dependence on collisions, so we can work out expressions for f for different situations, equilibrium and nonequilibrium.

The Boltzmann Equation

The situation pictured in the last paragraphs is somewhat analogous to the flow of a compressible fluid. The changes of density ρ and velocity $\mathbf{v}$ of such a fluid are governed by several equations; two of them are appropriate for our use. In the first place there is the *equation of continuity*, which states that the rate of creation $q(\mathbf{r}, t)$ of fluid at a certain point $\mathbf{r}$ is equal to the rate of change of density there, $\partial\rho/\partial t$, plus the net *outflow* of fluid from that region, $\text{div}(\rho\mathbf{v})$,

$$q(\mathbf{r}, t) = (\partial\rho/\partial t) + \text{div}(\rho\mathbf{v})$$
$$= (\partial\rho/\partial t) + \rho\,\text{div}(\mathbf{v}) + \mathbf{v}\cdot\text{grad}(\rho) \qquad (13\text{-}4)$$

The other equation states that the change of density $(d\rho/dt)$ of a portion of the fluid, *as it moves along*, is equal to the change $(\partial\rho/\partial t)$ in ρ *at the region in space* where the portion initially was, plus the change caused by the differences in density at different regions of space (see the discussion of Figure 8-4),

$$(d\rho/dt) = (\partial\rho/\partial t) + \mathbf{v}\cdot\text{grad}(\rho)$$

where $(d\rho/dt)$ is related to q; in fact combining the two we find

$$(d\rho/dt) = q(\mathbf{r}, t) - \rho\,\text{div}(\mathbf{v})$$

which says that the change in density of a portion of a fluid, as it moves along, is equal to the rate of creation of fluid in the portion, minus the reduction of density caused by the differential motions of the fluid.

In the case of the flow of points in six-dimensional phase space, the phase-space velocity $\mathbf{U}$ has six components, $\dot{x}$, $\dot{y}$, $\dot{z}$, $\dot{p}_x$, $\dot{p}_y$, $\dot{p}_z$ and the gradient operator has six components $(\partial\rho/\partial x)$, $(\partial\rho/\partial y)$, $(\partial\rho/\partial z)$, $(\partial\rho/\partial p_x)$, $(\partial\rho/\partial p_y)$, $(\partial\rho/\partial p_z)$. The density $\rho(\mathbf{r}, \mathbf{p}, t) = Nf(\mathbf{r}, \mathbf{p}, t)$ is thus subject to the six-dimensional equivalents of the equations written out above. But an important property of phase space, discussed at the beginning of this chapter, enables us to reduce the two equations to one. The property may be rephrased by stating that the six-dimensional divergence of the phase-space velocity $\mathbf{U}$ is *everywhere zero*. For

$$\text{div }\mathbf{U} = \frac{\partial\dot{x}}{\partial x} + \frac{\partial\dot{y}}{\partial y} + \frac{\partial\dot{z}}{\partial z} + \frac{\partial\dot{p}_x}{\partial p_x} + \frac{\partial\dot{p}_y}{\partial p_y} + \frac{\partial\dot{p}_z}{\partial p_z}$$
$$= \frac{1}{m}\left[\frac{\partial p_x}{\partial x} + \frac{\partial p_y}{\partial y} + \frac{\partial p_z}{\partial z}\right] - \frac{\partial}{\partial p_x}\left(\frac{\partial\phi}{\partial x}\right) - \frac{\partial}{\partial p_y}\left(\frac{\partial\phi}{\partial y}\right) - \frac{\partial}{\partial p_z}\left(\frac{\partial\phi}{\partial z}\right) = 0$$
$$(13\text{-}5)$$

since the components p_x, p_y, p_z are independent of the components x, y, z and the potential energy ϕ is independent of p_x, p_y, p_z. Therefore the change of density of the fluid, as it moves along in phase space, is exactly equal to the net rate of creation, $q = NQ$, of points in phase space by collision. If there were no collisions, a group of points, once started along its trajectories, would never change its density; wherever the trajectories spread apart the spacing along the trajectories would correspondingly decrease to maintain the density unchanged, as we pointed out in the two-dimensional case of Eq. (13-5).

Thus both equations for fluid flow reduce to the same equation, for phase space (grad_6 is the six-dimensional gradient),

$$(df/dt) = Q(\mathbf{r}, \mathbf{p}, t) = (\partial f/\partial t) + \mathbf{U} \cdot \mathrm{grad}_6(f)$$

$$= \frac{\partial f}{\partial t} + \dot{x}\frac{\partial f}{\partial x} + \dot{y}\frac{\partial f}{\partial y} + \dot{z}\frac{\partial f}{\partial z}$$

$$+ \dot{p}_x\frac{\partial f}{\partial p_x} + \dot{p}_y\frac{\partial f}{\partial p_y} + \dot{p}_z\frac{\partial f}{\partial p_z}$$

$$= (\partial f/\partial t) + (\mathbf{p}/m) \cdot \mathrm{grad}_r f - \mathrm{grad}_r\phi \cdot \mathrm{grad}_p f \qquad (13\text{-}6)$$

where grad_r is the gradient in position space x, y, z, grad_p is the gradient in momentum space p_x, p_y, p_z, and where $\phi(\mathbf{r})$ is the potential energy acting on every molecule ($\dot{\mathbf{p}} = -\mathrm{grad}_r\phi$). This is *Boltzmann's equation*. It is of basic importance in calculating the thermal and mechanical behavior of gases, as we shall demonstrate in the next chapters. The terms on the right-hand side control the time, space, and momentum dependence of the probability density f of molecules in phase space. The left-hand term, $Q(\mathbf{r}, \mathbf{p}, t)$, represents the net gain, in any region of phase space, of the number of molecular points entering the region over the number leaving the region because of collisions. It is a complicated integral involving the collision cross section σ_c [see Eq. (12-12)] and the values of f for those other regions of momentum space which can be connected to $\mathbf{p}$ by collision. As we have indicated earlier, it is zero for gases in equilibrium. We shall discuss its general properties in the next chapter.

The Maxwell-Boltzmann Distribution

Meanwhile we shall investigate the implications of Eq. (13-6) for equilibrium states, where Q is zero and $(\partial f/\partial t)$ is zero, and the

Boltzmann equation reduces to

$$(\mathbf{p}/m) \cdot \mathrm{grad}_r f = \mathrm{grad}_r \phi \cdot \mathrm{grad}_p f \qquad (13\text{-}7)$$

Our earlier, heuristic arguments regarding the dependence of the probability density f, for an equilibrium state, on momentum $\mathbf{p}$ should still be valid. In other words we would expect that $f(\mathbf{r}, \mathbf{p}) = f_r(\mathbf{r}) f_p(\mathbf{p})$, where $f_p(\mathbf{p})$ is the Maxwell distribution of Eq. (12-7). Therefore, in Eq. (13-7) $\mathrm{grad}_p f$ should be equal to $-(\mathbf{p}/mkT) f_r f_p$, so the equation becomes

$$(\mathbf{p}/m) \cdot [\mathrm{grad}_r f_r + (f_r/kT) \, \mathrm{grad}_r \phi] f_p = 0$$

or

$$\mathrm{grad}_r f_r = -(f_r/kT) \, \mathrm{grad}_r \phi \qquad \text{or} \qquad f_r(\mathbf{r}) = Be^{-\phi/kT} \qquad (13\text{-}8)$$

Thus the distribution function for this case is

$$f(\mathbf{r}, \mathbf{p}) = \frac{B}{(2\pi mkT)^{3/2}} \, \exp\left[-\frac{1}{kT}\left(\frac{p^2}{2m} + \phi\right)\right] \qquad (13\text{-}9)$$

This is known as the *Maxwell–Boltzmann distribution*. The formula states that the probability of presence of a molecule at a point $\mathbf{r}, \mathbf{p}$ in phase space is determined by its *total energy* $(\mathbf{p}^2/2m) + \phi(\mathbf{r}) = H(\mathbf{r}, \mathbf{p})$; the larger H is, the smaller is the chance that a molecule is present; the smaller T is, the more pronounced is this probability difference between points where H differs. Constant B is, of course, adjusted so that the integral of $f_r = Be^{-\phi/kT}$, over the volume of position space occupied by the gas, is unity. The factor f_p is already adjusted so its integral, over all momentum space, is unity. With such factors, the function $f(\mathbf{r}, \mathbf{p})$ is said to be *normalized*, and constant B is called a *normalization constant* for f_r.

This is the distribution function we spoke of at the beginning of this chapter. The discussion leading up to Eq. (13-9) does not pretend to be a logically satisfactory derivation; in fact it is no better than our "derivation" of the Maxwell distribution, on which it depends. Our excuse is twofold: a logical "derivation" of a physical law is less important than that the law correspond to nature, and the Maxwell and the Maxwell–Boltzmann distributions do thus correspond, as we have seen and will see. Second, we shall later demonstrate a more basic connection between these distributions and the general distribution functions of statistical mechanics which may be taken as a more logical "derivation" of these distributions, if this seems desirable. Meanwhile we should investigate the implications of Eq. (13-9).

A Simple Example

Suppose a gas at equilibrium at temperature T is confined to two interconnected vessels, one of volume V_1, at zero potential energy, the other of volume V_2 at the lower potential energy $\phi = -\gamma$. The connecting tube between the two containers should allow free passage of the gas molecules, although its volume should be negligible compared to V_1 or V_2. The potential difference between the vessels may be gravitational, the vessels being at different elevations above sea level; or it may be due to an electric potential difference, if the molecules possess electric charges.

In this simple case the factor $f_r = Be^{-\phi/kT}$ of the Maxwell–Boltzmann distribution is simply B throughout volume V_1 and $Be^{\gamma/kT}$ throughout V_2. Since the integral of f_r over the total volume must be unity, we must have $B = [V_1 + V_2 e^{\gamma/kT}]^{-1}$. The number of molecules per unit volume in the upper container V_1 is B times the total number N of molecules in the system; the density of molecules in the lower container V_2 is $NBe^{\gamma/kT}$. Since the distribution in momentum is Maxwellian in both vessels, the pressure in each container is kT times the density of molecules there.

$$\text{For } V_1: \qquad P_1 = \frac{NkT}{V_1 + V_2 e^{\gamma/kT}}$$

$$\text{In } V_2: \qquad P_2 = \frac{NkT}{V_1 e^{-\gamma/kT} + V_2} \qquad (13\text{-}10)$$

At temperatures high enough so that $kT \gg \gamma$ the exponentials in the denominators of these expressions are nearly unity and the pressures in the two vessels are both roughly equal to kT times the mean density $N/(V_1 + V_2)$ of molecules in the system. If the temperature is less than (γ/k), however, the pressures in the two vessels differ appreciably, being greater in the one at lower potential, V_2. When $kT \ll \gamma$ practically all the gas is in this lower container.

The mean energy of a point atom in the upper container is all kinetic, and thus is $\frac{3}{2}kT$; the mean energy of a point atom in the lower vessel is $\frac{3}{2}kT$ plus its potential energy there, $-\gamma$. Consequently, if the molecules are point atoms, having only translational kinetic energy, the total energy of the system is the sum of the number of molecules in each vessel times the respective mean energies,

$$U = NBV_1(\tfrac{3}{2}kT) + NBV_2 e^{\gamma/kT}(\tfrac{3}{2}kT - \gamma) \qquad \text{or}$$

$$U = \tfrac{3}{2}NkT - \frac{N\gamma V_2}{V_1 e^{-\gamma/kT} + V_2} \tag{13-11}$$

which changes from $N(\tfrac{3}{2}kT - \gamma)$ when $kT \ll \gamma$ and all the gas is in V_2 to $N\{\tfrac{3}{2}kT - [V_2\gamma/(V_1 + V_2)]\}$ when $kT \gg \gamma$ and the density is practically the same in both containers.

To compute the entropy and other thermodynamic potentials we must first recognize that there are three independent variables, T, V_1, and V_2. The appropriate partials of S can be obtained by the procedures of Chapter 8,

$$\left(\frac{\partial S}{\partial T}\right)_{V_1 V_2} = \frac{C_V}{T} = \frac{1}{T}\left(\frac{\partial U}{\partial T}\right)_{V_1 V_2};$$

$$\left(\frac{\partial S}{\partial V_1}\right)_{TV_2} = \left(\frac{\partial P_1}{\partial T}\right)_{V_1 V_2}; \qquad \text{etc.}$$

Therefore,

$$S = \tfrac{3}{2}Nk \ln\left(\frac{T}{T_0}\right) + \tfrac{3}{2}Nk + S_0 + Nk \ln\left(\frac{V_1 + V_2 e^{\gamma/kT}}{V_0}\right)$$
$$- \frac{(N\gamma V_2/T)}{V_1 e^{-\gamma/kT} + V_2} \tag{13-12}$$

$$F = U - TS = -TS_0 - \tfrac{3}{2}NkT \ln\left(\frac{T}{T_0}\right)$$
$$- NkT \ln\left(\frac{V_1 + V_2 e^{\gamma/kT}}{V_0}\right)$$

where S_0 and V_0 are constants of integration.

The heat capacity of the gas at constant V_1 and V_2,

$$C_{V_1 V_2} = \tfrac{3}{2}Nk + \frac{(N\gamma^2 V_1 V_2/kT^2)e^{-\gamma/kT}}{(V_1 e^{-\gamma/kT} + V_2)^2}$$

is $\tfrac{3}{2}kN$ both for $kT \ll \gamma$ and for $kT \gg \gamma$, but for intermediate temperatures C_v is larger than this. At low temperatures nearly all the molecules are in the lower vessel and additional heat merely speeds up the molecules; at temperatures near γ/k the added heat must push more molecules into the upper vessel as well as speed them all up; at very high temperatures the density in the two containers is nearly equal and additional heat again serves merely to increase kinetic energy. We also note that $P_1 = -(\partial F/\partial V_1)$, and similarly for P_2, as required by Eqs. (8-8).

A More General Distribution Function

The form of the Maxwell–Boltzmann distribution suggests some generalizations. In Eq. (13-4) the expression in the exponent is the total energy of position and of motion of the center of mass of the molecule, and f itself is the probability density of position and momentum of the center of mass. One obvious generalization is to put the total energy of the molecule in the exponent and to expect that the corresponding f is the probability density that each of the molecular coordinates and momenta have specified values. The position coordinates q need not be rectangular ones, they may be angles and radii, or other orthogonal curvilinear coordinates. We specify the nature of these coordinates in terms of their scale factors h, such that $h_i\, dq_i$ represents actual displacement in the q_i direction (as $r\, d\theta$ is displacement in the θ direction). For rectangular coordinates h is unity; for curvilinear coordinates h may be a function of the q's.

Suppose each molecule has v degrees of freedom; then it will need v coordinates $q_1, q_2, ..., q_v$ to specify its configuration and position in space. If the coordinates are mutually perpendicular and if h_i is the scale factor for coordinate q_i, the volume element for the q's is $dV_q = h_1\, dq_1\, h_2\, dq_2 \cdots h_v\, dq_v$ and the kinetic energy of the molecule is

$$\langle \text{K.E.} \rangle = \tfrac{1}{2} \sum_{i=1}^{v} m_i h_i^2 \dot{q}_i^2, \qquad \dot{q}_i = dq_i/dt \tag{13-13}$$

where m_i is the effective mass for the ith coordinate (total mass or reduced mass or moment of inertia, as the case may be).

Following the procedures of classical mechanics we define the momentum p_i, conjugate to q_i, as

$$p_i = \frac{\partial}{\partial \dot{q}_i} (\text{K.E.}) = m_i h_i^2 \dot{q}_i$$

We now define the *Hamiltonian function* for the molecule as the total energy of the molecule, expressed in terms of the p's and q's,

$$H(p, q) = \sum_{i=1}^{v} \frac{1}{2m_i} (p_i/h_i)^2 + \phi(q) \tag{13-14}$$

where ϕ is the potential energy of the molecule, expressed in terms of the q's. The h's may also be functions of the q's, but the only dependence of H on the p's is via the squares of each p_i, as written specifically in the sum. It can then be shown that the corresponding scale factors for the momentum coordinates, the other half of the $2v$-dimensional phase space for the molecule, are the reciprocals of the h's, so that the momentum volume element is $(dp_1/h_1)(dp_2/h_2) \cdots (dp_v/h_v) = dV_p$. For the kinetic energy is proportional to p^2 and the contribution of the ith momentum is $(p_i/h_c)^2$.

As an example, consider a diatomic molecule, with one atom of mass m_1 at position x_1, y_1, z_1 and another of mass m_2 at x_2, y_2, z_2. We can use, instead of these coordinates, the three coordinates of the center of mass, $x = [(m_1 x_1 + m_2 x_2)/(m_1 + m_2)]$ and similarly for y and z, plus the distance r between the two atoms and the spherical angles θ and φ giving the direction of r. Then the total kinetic energy of the molecule, expressed in terms of the velocities, is

$$\tfrac{1}{2}(m_1 + m_2)(\dot{x}^2 + \dot{y}^2 + \dot{z}^2) + \frac{\tfrac{1}{2}m_1 m_2}{m_1 + m_2}$$
$$\times (\dot{r}^2 + r^2\dot{\theta}^2 + r^2 \sin^2 \theta \, \dot{\varphi}^2)$$

so that the volume element $dV_q = dx \, dy \, dz \, dr \, r \, d\theta \, r \sin \theta \, d\varphi$. The momenta are

$$p_x = (m_1 + m_2)\dot{x} \quad \text{etc.,} \quad p_r = m_r \dot{r} \quad p_\theta = m_r r^2 \dot{\theta}$$
$$p_\varphi = m_r r^2 \sin^2 \theta \, \dot{\varphi}$$

where m_r is the reduced mass $[m_1 m_2/(m_1 + m_2)]$. The kinetic energy expressed in terms of the p's is

$$\frac{1}{2(m_1 + m_2)}(p_x^2 + p_y^2 + p_z^2) + \frac{1}{2m_r}\left(p_r^2 + \frac{p_\theta^2}{r^2} + \frac{p_\varphi^2}{r^2 \sin^2 \theta}\right)$$

and the volume element $dV_p = dp_x \, dp_y \, dp_z \, dp_r \, (dp_\theta/r)(dp_\varphi/r \sin \theta)$.

For a molecule with v degrees of freedom, the distribution function is

$$f(q, p) = \frac{1}{Z_q Z_p} e^{-H(p,q)/kT} \tag{13-15}$$

where

$$Z_q = \int \cdots \int e^{-\phi(q)/kT} \, dV_q$$

$$Z_p = \int \cdots \int \exp\left(-\sum_{i=1}^{v} \frac{p_i^2/h_i^2}{2m_i kT}\right) dV_p$$

$$= (2\pi kT)^{(1/2)v}\sqrt{m_1 m_2 \cdots m_v}$$

and where $f(q, p) \, dV_q \, dV_p$ is the probability that the first molecular coordinate is between q_1 and $q_1 + dq_1$, the second is between q_2 and $q_2 + dq_2$, and so on, that the first momentum coordinate lies between p_1 and $p_1 + dp_1$, and so on. Since the scale factors h_i are not functions of the p's, they enter as simple constants in the integration over the p's and thus the normalizing constant Z_p can be written out explicitly. We can also compute explicitly the mean total kinetic energy of the molecule, no matter what its position or orientation:

$$\langle K.E. \rangle_{total} = \frac{1}{Z_q} \int \cdots \int e^{-\phi/kT} \, dV_q$$

$$\times \frac{1}{Z_p} \int \cdots \int \sum_i \left(\frac{p_i^2}{2m_i h_i^2} \right) \exp\left(-\sum_i \frac{p_i^2/h_i^2}{2m_i kT} \right) dV_p$$

$$= \left(\frac{kT}{\pi^{(1/2)v}} \right) \int_{-\infty}^{\infty} \cdots \int \sum_i u_i^2 \exp\left(-\sum_i u_i^2 \right) du_1 \, du_2 \cdots du_v = \frac{v}{2} kT$$

$$(13\text{-}16)$$

Mean Energy per Degree of Freedom

Therefore each degree of freedom of the molecule has a mean kinetic energy $\frac{1}{2}kT$, no matter whether the corresponding coordinate is an angle or a distance and no matter what the magnitude of the mass m_i happens to be. The thermal energy of motion of the molecule is equally distributed among its degrees of freedom. The mean value of the potential energy of course depends on the nature of the potential function $\phi(q)$, although a comparison with the kinetic-energy terms indicates that if the sole dependence of ϕ on coordinate q_i is through a quadratic term $\frac{1}{2}m_i\omega_i^2 q_i^2$ then the mean potential energy for this coordinate is also $\frac{1}{2}kT$ (see below).

This brings us to an anomaly, the resolution of which will have to await our discussion of statistical mechanics. A diatomic molecule has six degrees of freedom (three for the center of mass, one for the interatomic distance, and two angles for the direction of r, as given above) even if we do not count the electrons as separate particles. We should therefore expect the total kinetic energy of a diatomic gas to be the number of molecules N times six times $\frac{1}{2}kT$ and thus U, the internal energy of the gas, to be at least $3NkT$ (it should be more, for there must be a potential energy, dependent on r, to hold the molecule together, and this should add something to U). However, measurements of heat capacity $C_v = (\partial U/\partial T)_v$ show that just above its boiling point (about 30°K) the U of H_2 is more nearly $\frac{3}{2}NkT$, that between about 100°K and about 500°K the U of H_2, O_2, and many other diatomic gases is roughly $\frac{5}{2}NkT$ and that only well above 1000°K is the U for most diatomic molecules equal to the expected value of $3NkT$ (unless the molecules have dissociated by then). The reasons for this discrepancy can only be explained in terms of quantum theory, as will be shown in Chapter 22.

A Simple Crystal Model

Hitherto we have tacitly assumed that the coordinates q of the molecules, as used in Eq. (13-14), are universal coordinates, referred to the same origin for all molecules. This need not be so; we can refer the coordinates of each molecule to its own origin, provided that the form of the Hamiltonian for each molecule is the same when expressed in terms of these coordinates. For example, each atom in a crystal lattice is bound elastically to its own equilibrium position, each oscillating about its own origin. The motion of each atom will affect the motion of its neighbors, but if we neglect this coupling, we can consider each atom in the lattice to be a three-dimensional harmonic oscillator, each with the same three frequencies of oscillation (in Chapter 20 we shall consider the effects of the coupling, which we here neglect). In a cubic lattice all three directions will be equivalent, so that (without coupling) the Hamiltonian for the jth atom is

$$H_j = \frac{1}{2m}(p_{xj}^2 + p_{yj}^2 + p_{zj}^2) + \frac{1}{2}m\omega^2(x_j^2 + y_j^2 + z_j^2) \qquad (13\text{-}17)$$

where x_j is the x displacement of the jth atom from its own equilibrium position and p_{xj} is the corresponding $m\dot{x}_j$. Thus expressed, H has the same form for any atom in the lattice.

In this case we can redefine the Maxwell–Boltzmann probability density as follows: $(1/Z_q Z_p)\exp(-H_j/kT)$ is the probability density that the jth atom is displaced x_j, y_j, z_j away from its own equilibrium position and has momentum components equal to p_{xj}, p_{yj}, p_{zj}. The normalizing constants are [using Eqs. (12-6)]

$$Z_q = \left(\int_{-\infty}^{\infty} e^{-m\omega^2 x^2/2kT}\,dx\right)^3 = \left(\frac{2\pi kT}{m\omega^2}\right)^{3/2};$$

$$Z_p = (2\pi mkT)^{3/2}$$

Therefore the probability density that *any* atom, chosen at random, has momentum **p** and is displaced a distance **r** from *its* position of equilibrium in the lattice is

$$f(\mathbf{r}, \mathbf{p}) = \left(\frac{\omega}{2\pi kT}\right)^3 \exp\left(-\frac{p^2}{2mkT} - \frac{m\omega^2 r^2}{2kT}\right) \tag{13-18}$$

for the simplified model of a crystal lattice we have been assuming.

We can now use this result to compute the total energy U, kinetic and potential, of the crystal of N atoms, occupying a volume V, at temperature T. In the first place we note that even at $T = 0$ the crystal has potential energy of over-all compression, when it is squeezed into a volume less than its equilibrium volume V_0. If the compressibility at absolute zero is κ, this additional potential energy is $(V - V_0)^2/2\kappa V_0$; the potential energy increases whether the crystal is compressed or stretched. We should also notice that the natural frequencies of oscillation of the atoms also are affected by compression; ω is a function of V.

The thermal energy of vibration of a typical atom in this crystal is given by the integral of Hf over all values of **r** and **p**. Again using Eqs. (12-6) we have

$$\int \cdots \int Hf\, dV_p\, dV_q = \frac{\frac{3}{2}}{m\sqrt{2\pi mkT}} \int_{-\infty}^{\infty} p^2 e^{-p^2/2mkT}\, dp$$

$$+ \tfrac{3}{2}m\omega^2 \sqrt{\frac{m\omega^2}{2\pi kT}} \int_{-\infty}^{\infty} x^2 e^{-m\omega^2 x^2/2kT}\, dx$$

$$= \tfrac{3}{2}kT + \tfrac{3}{2}kT = 3kT = 3(R/N_0)T \tag{13-19}$$

showing that the mean thermal kinetic energy per point particle is $\tfrac{3}{2}kT$ whether the particle is bound or free and that the mean thermal potential energy of a three-dimensional harmonic oscillator is also $\tfrac{3}{2}kT$, independent of the value of ω.

Thus, for this simple model of a crystal, with $N = nN_0$,

$$U = 3nRT + [(V - V_0)^2/2\kappa V_0]; \qquad C_v = 3nR = T(\partial S/\partial T)_v \tag{13-20}$$

One trouble with this simple model is that it does not provide us with enough to enable us to compute the entropy. We can obtain $(\partial S/\partial T)_v$ from Eq. (13-20) but we do not know the value of $(\partial S/\partial V)_T$

$= (\partial P/\partial T)_v$ [see Eq. (8-12)], unless we assume the equation of state (9-1) and set $(\partial P/\partial T)_v = \beta/\kappa$ from it. But an even more basic deficiency is that it predicts that $C_v = 3nR$ for all values of T. As shown in Figure 3-1 and discussed prior to Eq. (9-2), C_v is equal to $3nR$ for actual crystals only at high temperatures; as T goes to zero C_v goes to zero. We shall show in Chapters 18 and 20 that this behavior is a quantum-mechanical effect, related to the anomaly in the heat capacities of diatomic molecules, mentioned on page 191.

Magnetization and Curie's Law

Another example of the use of the generalized Maxwell–Boltzmann distribution is in connection with the magnetic equation of state of a paramagnetic substance (see page 21). Such a material has permanent magnetic dipoles of moment $\mathfrak{m}$ connected to its constituent molecules or atoms. If a uniform magnetic field $\mathfrak{B} = \mu_0 \mathfrak{H}$ (μ_0 is the permeability of vacuum) is applied to the material, each molecule will have an additional potential-energy term $-\mathfrak{m}\mathfrak{B} \cos \theta$ in its Hamiltonian, where θ is the angle between the dipole and the direction of the field (and $\mathfrak{m}$ is the dipole's moment). We note that the $\mathfrak{B}$ and $\mathfrak{H}$ used here are those acting on the individual molecules, which differ from the fields outside the substance.

The distribution function, giving the probability density that the molecule has a given momentum, position, and orientation, can be written

$$f = f_p f_q f_\theta; \qquad f_\theta = (1/Z_\theta) \exp(\mathfrak{m}\mathfrak{B} \cos \theta/kT)$$

where f_p is the momentum distribution, so normalized that the integral of f_p over all momenta is unity; f_q is the position distribution for all position coordinates except θ, the orientation angle of the magnetic dipole (f_q is also normalized to unity). Thus the factor f_θ gives the probability distribution for orientation of the magnetic dipole; each dipole is most likely to be oriented along the field ($\theta = 0$) and least likely to be pointed against the field ($\theta = \pi$). When $kT \gg \mathfrak{m}\mathfrak{B}$ the difference is not great; the thermal motion is so pronounced that the dipoles are oriented almost at random. When $kT \ll \mathfrak{m}\mathfrak{B}$ the difference is quite pronounced. Nearly all the dipoles are lined up parallel to the magnetic field; the thermal motion is not large enough to knock many of them askew.

The factor f_θ can be normalized separately,

$$Z_\theta = \int_0^\pi \exp(\mathfrak{m}\mathfrak{B} \cos \theta/kT) \sin \theta \, d\theta = \left(\frac{2kT}{\mathfrak{m}\mathfrak{B}}\right) \sinh\left(\frac{\mathfrak{m}\mathfrak{B}}{kT}\right)$$

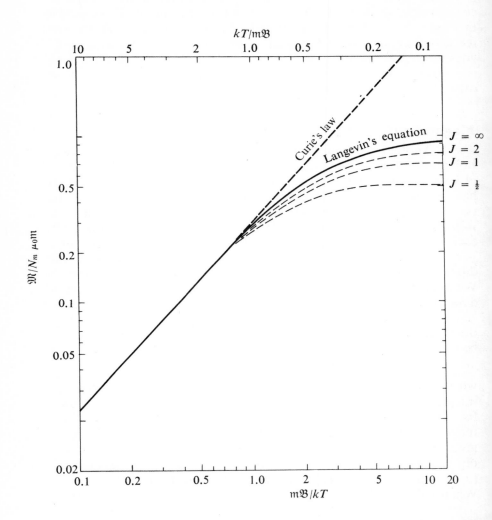

FIGURE 13-4

Magnetization curve for paramagnetic substances. Solid curve is Langevin's equation for classical, magnetic dipoles; dashed curves are for quantized dipoles for different values of angular momentum quantum number J.

and can be used separately to find the mean value of the component $\mathfrak{m} \cos \theta$ of the dipole moment along the magnetic field. This, times the number of dipoles per unit volume, is, of course, the magnetic polarization $\mathfrak{P}$ of the material, as defined in Chapter 3. And, since the magnetization is $\mathfrak{M} = \mu_0 V \mathfrak{P} = V\chi\mathfrak{B}$, we have Langevin's equation of state,

$$\mathfrak{M} = \frac{\mu_0 N}{Z_\theta} \int_0^\pi (\mathfrak{m} \cos \theta) \exp(\mathfrak{m}\mathfrak{B} \cos \theta / kT) \sin \theta \, d\theta \qquad (13\text{-}21)$$

$$= N\mathfrak{m}\mu_0 \frac{(2/x)\cosh x - (2/x^2)\sinh x}{(2/x)\sinh x}, \qquad x = \frac{\mathfrak{m}\mathfrak{B}}{kT}$$

$$= N\mathfrak{m}\mu_0 \left[\coth\left(\frac{\mathfrak{m}\mathfrak{B}}{kT}\right) - \left(\frac{kT}{\mathfrak{m}\mathfrak{B}}\right) \right]$$

$$\rightarrow \begin{cases} (N\mu_0\mathfrak{m}^2\mathfrak{B}/3kT), & kT \gg \mathfrak{m}\mathfrak{B} \\ N\mu_0\mathfrak{m}, & kT \ll \mathfrak{m}\mathfrak{B} \end{cases}$$

Thus we see that at low fields (or high temperatures) the dipoles tend only slightly to line up with the field and the magnetization $\mathfrak{M}$ is proportional to $\mathfrak{B} = \mu_0\mathfrak{H}$, but that at high magnetic intensities (or low temperatures) all the dipoles line up and the magnetization reaches its asymptotic value $N\mu_0\mathfrak{m}$ (i.e., it is saturated), as shown in Figure 13-4. Since the magnetic moment of an oxygen molecule is roughly 3×10^{-23} mks units, $\mathfrak{m}\mathfrak{B}$ is no larger than 3×10^{-23} joule for $\mathfrak{B} = 1$ weber per m^2 ($= 10{,}000$ gauss, a quite intense field). Since $kT = 3 \times 10^{-23}$ joule for $T \simeq 2°K$, the parameter $x = \mathfrak{m}\mathfrak{B}/kT$ is considerably less than unity for O_2 (for example) for temperatures greater than $30°K$ and/or $\mathfrak{B}$ less than 1 weber/m^2. In such cases, where $x \ll 1$, the polarization $\mathfrak{P}$ is much smaller than $\mathfrak{H}$, so that the $\mathfrak{H}$ acting on the molecule is not much different from the $\mathfrak{H}$ outside the material.

Thus for most temperatures and field strengths, for paramagnetic materials like O_2, x is very small and Curie's law

$$\mathfrak{M} = nD\mathfrak{H}/T; \qquad D = N_0\mu_0^2\mathfrak{m}^2/3k \qquad (13\text{-}22)$$

is a good approximation for the magnetic equation of state. Kinetic theory has thus not only derived Curie's law [see Eq. (3-8)] and obtained a relation between the Curie constant D and the molecular characteristics of the material (such as $\mathfrak{m}$ and the basic constants N_0, μ_0, and k) but has also determined the limits beyond which

Curie's law is no longer valid, and the equation of state which then holds.

For example, for the paramagnetic perfect gas of Chapter 7, the more accurate equation for $T\,dS$ and the adiabatic formula (7-14) is

$$T\,dS = \left[\tfrac{5}{2}nR - \frac{3nD}{\alpha^2}(x^2\operatorname{csch}^2 x - 1)\right]dT - \frac{nRT}{P}\,dP$$

$$+ \frac{3nDT}{\alpha^2\mathfrak{H}}(x^2\operatorname{csch}^2 x - 1)\,d\mathfrak{H}$$

and

$$\frac{T^{5/2}}{P}\left[\frac{T}{\alpha\mathfrak{H}}\sinh\left(\frac{\alpha\mathfrak{H}}{T}\right)\right]^{3D/R\alpha^2}\exp\left[-\frac{3D\mathfrak{H}}{R\alpha T}\coth\left(\frac{\alpha\mathfrak{H}}{T}\right)\right] = \text{const.}$$

$$(13\text{-}23)$$

where $x = \alpha\mathfrak{H}/T$ and $\alpha = \mathfrak{m}\mu_0/k$. This reduces to Eq. (7-14) when x is small.

The foregoing theoretical model is still only an approximate one. In the first place it has neglected the interaction between each atomic magnet and its neighbors; we assumed that the only magnetic field acting on each magnet is the externally applied field $\mathfrak{H}_{ap}$. This is not a bad approximation for paramagnetic gases, where the atomic magnets are far enough apart so that their interactions are quite small. But it is not a good approximation for paramagnetic solids. In these materials the field $\mathfrak{H}$ acting on the individual magnet differs from the applied field $\mathfrak{H}_{ap}$ because of the polarization $\mathfrak{P}$ of the rest of the material. For solids with simple crystal structure the correction is $\mathfrak{H} = \mathfrak{H}_{ap} + \tfrac{4}{3}\pi\mathfrak{P}$. But the polarization $\mathfrak{P} = (\mathfrak{M}/\mu_0 V)$ is proportional to the acting field $\mathfrak{H}$, the susceptibility χ of the material being the proportionality factor (see page 21). Therefore the field acting on the magnets is related to the applied field by the equation

$$\mathfrak{H}(1 - \tfrac{4}{3}\pi\chi) = \mathfrak{H}_{ap} \qquad \text{or} \qquad \mathfrak{P} = \chi\mathfrak{H} = \frac{\chi}{1 - \tfrac{4}{3}\pi\chi}\mathfrak{H}_{ap}$$

The measured magnetic susceptibility χ_{meas}, the ratio between the polarization $\mathfrak{P}$ and the applied field $\mathfrak{H}_{ap}$, is thus

$$\chi_{\text{meas}} = (\mathfrak{P}/\mathfrak{H}_{ap}) = \frac{\chi}{1 - \tfrac{4}{3}\pi\chi}$$

where χ is equal to the $\mathfrak{M}$ of Eq. (13-21) divided by $V\mathfrak{B} = V\mu_0\mathfrak{H}$.

This correction is satisfactory for substances with a low density of atomic magnets, such as iron ammonium alum, which is paramagnetic down to a fraction of a degree Kelvin. In this case $\frac{4}{3}\pi\chi$ is never larger than unity. In materials such as iron, however, the interaction effect, tending to line the magnets parallel to each other, is enhanced by quantum-mechanical effects, which have the effect of making the coefficient of χ in the denominator very much larger than $\frac{4}{3}\pi$. Hence χ_{meas} becomes infinite for T less than some value T_c, called the Curie temperature. Below T_c the interaction between magnets is so strong that they line up parallel to each other whether an applied field is present or not. These materials are appropriately called ferro-magnetic materials; they can make permanent magnets below the Curie temperature. Other materials have interactions such that, below the Curie temperature, the magnets oppose each other in pairs, so the material becomes magnetically neutral; these are called anti-ferromagnetic materials. In either case, when the temperature is above the Curie temperature the material is paramagnetic, though with an equation of state $\mathfrak{M} = [nD\mathfrak{H}/(T - T_c)]$, instead of the simple Curie law $(nD\mathfrak{H}/T)$. Equation (13-21), with $(T - T_c)$ sub-stituted for T, is a fairly good approximation to the behavior of these substances when $T > T_c$.

In still other cases the interaction is not directly between the atomic magnets but is with the crystal lattice itself. These forces are much smaller than those responsible for ferro- and antiferro-magnetism, and the equation of state (13-21) or its quantum analog, Eq. (13-26), is valid down to less than $1°K$. Substances of this kind are the paramagnetic salts, such as iron ammonium alum, which are used to reach very low temperatures by adiabatic demagnetization [see Eq. (8-29)]. We postpone further discussion of these phenomena until Chapter 20, when we can use the machinery of statistical mechanics to work it all out.

Quantum Theory of Paramagnetism

The third deficiency of Eqs. (13-21) is the neglect of quantum effects. The electrons and nucleons which are the atomic magnets in a paramagnetic material have spins and orbital angular momenta which are quantized in magnitude and in orientation with respect to the magnetic field. The magnitude of their magnetic moment can be $\mathfrak{m} = \sqrt{[J(J + 1)]}\mathfrak{m}_B$, where J is an integer or a half-integer and where

$$\mathfrak{m}_B = (eh/4\pi m) = 0.927 \times 10^{-23} \text{ amp} - m^2 \qquad \text{(for an electron)}$$

is the Bohr magneton. The allowed values of the component of $\mathfrak{m}$ along the magnetic field are then $\mathfrak{m}_z = m\mathfrak{m}_B$, where the magnetic quantum number m can take on the $2J + 1$ values J, $J - 1$,..., $-J + 1$, $-J$.

As with the classical case, the energy of interaction between field and magnet is

$$E_m = -\mathfrak{m}_z \mathfrak{B} = -m\mathfrak{m}_B \mathfrak{B} \qquad (13\text{-}24)$$

The Maxwell–Boltzmann distribution holds as well for quantized as for classical systems, so the factor f_θ, giving the probability distribution for the orientation of the atomic magnets, is now the probability that the component of $\mathfrak{m}$ along $\mathfrak{H}$ is $m\mathfrak{m}_B$,

$$f_\theta(m) = (1/Z_\theta)e^{m\mathfrak{m}_B \mathfrak{B}/kT}$$

where the normalization factor

$$Z_\theta = \sum_{m=-J}^{J} e^{m\mathfrak{m}_B \mathfrak{B}/kT} = \sinh\left[\frac{(J + \frac{1}{2})\mathfrak{m}_B \mathfrak{B}}{kT}\right] \bigg/ \sinh\left(\frac{\mathfrak{m}_B \mathfrak{B}}{2kT}\right)$$

$$(13\text{-}25)$$

The magnetization is then

$$\mathfrak{M} = \mu_0 N \sum_{m=-J}^{J} m\mathfrak{m}_B f_\theta(m) = \mu_0 NkT\frac{\partial}{\partial\mathfrak{B}} \ln Z_\theta$$

$$= \mu_0 N\mathfrak{m}_B\left\{(J + \tfrac{1}{2})\coth\left[\frac{(J + \frac{1}{2})\mathfrak{m}_B \mathfrak{B}}{kT}\right] - \tfrac{1}{2}\coth\left(\frac{\mathfrak{m}_B \mathfrak{B}}{2kT}\right)\right\}$$

$$\rightarrow [\mu_0 N\mathfrak{m}_B^2 J(J + 1)\mathfrak{B}/3kT] \qquad (kT \gg J\mathfrak{m}_B \mathfrak{B}) \qquad (13\text{-}26)$$

$$\rightarrow (\mu_0 JN\mathfrak{m}_B) \qquad (kT \ll \tfrac{1}{2}\mathfrak{m}_B \mathfrak{B})$$

which is to be compared with Eqs. (13-21). We note that when $J = \frac{1}{2}$, the second line becomes $\frac{1}{2}\mu_0 N\mathfrak{m}_B \tanh(\mathfrak{m}_B \mathfrak{B}/2kT)$. The approximate formula for $kT \gg J\mathfrak{m}_B \mathfrak{B}$ (the Curie equation of state) is the same for either a quantum or a classical system, with $\sqrt{[J(J + 1)]}\mathfrak{m}_B$, in the quantum case, equal to the moment $\mathfrak{m}$ of the classical magnet, which can have any orientation. The exact formulas are not quite identical, however, as is shown in Figure 13-4. The dashed lines are for different values of the angular momentum quantum number J. We see that for small values of J the saturation value of $\mathfrak{M}$ is smaller than $\mu_0\sqrt{[J(J + 1)]}\mathfrak{m}_B = \mu_0 \mathfrak{m}$. As we see from the figure, the classical (Langevin) and quantum formulas are nearly equal over most of the range of values of magnetization.

We postpone until Chapters 18 and 20 any discussion of entropy and other thermodynamic properties of paramagnetic substances. There we shall complete the discussion that we started in Chapter 8.

CHAPTER

14

TRANSPORT
PHENOMENA

In a gas near equilibrium the situation is that discussed in the last half of Chapter 8. Some or all of the extensive variables (internal energy or net momentum or net particle density, etc.) will flow from one region to another, in the direction which will increase the entropy and thus get the system closer to equilibrium. The distribution function $f(\mathbf{r}, \mathbf{p}, t)$ will approach its equilibrium form $f_0(\mathbf{r}, \mathbf{p})$ as time goes on, and the various net fluxes, which can be computed from f, will correspond to the fluxes $\mathbf{J}_n$, defined in Eqs. (8-32) to (8-34). Thus the approximate form of the Boltzmann Eq. (13-6) is the kinetic-theory counterpart of the nonequilibrium thermodynamics of Chapter 8. We should be able to compute the conductivities L_{mn} of Eq. (8-34) in terms of molecular properties, such as mean velocities, mean free time, etc. Before we do this, however, we should examine, in a bit more detail, the nature of the collision term Q in the Boltzmann equation.

The Effect of Collisions

Having studied the case of equilibrium, where the collision term Q is zero, we must now return to this term, which becomes important in situations of near-equilibrium, of the sort discussed in the last half of Chapter 8. As stated earlier, the function $Q(\mathbf{r}, \mathbf{p}, t)$ of Eq. (13-6) is the net increase of probability density f at point

$\mathbf{r}$, $\mathbf{p}$ in phase space because of collisions. Thus it must be the difference of two terms, $Q_+ - Q_-$, where Q_- represents the rate of leaving volume element $dV_r\, dV_p$ and Q_+ the rate of entering, because of collisions.

Because every collision involves two molecules, Q must relate the density of two regions of phase space, the position $\mathbf{r}$, $\mathbf{p}$ of the incident molecule, which is leaving or entering the volume element $dV_r\, dV_p$ and the position $\mathbf{r}$, $\mathbf{p}_t$ of the target molecule, which it hits (position $\mathbf{r}$ must be the same of course). After the collision, if the incident molecule is leaving $dV_r\, dV_p$, the two molecules have final momenta $\mathbf{p}'$ and $\mathbf{p}'_t$, elsewhere in momentum space. In the term Q_+, where the incident molecule is entering $dV_r\, dV_p$ then $\mathbf{p}$, $\mathbf{p}_t$ are the final momenta, after collision, and $\mathbf{p}'$, $\mathbf{p}'_t$ are the initial momenta, before collision.

To see how the primed momenta are related to the unprimed ones, let us return to the definition of collision cross section, given in the paragraph preceding Eq. (12-12). Here we likened a collision to the impact of a bullet on a target of area σ, the collision cross section, and deduced that in a distance dx the incident molecule would strike $(N/V)\sigma\, dx$ such targets, where N/V is the number of target molecules per unit volume. This is pictured in Figure 14-1a, where the shaded circle σ is on the center of mass O of the two molecules which are going to collide. In the coordinate system where the center of mass is at rest the two molecules are approaching σ with equal and opposite velocities, $\mathbf{v}_m$ and $-\mathbf{v}_m$. Area σ is placed perpendicular to these velocities and with its center on the projected track of the target molecule, number 2. Then if the projected track of the incident molecule, number 1, penetrates the target there will be collision and the motions of both molecules will be deflected, as shown in the figure. After the collision, if the impact is elastic, the magnitudes of the final velocities, $\mathbf{v}'_m$ and $-\mathbf{v}'_m$, will equal the magnitudes of the initial velocities $\mathbf{v}_m$ and $-\mathbf{v}_m$ (in coordinates at rest with the center of mass), but will be deflected by the angle ϑ from their initial direction, in a plane denoted by the angle φ.

The exact values of the angles ϑ and φ are determined by the details of the collision. If the two molecules hit head-on they will rebound back in their tracks and ϑ will equal π; if their paths are too far apart they will not be deflected at all. The probability that the collision will change $\mathbf{v}_m$ into vector $\mathbf{v}'_m$, pointed within the solid angle $\sin\vartheta\, d\vartheta\, d\varphi$ is determined by the nature of the force field between the molecules and by where on the target the projected path of the incident molecule strikes. Examination of the figure

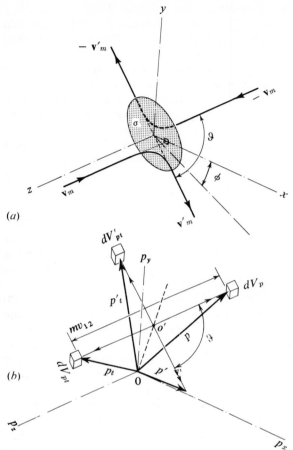

FIGURE 14-1

(a) *Paths of molecules during collision, in coordinates at rest with respect to center of mass, (b) Momenta* **p**, **p**$_t$, *prior to collision and* **p**′, **p**′$_t$ *after collision, in coordinates at rest with respect to laboratory.*

indicates that it can be set proportional to the element of target area $d\sigma$, within which the projected incident track strikes σ. For then the number of collisions per length of path dx which result in deflections by ϑ, φ will be

$$(N_t/V)\, d\sigma\, dx = (N_t/V)\, dx\, g(\vartheta, \varphi) \sin\vartheta\, d\vartheta\, d\varphi \qquad (14\text{-}1)$$

where $g(\vartheta, \varphi)$ is determined by the nature of the intermolecular force and where (N_t/V) is the number of target molecules of its momentum class per unit volume. Then an integration over all possible angles of deflection results in the number $(N_t\sigma/V)\,dx$ of collisions per path length dx with target molecules of this class, as stated in the discussion preceding Eq. (12-12).

To translate this to stationary coordinates (at rest with respect to the laboratory) we first change collisions per length dx to collisions per interval dt of time, by multiplying by the mutual speed $v_{12} = 2v_m$ of the colliding molecules. The transformation back to the coordinates p_x, p_y, p_z (at rest with the laboratory) is illustrated in Figure 14-1b. Origin O is at rest in the laboratory and O' is at rest at the center of mass. The initial momentum of the incident molecule is $\mathbf{p}$, within momentum element dV_p; its final momentum is $\mathbf{p}'$, related to $\mathbf{p}$ and ϑ, and $\mathbf{p}_m = m\mathbf{v}_m$ as shown. The initial momentum of the target molecule is $\mathbf{p}_t$, in element dV_{pt}. The number per unit volume of such target molecules is $(N_t/V) = Nf(\mathbf{r}, \mathbf{p}_t, t)\,dV_{pt}$ according to the definition of the probability density f; N is the total number of molecules in the system.

Therefore the number of collisions in time dt between the incident molecule and the target molecules that are in element dV_{pt}, which result in deflections by angles ϑ, φ is

$$d\sigma\,(N_t/V)v_{12}\,dt = Nv_{12}f(\mathbf{r}, \mathbf{p}_t, t)\,dV_{pt}\,d\sigma\,dt$$

This, times the probability of finding an incident molecule in the element $dV_r\,dV_p$, $f(\mathbf{r}, \mathbf{p}, t)\,dV_r\,dV_p$, is the mean number of collisions per second, in element $dV_r\,dV_p$, between molecules of momentum $\mathbf{p}$ with those of momentum $\mathbf{p}_t$. Integration over $d\sigma$ and over the momenta of the target molecules gives us

$$N\,dV_r\,dV_p\,dt \int d\sigma \int dV_{pt}v_{12}f(\mathbf{r}, \mathbf{p}, t)f(\mathbf{r}, \mathbf{p}_t, t)$$

$$= dV_r\,dV_p\,dt\,Q_-(\mathbf{r}, \mathbf{p}, t) \qquad (14\text{-}2)$$

as the chance of molecules *being lost* to element $dV_r\,dV_p$ in time dt, because of collision with a target molecule of any velocity.

Now $d\sigma\,v_{12}(N_t/V)$ is, according to Eq. (13-13), the reciprocal of the mean free time, the *frequency* of collisions between molecules of momentum $\mathbf{p}$ and those of momentum $\mathbf{p}_t$. Therefore

$$N\int d\sigma \int v_{12}\,dV_{pt}\,f(\mathbf{r}, \mathbf{p}_t, t)$$

can be considered to be the frequency of collisions of all sorts, suffered by molecules of momentum $\mathbf{p}$; it is a *mean, reciprocal, mean free time*, and can be written as $1/t_c$. It is, of course, a function of the magnitude p of momentum of the incident molecules in $dV_r\,dV_p$. Therefore

$$Q_-(\mathbf{r}, \mathbf{p}, t) = N \int d\sigma \int dV_{pt} v_{12} f(\mathbf{r}, \mathbf{p}, t) f(\mathbf{r}, \mathbf{p}_t, t)$$

$$= [f(\mathbf{r}, \mathbf{p}, t)/t_c(p)] \tag{14-3}$$

since the probability density $f(\mathbf{r}, \mathbf{p}, t)$ of the incident molecules is independent of ϑ, φ, and $\mathbf{p}_t$, and can be taken outside the integral. Thus $Q_-(\mathbf{r}, \mathbf{p}, t)$ measures the rate of loss of molecules in $dV_r\,dV_p$ because of collisions.

But what about $Q_+(\mathbf{r}, \mathbf{p}, t)$, which measures the reverse process, in which molecules, originally with momenta $\mathbf{p}'$, happen to collide so they end up in element $dV_r\,dV_p$? Here the labels of Figures 14-1a and b should be reversed; the incident molecule has initially the momentum $\mathbf{p}'$ and only finally is in $dV_r\,dV_p$; the target molecule initially has momentum $\mathbf{p}'_t$ and ends up with momentum $\mathbf{p}_t$. Because elastic collisions are symmetric with respect to time, the paths of the colliding molecules in 14-1a would be the same if the arrows were reversed; so g of Eq. (14-1) is unchanged and the geometry of Figure 14-1b is unchanged. However, the *initial* momenta in the coordinates at rest with the laboratory is the primed pair $\mathbf{p}'$, $\mathbf{p}'_t$, instead of the unprimed pair $\mathbf{p}$, $\mathbf{p}_t$. Therefore the numbers per unit volume, which must be inserted in the formula for the collision frequency, are the probability densities for the primed momenta, and

$$Q_+(\mathbf{r}, \mathbf{p}, t) = N \int d\sigma \int dV'_{pt} v_{12} f(\mathbf{r}, \mathbf{p}', t) f(\mathbf{r}, \mathbf{p}'_t, t) \tag{14-4}$$

where $\mathbf{p}'$ and $\mathbf{p}'_t$ are related to $\mathbf{p}$ and the ϑ of $d\sigma = g(\vartheta, \varphi) \cdot \sin\vartheta \cdot d\vartheta\,d\varphi$ by the geometry of Figure 14-1b.

The Boltzmann Equation

Rederiving the Boltzmann equation, for practice, we write out the equation that equates the difference between the number of molecules in $dV_r\,dV_p$ at time $t + dt$ and the number at time t (letting the volume element move with the molecular cloud in phase space) with the difference between the number scattered into the element $dV_r\,dV_p$ and that scattered out in time dt,

$$N\,dV_r\,dV_p\left[f\left(\mathbf{r} + \frac{\mathbf{p}}{m}dt, \mathbf{p} - dt\,\text{grad}\,\phi, t + dt\right) - f(\mathbf{r}, \mathbf{p}, t)\right]$$

$$= N\,dV_r\,dV_p[Q_+(\mathbf{r}, \mathbf{p}, t) - Q_-(\mathbf{r}, \mathbf{p}, t)]\,dt$$

or

$$\frac{\partial}{\partial t} f(\mathbf{r}, \mathbf{p}, t) + \frac{\mathbf{p}}{m} \cdot \text{grad}_r \, f - \text{grad} \, \phi \cdot \text{grad}_p \, f$$

$$= Q_+(\mathbf{r}, \mathbf{p}, t) - Q_-(\mathbf{r}, \mathbf{p}, t)$$

$$= N \int d\sigma \int v_{12} f(\mathbf{r}, \mathbf{p}', t) f(\mathbf{r}, \mathbf{p}'_t, t) \, dV'_{pt} - \frac{1}{t_c(p)} f(\mathbf{r}, \mathbf{p}, t)$$

$$(14\text{-}5)$$

where Eq. (13-5) ensures that $N \, dV_r \, dV_p$ is the same for $t + dt$ as for t.

This is the usual form of the Boltzmann equation, for the distribution function $f(\mathbf{r}, \mathbf{p}, t)$ for single-constituent gases having elastic collisions only. We have used Eq. (14-3) to simplify the Q_- term; the Q_+ term cannot be similarly simplified. The Q_- term is proportional to $f(\mathbf{r}, \mathbf{p}, t)$ since the molecules scattered *out* of $dV_r \, dV_p$ all have the same initial momentum. This is not the case with the Q_+ integral; the molecules scattered *into* $dV_r \, dV_p$ come from all varieties of initial momenta, their relative proportions depending on the scattering function $g(\vartheta, \varphi)$ of Eq. (14-1) and the relative velocity v_{12}. In a very real sense the Q_+ term links the probability density of one momentum with that of another, and thus controls the shape of $f(\mathbf{r}, \mathbf{p}, t)$.

In states of thermodynamic equilibrium the "detailed balance" of transitions from one momentum to another, caused by collisions, ensures that $Q_+ = Q_-$, so that an equilibrium distribution, which can be written as $f_0(\mathbf{r}, \mathbf{p})$, satisfies

$$(\mathbf{p}/m) \cdot \text{grad}_r f_0 = \text{grad} \, \phi \cdot \text{grad}_p f_0$$

which is Eq. (13-7) for the Maxwell–Boltzmann distribution. For states close to equilibrium, when the system is slowly and spontaneously reverting to equilibrium, the term Q_+ cannot be much different from $Q_- = [f(\mathbf{r}, \mathbf{p}, t)/t_c(p)]$. Since it is the term that ensures that f eventually reverts to an equilibrium distribution $f_0(\mathbf{r}, \mathbf{p})$, we might expect that, for situations such as those discussed in the latter half of Chapter 8, Q_+ would be approximately equal to $[f_0(\mathbf{r}, \mathbf{p})/t_c]$. In other words, we shall assume that, for situations near equilibrium, the Boltzmann equation has the approximate form

$$\frac{\partial}{\partial t} f(\mathbf{r}, \mathbf{p}, t) + \frac{\mathbf{p}}{m} \cdot \text{grad}_r \, f - \text{grad}_r \, \phi \cdot \text{grad}_p \, f$$

$$= \frac{1}{t_c} [f_0(\mathbf{r}, \mathbf{p}) - f(\mathbf{r}, \mathbf{p}, t)] \qquad (14\text{-}6)$$

where f_0 is the equilibrium distribution closest to the nonequilibrium distribution $f(\mathbf{r}, \mathbf{p}, t)$. This is the form of the Boltzmann equation we shall use in the next chapter. All the effects of collisions have been condensed into the single term $t_c(p)$, the average, mean free time of molecules with speed p/m, evaluated for the equilibrium distribution $f_0(\mathbf{r}, \mathbf{p})$, in accordance with Eq. (14-3).

Near equilibrium the terms on the right-hand side of Eq. (14-6) are large compared to those on the left; the rate $Q_- \simeq (f/t_c)$ of scattering of molecules out of $dV_r\, dV_p$ is nearly equal to the rate $Q_+ \simeq (f_0/t_c)$ of scattering into $dV_r\, dV_p$, so that the difference is much smaller than either term separately. In such a case a good approximation to Eq. (14-6) is

$$f + t_c(\partial f/\partial t) \simeq f_0 - t_c[(\mathbf{p}/m) \cdot \text{grad}_r\, f_0 - \text{grad}\, \phi \cdot \text{grad}_p\, f_0]$$
$$= f_0 - t_c(p)[(\mathbf{p}/m) \cdot \text{grad}_r\, f_0 + \mathbf{F} \cdot \text{grad}_p\, f_0] \qquad (14\text{-}7)$$

for near-equilibrium conditions. We can write the term in brackets either in terms of the gradient of the potential energy ϕ or directly in terms of the force $\mathbf{F}$ which acts on each molecule. In general t_c is a function of the magnitude of the momentum p, but we often shall assume that it is constant, roughly equal to the mean free time $\tau = \langle m\lambda/p \rangle$, from Eq. (12-13).

To demonstrate the workings of this equation, we consider the case of a gas of uniform density, having initially a distribution in momentum $f_i(\mathbf{p})$ of its molecules which differs from the Maxwell distribution (12-7); there may be more fast particles in relation to slow ones than (12-7) requires, or there may be more going in the x direction than in the y, or some other asymmetry of momentum which is still uniform in density. Such a distribution f_i, since it is not Maxwellian, is not an equilibrium distribution. However, since the lack of equilibrium is entirely in the "momentum coordinate" part of phase space, the distribution can return to equilibrium in one collision time; we would expect that at the next collision each molecule would regain its place in a Maxwell distribution, so to speak. The molecules do not all collide at once, the chance that a given molecule has not had its next collision, after a time t, is $e^{-t/\tau}$, where τ is the mean free time between collisions [see Eqs. (11-17) and (12-13)]. Thus we would expect that our originally anisotropic distribution would "relax" from f_i back to f_0 with an exponential dependence on time of $e^{-t/\tau}$ (note that τ is proportional to λ, the molecular mean free path).

But if f is independent of $\mathbf{r}$ and there is no force $\mathbf{F}$ acting, a solution of Eq. (14-7) which starts as $f = f_i$ at $t = 0$ is

$$f = f_0 + (f_i - f_0)e^{-t/t_c} \qquad (14\text{-}8)$$

which has just the form we persuaded ourselves it should have, except that the relaxation time t_c is in the exponent, rather than the mean free time $\tau = \langle m\lambda/p \rangle$. We would thus expect that the relaxation time t_c, entering Eq. (14-7) would be approximately equal to $\langle m\lambda/p \rangle$. Detailed calculations for the few cases which can be carried out, plus indirect experimental checks (described later) indicate that it is not a bad approximation to set $t_c = \langle m\lambda/p \rangle = \tau$.

Transport of Momentum; Viscosity

The general results of our investigations of the effects of collisions may be put into words as follows. Each molecule, on the average, forgets its past history whenever it suffers a collision. After each collision it begins again to be affected by any non-equilibrium conditions that are present; it loses the effects at the next collision. Thus we can picture each molecule as transporting various quantities (momentum, energy and, of course, itself) from one collision to the next. At each collision, on the average, it reverts to the mean status of the molecules in the region near the collision. Thus, as it travels along the next mean free path, it carries a sample of the properties of the gas in the neighborhood of its last collision. If the gas is not in equilibrium, the properties of the gas in the neighborhood of its next collision may not be the same as the sample carried by the molecule. During the next collision the molecule, on the average, will give up this difference to the gas in this new neighborhood. Thus the molecules, in their motion from one collision to the next, will couple the properties of the gas at one point with those at another, tending to equalize the properties at the two points. The longer the mean free path λ the more long-range is this coupling.

For example, suppose the gas has a net flow [a drift velocity, see Eq. (12-3)] in the z direction ($\langle p_z \rangle \neq 0$) and suppose that this flow differs from point to point ($\langle p_z \rangle$ is a function of x, for example) so there is a rate of shear in the gas, as shown in Figure 14-2. Then all the molecules going to the right between collisions will tend to carry an excess of p_z and all those going to the left will tend to carry a debit, both tending to equalize p_z at the two values of x. The rate of transport of z momentum, by all molecules traveling to the right, will constitute a force, the force of *viscosity*, tending to diminish the rate of shear of the gas.

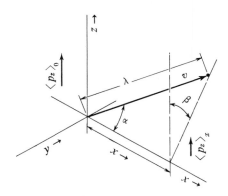

FIGURE 14-2

Transport of momentum $\langle p_z \rangle$ from one collision to the next.

Viewed macroscopically, each surface of each element of volume $dx\,dy\,dz$ of a fluid is subject to a stress force, transmitted to it from the rest of the fluid. The normal components of these forces are proportional to the pressure P; for example the x component F_{xx} of the force $\mathbf{F}_x$ on the right-hand face $dy\,dz$, perpendicular to the x axis, is equal to $-P\,dy\,dz$, as long as the fluid is at rest (the minus sign indicates that the force is into the element). But if one part of the fluid is in motion relative to the rest, the stress forces $\mathbf{F}_x$, $\mathbf{F}_y$, $\mathbf{F}_z$ may not all be normal to their respective elementary surfaces. In general the nine components of stress are related to the three components of mean fluid velocity by the equations

$$F_{xx} = -P + (\lambda - \tfrac{2}{3}\eta)\,\operatorname{div}\mathbf{u} + 2\eta(\partial u_x/\partial x),\ \text{etc.}$$
$$F_{xy} = F_{yx} = \eta[(\partial u_x/\partial y) + (\partial u_y/\partial x)],\ \text{etc.}$$

where λ is the coefficient of compressive friction and η is the coefficient of viscosity, the value of which we are about to compute, for a gas, in terms of the microstructure of the gas.

In particular, if one portion of the fluid slides past its neighbor, as would be the case if the mean fluid velocity $\mathbf{u} = \langle \mathbf{v} \rangle$ were all in the z direction and u_z were a function of x, the shear stress F_{xz} would equal $\eta(\partial u_z/\partial x)$, where η is the coefficient of viscosity. The shear flow is opposed because the gas molecules' wanderings tend to even out the motion, slowing down the faster parts and speeding up the slower.

To estimate the effect we need to compute the mean number of molecules per second, crossing the y, z plane in the positive x direction, and then to calculate the mean distance they travel in the x direction before their next collision. The number of molecules, per unit volume of position space, having their momentum vector in the element $dV_p = p^2\, dp\, d\beta \sin\alpha\, d\alpha$ (see Figure 14-2) is $(N/V)f(\mathbf{p})\, dV_p$, where we have assumed that the density N/V of molecules in position space is uniform and thus that f depends only on $\mathbf{p}$. The number of these crossing a unit area of the y, z plane per second, in the positive x direction is then

(flow to right through y, z plane)

$$= \frac{N}{V}\int\limits_{0}^{\infty} p^2\, dp \int\limits_{0}^{2\pi} d\beta \int\limits_{0}^{\pi/2} \sin\alpha\, d\alpha\, f(\mathbf{p}) \frac{p}{m}\cos\alpha$$

$$= \frac{N}{V}\left\langle \frac{p}{m}\right\rangle \frac{1}{4\pi}\int\limits_{0}^{2\pi} d\beta \int\limits_{0}^{\pi/2} \cos\alpha\sin\alpha\, d\alpha = \frac{N}{4V}\langle v\rangle \qquad (14\text{-}9)$$

where we have integrated over α only in the positive directions ($\alpha < \tfrac{1}{2}\pi$). If f is not much different from the Maxwell distribution the mean speed $\langle v\rangle$ will be given by Eq. (12-10). The mean flow through the y, z plane to the right is thus at a quarter the mean speed $\langle v\rangle$; an equal number of molecules flow to the left through the y, z plane, so the drift velocity in the x direction is zero.

Those molecules having a rightward motion will, on the average, travel a distance λ before their next collision. Their mean change in x component from one collision to the next is then

$$\frac{N}{V}\int\limits_{0}^{\infty} p^2\, dp \int\limits_{0}^{2\pi} d\beta \int\limits_{0}^{\pi/2} \sin\alpha\, d\alpha\, f(\mathbf{p}) \frac{p}{m}\cos\alpha\left[(\lambda\cos\alpha)\Big/(N/4V)\langle v\rangle\right]$$

$$= \frac{4\lambda}{\langle v\rangle}\frac{\langle p/m\rangle}{4\pi}\int\limits_{0}^{2\pi} d\beta \int\limits_{0}^{\pi/2} \cos^2\alpha\sin\alpha\, d\alpha = \tfrac{2}{3}\lambda$$

two-thirds of a mean free path λ. Each of these molecules, therefore, transports an average z momentum from the region $x = 0$ to the region $x = (2\lambda/3)$ and, if $\langle p_z\rangle$ is a function of x, the net rate of transport difference in the positive x direction is

$$(N/4V)\langle v\rangle\tfrac{2}{3}\lambda\frac{\partial}{\partial x}\langle p_z\rangle = (mN\lambda\langle v\rangle/6V)\frac{\partial}{\partial x}\langle v_z\rangle$$

A like difference is carried to the left, by molecules going to the left, but with the difference having an opposite sign, so that the two effects add.

This rate of transport of momentum is the viscosity force, tending to reduce the rate of shear, proportional to the rate of shear $[\partial \langle v_z \rangle / \partial x]$ of the gas. The proportionality constant,

$$\eta = \frac{mN\lambda}{3V} \langle v \rangle = \frac{\langle p \rangle}{3\sigma_c} \simeq \frac{N}{V} \tau kT \tag{14-10}$$

is the *coefficient of viscosity* of the gas. As noted earlier, this coefficient is proportional to the mean free path λ, times the molecular density N/V, times the mean molecular speed $\langle v \rangle = \sqrt{(8kT/\pi m)}$. Since the mean free path is inversely proportional to the molecular density, we see that the coefficient of viscosity (at least to the degree of approximation of this calculation) is *independent* of the molecular density, simply proportional to the average momentum magnitude $\langle p \rangle$ and inversely proportional to the molecular collision cross section σ_c. As the density of the gas is reduced, the number of collisions is reduced, but the mean distance $\frac{2}{3}\lambda$, over which the molecule carries excess momentum (or debit) between collisions is *increased* and the two effects just cancel each other out.

The Boltzmann equation is a more accurate way of deriving this coefficient. The expression for η, obtained from Eq. (14-7), differs from that of (14-10) by the factor τ/t_c because of the differences in the averaging process. If details of the variation of t_c with p are taken into account, the derivation via the Boltzmann equation is much more accurate than the rather heuristic derivation just worked out. But rather than going through the viscosity calculations again (it is given as a problem) we shall demonstrate the power of the Boltzmann equation on other transport problems.

Electric Conductivity in a Gas

Suppose that a certain number N_i of the molecules of a gas are ionized (N_i being small compared to the total number N of molecules) and suppose that initially the gas is at equilibrium at temperature T. At $t = 0$ a uniform electric field $\mathfrak{E}$, in the positive x direction, is turned on. The ions will then experience a force $e\mathfrak{E}$ in the x direction, where e is the ionic charge. Imposed on the random motion of the ions between collisions will be a "drift" in the x direction. This is not an equilibrium situation, since the drift velocity of the ions will heat up the gas. But if N_i/N is small, and if

$\mathfrak{E}$ is small enough, the heating will be slow and we can neglect the term $\partial f/\partial t$ in Eq. (14-7) in comparison with the other terms.

Since the ions are initially uniformly distributed in space and since the ionic drift is slow, we can assume that f is more-or-less independent of $\mathbf{r}$. Thus Eq. (14-7) for the ions becomes

$$f \simeq f_0 - t_c \mathbf{F} \cdot \text{grad}_p \, f_0 = f_0 - et_c \mathfrak{E}(\partial f_0/\partial p_x)$$
$$\simeq [1 + et_c(\mathfrak{E}/mkT)p_x] f_0 \qquad (14\text{-}11)$$

where f_0 is the Maxwell distribution,

$$f_0 = [1/(2\pi mkT)^{3/2}] \exp\left(\frac{-p_x^2 - p_y^2 - p_z^2}{2mkT}\right)$$

of the neutral molecules. Function f will be a good approximation to the correct momentum distribution of the ions if the second term in the brackets of Eq. (14-11) is small compared to the first term, unity, over the range of values of p_x for which f_0 has any appreciable magnitude. The term $et_c \mathfrak{E} p_x/mkT$ can be written $(e\lambda\mathfrak{E}/kT)(p_x \langle 1/p \rangle)$ if we assume that $t_c = \tau = \langle m\lambda/p \rangle$, λ being the mean free path of the molecule [see Eq. (12-11)]. Since $e\lambda\mathfrak{E}$ is the energy that would be gained by the ion (in electron volts, if the ion is singly ionized) by falling through a mean free path in the direction of $\mathfrak{E}$, and since kT in electron volts is $T/7500$, then for a gas (such as O_2) at standard conditions, where $\lambda \simeq 10^{-7}$ m [see the discussion following Eq. (12-10)] and $T \simeq 300$, the factor $e\lambda\mathfrak{E}/kT \simeq \mathfrak{E}/40,000$, $\mathfrak{E}$ being in volts per meter. Thus if $\mathfrak{E}$ is as large as 4000 volts per meter, the second term will not equal the first in Eq. (14-11) until p_x is 10 times the mean momentum $\langle p \rangle$ and by this time the exponential factor of f_0 will equal about e^{-50}. Thus, for a wide range of values of T and of $\mathfrak{E}$, either f is vanishingly small or else the second term in brackets of Eq. (14-11) is small compared to the first.

What Eq. (14-11) indicates is that the momentum distribution of the ions, in the presence of the electric field, is slightly nonisotropic; somewhat more of them are going in the direction of the field (p_x positive) than are going in the opposite direction (p_x negative). There is a net *drift velocity* of the ions in the x direction:

$$V_x = \int\int\int_{-\infty}^{\infty} (p_x/m) f \, dV_p \simeq \int\int\int_{-\infty}^{\infty} \left(\frac{p_x}{m} + \frac{2et_c\mathfrak{E}}{mkT}\frac{p_x^2}{2m}\right) f_0(\mathbf{p}) \, dV_p$$

$$\simeq \frac{et_c\mathfrak{E}}{m} \simeq \frac{e\tau\mathfrak{E}}{m} = \lambda e\mathfrak{E}\left(\frac{2}{\pi mkT}\right)^{1/2} \simeq \frac{\lambda e\mathfrak{E}}{m\langle v \rangle} = \Omega\mathfrak{E} \qquad (14\text{-}12)$$

where we have used the fact that for a Maxwell distribution f_0, $\langle p_x \rangle = 0$ and $\langle p_x^2/2m \rangle = \frac{1}{2}kT$ and we have also used Eq. (12-13) for the mean free time τ.

We see that the drift velocity $\mathbf{V}$ of the ion is proportional to the electric intensity, as though the ion were moving through a viscous fluid. The proportionality factor $\Omega \simeq et_c/m \simeq \lambda e/\langle v \rangle$ is called the *mobility* of the ion. The current density $\mathbf{I} = (N_i e \mathbf{V}/V)$ (in amperes per square meter) is

$$\mathbf{I} \simeq (N_i \lambda e^2/V)(2/\pi m k T)^{1/2} \mathfrak{E} \tag{14-13}$$

obeying Ohm's law, with a conductivity $N_i e\Omega/V = N_i e^2 t_c/mV$.

Drift Velocity

It is interesting to see that the drift velocity, and therefore the current density, is proportional to the mean free time between collisions and is thus *inversely* proportional to the square root of the temperature T. As T increases, the *random* velocity $\langle v \rangle$ of the ions (and neutral molecules) increases, but the *drift* velocity $\mathbf{V}$ of the ions decreases. One can visualize the process by imagining the flight of an ion from one collision to its next. Just after each collision the ion comes away in a random direction, with no initial preference for the direction of $\mathfrak{E}$. But during its free flight the electric field acts on it, turning its motion more and more in the positive x direction (its path is a portion of a parabola) and thus adding more and more positive x component to its velocity. This accentuation of the positive x motion is completely destroyed by the next collision (on the average) and the molecule starts on a new parabolic path. If the mean free time is long, the molecule has plenty of time to add quite a bit of excess v_x; if τ is small, the molecule hardly has time to get acted on by the field before it collides again. Thus the higher the temperature, the greater the random velocity $\langle v \rangle$, the shorter the mean free time τ and the smaller the drift velocity and current density. This, of course, checks with the measurements of gaseous conduction.

In most ionized gases, free electrons will be present as well as positive ions. The electrons will also have a drift velocity, mobility, and current density, given by Eqs. (14-12) and (14-13), only with a negative value of charge e, a different value of λ, and a much smaller value of mass m. Thus the drift velocity will be opposite in direction to that of the ions but, since the charge e is negative, the *current* density is in the same direction as that of the ions. Since the electronic mean free path is roughly 2 to 4 times that of the ions and since the electronic mass is several thousand times smaller, the electronic mobility is 500 to 1000 times greater than that of the positive ions and therefore most of the current in an ionized gas is carried by the electrons.

Diffusion

Another nonequilibrium situation is one in which different kinds of molecules mix by diffusion. To make the problem simple, suppose we have a small number N_i of radioactive "tagged" molecules in a gas of N nonradioactive molecules of the same kind. Suppose, at $t = 0$, the distribution in space of the tagged molecules is not uniform (although the density of the mixture *is* uniform). Thus the distribution function for the tagged molecules is a function of $\mathbf{r}$, and we have to write our "0-th approximation" as

$$f_0 = f_r(\mathbf{r}) \cdot f_p(\mathbf{p}); \qquad f_p = \frac{e^{-p^2/2mkT}}{(2\pi mkT)^{3/2}}; \qquad \int \int \int f_r \, dV_r = 1$$

$$(14\text{-}14)$$

The distribution function f for the diffusing molecules will change with time, but we shall find that the rate of diffusion is slow enough so that the term $\partial f/\partial t$ in Eq. (14-1) is negligible compared to other terms.

In the case of diffusion there is no force $\mathbf{F}$, but f_0 does depend on $\mathbf{r}$, so the approximate solution of Eq. (14-7) is

$$f \simeq f_0 - t_c(\mathbf{p}/m) \cdot \text{grad}_r \, f_0 = [f_r - (t_c/m)\mathbf{p} \cdot \text{grad}_r \, f_r] f_p \quad (14\text{-}15)$$

where the vector $\text{grad}_r \, f_r$ points in the direction of increasing density of the tagged molecules. The anisotropy is again in the momentum distribution, but here the preponderance is opposite to $\text{grad}_r \, f_r$; there is a tendency of the tagged molecules to flow *away* from the region of highest density. The conditional probability density that a molecule, if it is at point $\mathbf{r}$, has a momentum $\mathbf{p}$, is [see Eq. (11-2)]

$$(f/f_r) \simeq [1 - (t_c/m)\mathbf{p} \cdot \mathbf{g}] f_p; \qquad \mathbf{g} = (1/f_r) \, \text{grad}_r \, f_r$$

From this we can compute the mean drift velocity of the tagged molecules which are at point $\mathbf{r}$ (for convenience we point the x axis in the direction of $\mathbf{g}$):

$$\mathbf{V} \simeq \int\limits_{-\infty}^{\infty} \int \int (\mathbf{p}/m)[1 - (t_c/m)p_x g]f_p \, dV_p$$

$$= -2(t_c/m)\mathbf{g} \int\limits_{-\infty}^{\infty} \int \int (p_x^2/2m)f_p \, dV_p \simeq -(t_c kT/m)\mathbf{g}$$

$$\simeq -\lambda(2kT/\pi m)^{1/2} \, \mathbf{g} \simeq -\tfrac{1}{2}\lambda\langle v\rangle(1/f_r)\,\text{grad}_r \, f_r$$

We see that in this case the drift velocity increases as T increases.

The density ρ_i of tagged molecules at $\mathbf{r}$ is $N_i f_r$ molecules per unit volume, so the flux $\mathbf{J}$ of tagged particles at $\mathbf{r}$, the net diffusive flow caused by the uneven distribution of these particles, is

$$\mathbf{J} = N_i f_r \mathbf{V} \simeq -D \, \text{grad}_r \, \rho_i; \qquad D = t_c kT/m \simeq \lambda(2kT/\pi m)^{1/2}$$

$$(14\text{-}16)$$

where constant D is called the *diffusion constant* of the tagged molecules. A density gradient of tagged molecules produces a net flow away from the regions of high density, the magnitude of the flow being proportional to the diffusion constant D. We note that there is a simple relationship between D and the mobility Ω of the same molecule when ionized and in an electric field, as given in Eq. (14-12),

$$D = (kT/e)\Omega \qquad\qquad (14\text{-}17)$$

which is more accurate than our approximation for t_c. This simple relationship between diffusion constant, mobility, and temperature, is observed experimentally.

Thus a measurement of diffusion in a gas enables us to predict the electrical conductivity of the gas or vice versa. Equation (14-16) is the basic equation governing diffusion. By adding to it the equation of continuity, we obtain

$$\partial\rho_i/\partial t = -\text{div}\,\mathbf{J} \simeq D\nabla^2\rho_i \qquad\qquad (14\text{-}18)$$

which is called the *diffusion equation*.

There are a number of other transport problems, heat flow and viscosity, for example, which can be worked out by use of the Boltzmann equation. These will be given as problems.

The Chemical Potential

But we have not yet made the connection between the entropy parameters F_n and the fluxes $\mathbf{J}_n$ of extensive quantities X_n, discussed in the latter half of Chapter 8, and the Boltzmann equation, of this chapter. The connection comes through the chemical potential μ, of Eqs. (4-1), (6-3), and (6-6), which will play a more and more important role in our subsequent discussions.

To show the relationship between the chemical potential and the Maxwell–Boltzmann distribution, we write the distribution in terms of the particle density function $\rho_0(\mathbf{r}, \mathbf{p}) = Nf_0(\mathbf{r}, \mathbf{p})$,

$$\rho_0(\mathbf{r}, \mathbf{p}) = \frac{NB}{(2\pi mkT)^{3/2}} \exp\left[-\frac{(p^2/2m) + \phi(\mathbf{r})}{kT}\right] \tag{14-19}$$

where

$$\int \cdots \int \rho_0(\mathbf{r}, \mathbf{p}) \, dV_r \, dV_p = N$$

For conditions of complete equilibrium $\rho(\mathbf{r}) = NBe^{-\phi/kT}$ is the particle density in position space; when $\phi = 0$, $\rho(\mathbf{r}) = (N/V)$, where V is the volume occupied by the N molecules. The integral of ρ_0 times the energy of one molecule at $\mathbf{r}, \mathbf{p}$, taken over all phase space occupied by the molecules, is then the total energy U of the gas and the integral of $\mathbf{v} = (\mathbf{p}/m)$ times $\rho_0(\mathbf{r}, \mathbf{p})$ is the particle flux $\mathbf{J}_n = \rho(\mathbf{r})\langle \mathbf{v} \rangle$, where $\langle \mathbf{v} \rangle$ is the drift velocity.

Purely formally let us simplify the expression for ρ_0 by writing

$$\rho_0(\mathbf{r}, \mathbf{p}) = \exp\left[\frac{\mu - (p^2/2m)}{kT}\right]$$

where

$$\mu + \phi = kT \ln[NB/(2\pi mkT)^{3/2}] \tag{14-20}$$

To see why we write it this way and, in particular, why we write it with the symbol μ, we manipulate the integral for U, using the fact that $kT \ln \rho_0 = [\mu - (p^2/2m)]$,

$$U = \int \cdots \int \left(\frac{p^2}{2m} + \phi\right) \rho_0 \, dV_r \, dV_p$$

$$= \int \cdots \int \rho_0[-kT \ln(\rho_0/e) - kT + (\mu + \phi)] \, dV_r \, dV_p$$

$$= \left\{-kT \int \cdots \int \rho_0 \ln(\rho_0/e) \, dV_r \, dV_p - NkT + N(\mu + \phi)\right\}$$

where e is here the base of the natural logarithms ($\ln e = 1$). This begins to look somewhat like Euler's Eq. (6-5), since $NkT = PV$. If we define

$$s(\mathbf{r}, \mathbf{p}) = -k\rho_0(\mathbf{r}, \mathbf{p}) \ln[\rho_0(\mathbf{r}, \mathbf{p})/e] \tag{14-21}$$

as the *entropy density* in phase space of the gas at the point $\mathbf{r}, \mathbf{p}$, we can write

$$U = ST - PV + N\langle \mu + \phi \rangle \tag{14-22}$$

where S is the integral of $s(\mathbf{r}, \mathbf{p})$ over all occupied regions of phase space. Thus the symbol we used in the form for $\rho_0(\mathbf{r}, \mathbf{p})$ of Eq. (14-20) turns out to be the chemical potential *per molecule*, to which must be added the mechanical potential energy ϕ to obtain the total potential energy $(\mu + \phi)$ of the molecule at $\mathbf{r}, \mathbf{p}$.

When the system is not quite in equilibrium, we can keep the form (14-20) for $\rho_0(\mathbf{r}, \mathbf{p})$, only now μ and T may be functions of position. The chemical potential $\mu(\mathbf{r})$ of a molecule at point $\mathbf{r}$ of position space, is then related to the particle density $\rho(\mathbf{r}) = NBe^{-\phi/kT}$ in position space by the equation

$$\mu(\mathbf{r}) = kT \ln[\rho(\mathbf{r})/(2\pi mkT)^{3/2}] \tag{14-23}$$

The parameter μ can thus be thought of as a contribution to the potential energy of the molecule, because of the thermal motion of its neighbors. If $\rho(\mathbf{r})$ increases, μ increases; also μ increases if T increases.

Returning to the Boltzmann Eq. (14-7), multiplying by N to change it to one for the first approximation to $\rho(\mathbf{r}, \mathbf{p})$, the non-equilibrium density of particles in phase space (assuming that $\partial\rho_0/\partial t$ is zero),

$$\rho(\mathbf{r}, \mathbf{p}) \simeq \rho_0 - t_c\{(\mathbf{p}/m) \cdot \mathrm{grad}_r\, \rho_0 - \mathrm{grad}_r\, \phi \cdot \mathrm{grad}_p\, \rho_0\}$$

$$= \rho_0 - t_c \frac{\mathbf{p}}{mk} \cdot \left\{ \frac{p^2}{2m} \mathrm{grad}_r\left(\frac{1}{T}\right) - \mathrm{grad}_r\left(\frac{\mu}{T}\right) \right.$$

$$\left. - \frac{1}{T} \mathrm{grad}_r\, \phi \right\} \rho_0(\mathbf{r}, \mathbf{p})$$

$$= \rho_0 - t_c \frac{\mathbf{p}}{mk} \cdot \left\{ \left(\frac{p^2}{2m} + \phi\right) \mathrm{grad}_r\left(\frac{1}{T}\right) \right.$$

$$\left. - \mathrm{grad}\left(\frac{\mu + \phi}{T}\right) \right\} \rho_0 \tag{14-24}$$

We define the energy and particle fluxes at **r** (see Eq. 14-12) as

$$\mathbf{J}_u = \int \cdots \int \left(\frac{p^2}{2m} + \phi \right) (\mathbf{p}/m)\rho(\mathbf{r}, \mathbf{p}) \, dV_p$$

$$\mathbf{J}_n = \int \cdots \int (\mathbf{p}/m)\rho(\mathbf{r}, \mathbf{p}) \, dV_p \qquad (14\text{-}25)$$

where we assume that the gas molecules are point masses, since we have included potential energy and only kinetic energy of translation in the term $[(p^2/2m) + \phi]$ in the integral for $\mathbf{J}_u$.

Multiplying Eq. (14-24) by $(\mathbf{p}/m)[p^2/2m] + \phi]$ and integrating produces $\mathbf{J}_u$; multiplying by $(\mathbf{p}/m)$ and integrating produces $\mathbf{J}_n$;

$$\mathbf{J}_u = L_{uu} \, \mathrm{grad}(1/T) - L_{un} \, \mathrm{grad} \left(\frac{\mu + \phi}{T} \right)$$

$$(14\text{-}26)$$

$$\mathbf{J}_n = L_{nu} \, \mathrm{grad}(1/T) - L_{nn} \, \mathrm{grad} \left(\frac{\mu + \phi}{T} \right)$$

where

$$L_{uu} = \int \cdots \int \left(\frac{t_c}{k} \right)\left(\frac{p_x}{m} \right)^2 \left(\frac{p^2}{2m} + \phi \right)^2 \rho_0(\mathbf{r}, \mathbf{p}) \, dV_p$$

$$\rightarrow L_{uu}^0 = \frac{35 t_c}{4mk} (kT)^3 \rho(\mathbf{r})$$

$$L_{un} = L_{nu} = \int \cdots \int \left(\frac{t_c}{k} \right)\left(\frac{p_x}{m} \right)^2 \left(\frac{p^2}{2m} + \phi \right) \rho_0(\mathbf{r}, \mathbf{p}) \, dV_p$$

$$\rightarrow L_{un}^0 = \frac{5 t_c}{4mk} (kT)^2 \rho(\mathbf{r})$$

$$L_{nn} = \int \cdots \int \left(\frac{t_c}{k} \right)\left(\frac{p_x}{m} \right)^2 \rho_0(\mathbf{r}, \mathbf{p}) \, dV_p = L_{nn}^0 = \frac{t_c}{mk} kT \rho(\mathbf{r})$$

where the coefficients L_{uu}^0, $L_{un}^0 = L_{nu}^0$ are the values for $\phi = 0$, obtained by using Eq. (14-20) and by assuming that $t_c(p)$ varies slowly enough with the magnitude of the momentum, p, so that it can be replaced by its constant, average value t_c, and taken out of the integral. It is not difficult to see that

$$L_{uu} = L_{uu}^0 + 2\phi L_{un}^0 + \phi^2 L_{nn}^0 \qquad \text{and} \qquad L_{un} = L_{un}^0 + \phi L_{nn}^0$$

But Eqs. (14-26) are just the flow Eqs. (8-36) we set up by thermodynamic arguments, for the nonequilibrium flow of energy and matter. Our assumption, embodied in Eqs. (14-20) and (14-23), relating the Maxwell–Boltzmann distribution function to the chemical potential μ, has resulted in a formal identity between the results of the Boltzmann equation, in terms of atomic constants and the equations derived by Onsager from thermodynamics, for example, the symmetry requirements of Eq. (8-35), that $L_{un} = L_{nu}$ comes directly from Eq. (14-24) and ultimately from the form of the Maxwell–Boltzmann distribution. We thus can use our knowledge of atomic behavior to compute the values of the coefficients L, rather than obtaining their values experimentally; or else we can use the empirical values of the L's to assist us in determining various atomic properties, such as the value of t_c.

Heat Conduction

Let us apply these results to the problem of heat conduction through a gas placed between two parallel plates, with a temperature difference maintained between them, as shown in Figure 14-3. If the gas is free to move anywhere between the plates it will interact with the gravitational field and carry most of the heat from one plate to the other by *convection* currents. The gas in contact with the warmer plate will expand, have less density than the rest of the gas and thus will be displaced by the cooler gas from the cooler plate and a convection loop of gas motion, as indicated by the dashed-line loops, will carry most of the heat. If, however, the region between the plates is divided into a large number of prism-like

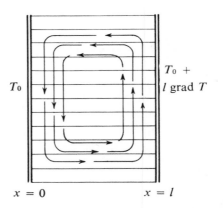

T_0 $T_0 +$ $l\ \text{grad}\ T$

$x = 0$ $x = l$

FIGURE 14-3

Conduction and convection of heat by gas between plates at different temperatures.

compartments, as indicated by the solid lines, stretching from one plate to the other, the convection current can be restrained and the heat will be carried from one plate to the other by the more disorganized sort of motion called heat *conduction*. The compartment walls should restrain the motion but must not conduct the heat themselves.

In such a case the temperature of the gas will not be uniform, but will change linearly from T_0 to $T_0 + l\operatorname{grad} T$ as x goes from 0 to l. The situation is thus one which can be described by Eqs. (14-24) and (14-26). We measure the heat conductivity of the gas by measuring the rate of flow of heat through the gas when the gas is kept at rest; in other words we measure the ratio between $\mathbf{J}_q$ [see Eq. (8-40)] and $-\operatorname{grad} T$, when $\mathbf{J}_n$ is kept zero. According to Eqs. (14-26) this will occur when

$$(L_{un}^0 + \phi L_{nn}^0)\operatorname{grad}(1/T) - L_{nn}^0\operatorname{grad}[(\mu + \phi)/T] = 0$$

or

$$(L_{un}^0 - \mu L_{nn}^0)\operatorname{grad}(1/T) = (L_{nn}^0/T)\operatorname{grad}(\mu + \phi)$$

or

$$\operatorname{grad}(\mu + \phi) = -(1/T)[(L_{un}^0/L_{nn}^0) - \mu]\operatorname{grad} T$$
$$= (\tfrac{5}{2}kT - \mu)(1/T)\operatorname{grad} T \tag{14-27}$$

or

$$\operatorname{grad}[(\mu + \phi)/T] = -[\tfrac{5}{2}k + (\phi/T)]\operatorname{grad}(\ln T)$$

This means that, if no force field is present, and $\phi = 0$, in order that $\mathbf{J}_n$ be zero μ will have to adjust itself so that $\operatorname{grad}(\mu/T)$ is equal to $-\tfrac{5}{2}kT\operatorname{grad}(\ln T)$. Therefore, from Eq. (14-23), the gas density $\rho(\mathbf{r})$ will have to adjust itself so that

$$\operatorname{grad}\{\ln[\rho(\mathbf{r})/(2\pi mkT)^{3/2}]\} = -\tfrac{5}{2}\operatorname{grad}(\ln T)$$

or

$$\operatorname{grad}[\ln \rho(\mathbf{r})] = -\operatorname{grad}(\ln T) \quad\text{or}\quad \rho(\mathbf{r}) = [C/T(\mathbf{r})] \tag{14-28}$$

The density $\rho(\mathbf{r})$ of the gas, in the region between the plates, must vary inversely as the absolute temperature of the gas; otherwise the gas pressure $P = \rho(\mathbf{r})kT(\mathbf{r})$ would vary from point to point, and the gas could not be at rest. Of course we could have achieved the same result by applying a force field, equal to

$$[\tfrac{5}{2}k - (\mu/T)]\operatorname{grad} T = -\{\ln[\rho(\mathbf{r})/(2\pi mkT)^{3/2}] - \tfrac{5}{2}\}\operatorname{grad} T$$

to hold the gas stationary against the pressure gradient produced by the temperature gradient.

To find the flow of heat when $\mathbf{J}_n = 0$, we must use the relationship $\mathbf{J}_s = (1/T)\mathbf{J}_u - (\mu/T)\mathbf{J}_n$ of Eq. (8-33). Since the flow of heat, $\mathbf{J}_q$, is equal to T times the flow of entropy, we have $\mathbf{J}_q = \mathbf{J}_u - \mu\mathbf{J}_n$. Therefore, from Eqs. (14-26),

$$\mathbf{J}_q = \left[L_{uu}^0 - 2\mu L_{un}^0 + \mu^2 L_{nn}^0\right]\mathrm{grad}(1/T)$$

$$- (1/T)(L_{un}^0 - \mu L_{nn}^0)\,\mathrm{grad}(\mu + \phi) \tag{14-29}$$

When $\mathbf{J}_n$ is kept zero, so that Eq. (14-27) is satisfied, we can eliminate $\mathrm{grad}(\mu + \phi)$ between the two equations and obtain

$$\mathbf{J}_q = \frac{L_{uu}^0 L_{nn}^0 - (L_{un}^0)^2}{L_{nn}^0}\,\mathrm{grad}\left(\frac{1}{T}\right) = -\frac{5}{2}\frac{k}{m}t_c\rho(\mathbf{r})\,\mathrm{grad}\,T$$

Therefore the thermal conductivity $\kappa = -[\mathbf{J}_q/\mathrm{grad}\,T]$ of the gas of point particles of mass m and density $\rho(\mathbf{r})$ particles per unit volume is

$$\kappa = \tfrac{5}{2}t_c\frac{k}{m}kT\rho(\mathbf{r}) \simeq \tfrac{5}{2}k\lambda\frac{kT}{\langle p\rangle}\rho(\mathbf{r}) \simeq \frac{5}{6}\frac{k}{\sigma_c}\langle v\rangle \tag{14-30}$$

But from Eq. (3-2), $\tfrac{3}{2}(k/m) = \tfrac{3}{2}(N_0 k/N_0 m) = \tfrac{3}{2}(R/M)$, where N_0 is Avogadro's number, the number of molecules in a mole, and M is the mass of a mole, i.e., the molecular weight of the molecules of the gas. Also $\tfrac{3}{2}R$ is the specific heat c_v of a gas of point atoms, the heat capacity of a mole of gas. Therefore the heat conductivity of the gas of point atoms is $\tfrac{5}{3}(N/V)t_c(kT)(c_v/M)$, where we have used (N/V) instead of $\rho(\mathbf{r})$ for the particle density. This equation also holds for other gases, with other values of c_v.

Referring to Eq. (14-10) (or to Problem 14-2) for the coefficient of viscosity η, we see that, to the degree of approximation of our calculations, there is a close relationship between the coefficient of viscosity of a gas, its specific heat c_v, and its molecular weight M,

$$\kappa = \frac{5}{3}\frac{N}{V}t_c kT(c_v/M) = \tfrac{5}{3}\eta(c_v/M) \tag{14-31}$$

This relationship also holds for polyatomic gases, where both κ and c_v are increased because the larger number of degrees of freedom of the molecules enables them to carry more thermal energy. The equation checks quite well with experimental measurements, even better than Eq. (14-17) between diffusion and mobility, over a wide range of temperatures and densities. Its derivation and experimental confirmation was one of the early triumphs of kinetic theory.

CHAPTER

FLUCTUATIONS

Any system in thermal equilibrium with its surroundings undergoes fluctuations in position and velocity because of the thermal motion of its own molecules as well as of any molecules that may surround it. Kinetic theory, which enables us to compute the mean thermal kinetic and potential energy inhering in each degree of freedom of the system, makes it possible to compute the variance (i.e., the mean-square amplitude of the fluctuations) of each coordinate and momentum of the system. It is often useful to know the size of these variances, for they tell us the lower bound to the accuracy of a piece of measuring equipment and they sometimes give us a chance to measure, indirectly, the magnitude of some atomic constants, such as Avogadro's number N_0.

Equipartition of Energy

Referring to Eqs. (13-14) and (13-15), we see that if the Hamiltonian function for a system can be separated into a sum of terms $[(1/2m_i)(p_i/h_i)^2 + \phi_i(q_i)]$, each of which is a function of just one pair of variables, p_i and q_i; then the Maxwell–Boltzmann probability density can be separated into a product of factors, $(1/Z_i) \exp[-(1/2m_i kT)(p_i/h_i)^2 - (1/kT)\phi_i(q_i)]$, each of which gives the distribution in momentum and position of one separate degree of freedom. Even if the potential energy, or the scale factors h, cannot be completely separated for all the degrees of freedom of the system, if the potential energy does not depend on some coordinates q_j (such as the x coordinates of the center of mass of a dust particle floating freely in the air), then all values of that coordinate are equally likely (the dust particle can be anywhere in the gas) and its momentum will be distributed according to the probability density

$$f_{pj}(p_j) = (1/Z_j)\exp\left[-\frac{1}{2m_jkT}\left(\frac{p_j}{h_j}\right)^2\right]; \qquad Z_j = (2\pi m_j kT)^{1/2} \tag{15-1}$$

The mean thermal kinetic energy of the jth degree of freedom is thus

$$\langle K.E.\rangle = \int_{-\infty}^{\infty}(p_j^2/2m_jh_j^2)f_{pj}(dp_j/h_j) = \tfrac{1}{2}kT \tag{15-2}$$

whether the coordinate is an angle or a distance or some other kind of curvilinear coordinate. Therefore the kinetic energy of thermal motion is equally apportioned, on the average, over all separable degrees of freedom of the system, an energy $\tfrac{1}{2}kT$ going to each. If the potential energy is independent of q_j, then $\tfrac{1}{2}kT$ is the total mean energy possessed by the jth degree of freedom. On the average the energy of rotation of a diatomic molecule (described in terms of two angles) would be kT and the average energy of translation of its center of mass would be $\tfrac{3}{2}kT$. A light atom (helium for example) will have a higher mean speed than does a heavy atom (xenon for example) at the same temperature, in order that the mean kinetic energy of the two be equal. In fact the mean-square value of the jth velocity, when q_j is a rectangular coordinate (i.e., when $h_j = 1$), is

$$\langle \dot{q}_j^2\rangle = \langle p_j^2/m_j^2\rangle = kT/m_j \tag{15-3}$$

Mean-Square Velocity

For example, the x component of the velocity of a dust particle in the air fluctuates irregularly, as the air molecules knock it about. The average value of $\dot{x}$ is zero (if the air has no gross motion) but the mean-square value of $\dot{x}$ is just kT divided by the mass of the particle. We note that this mean-square value is independent of the pressure or density of the air the particle is floating in, and is thus independent of the number of molecules which hit it per second. If the gas is rarefied only a few molecules hit it per second and the value of $\dot{x}$ changes only a few times a second; if the gas is dense the collisions occur more often and the velocity *changes* more frequently per second (as shown in Figure 15-1), but

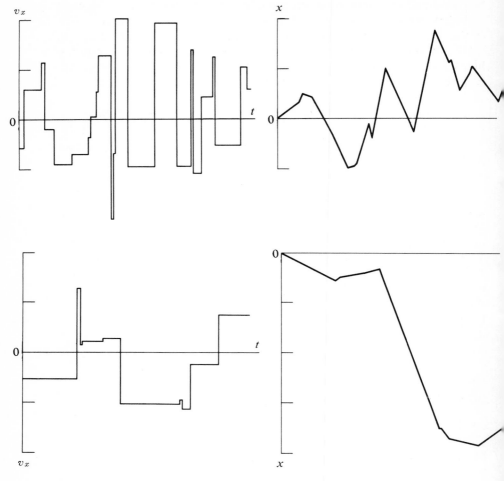

FIGURE 15-1

*Variation with time of x component of velocity and displacement of Brownian
motion. Lower curves for mean time between collisions three times that for
upper curves.*

the mean-square value of the velocity is the same in both cases if the temperature is the same.

Even if the potential energy does depend on the coordinate q_j, the mean kinetic energy of the jth degree of freedom is still $\frac{1}{2}kT$. If the potential energy can be separated into a term $\phi_j(q_j)$ and another term which is independent of q_j, then the probability density that the jth coordinate has a value q_j is

$$f_{qj}(q_j) = (1/Z_{qj})e^{-\phi_j/kT}; \qquad Z_{qj} = \int e^{-\phi_j/kT} h_j \, dq_j \qquad (15\text{-}4)$$

where the integration is over the allowed range of q_j. The mean value of the potential energy turns out to be a function of kT, but the nature of the function depends on how ϕ_j varies with q_j.

The usual case is the one where the scale constant is unity and where the potential energy $\phi_j = \frac{1}{2}m_j\omega_j^2 q_j^2$ has a quadratic dependence on q_j, so that in the absence of thermal motion the displacement q_j executes simple harmonic motion with frequency $\omega_j/2\pi$. In this case $Z_{qj} = (2\pi kT/m_j\omega_j^2)^{1/2}$, and the mean value of potential energy when thermal fluctuations are present is

$$\langle \phi_j \rangle = \frac{1}{2}m_j\omega_j^2 \int_{-\infty}^{\infty} q_j^2 \, f_{qj} \, dq_j = \frac{1}{2}kT \qquad (15\text{-}5)$$

Thus when the potential energy per degree of freedom is a quadratic function of each q, the mean potential energy per q is $\frac{1}{2}kT$, independent of ω_j, and equal to the mean kinetic energy, so that the total mean energy for this degree of freedom is kT. Also the mean-square displacement of the coordinate from its equilibrium position is

$$\langle q_j^2 \rangle = kT/m_j\omega_j^2$$

for coordinates with quadratic potentials.

As an example of a case where ϕ_j is not quadratic, we recall the case of the magnetic dipoles, where $\phi_j = -\mathfrak{m}\mathfrak{B}\cos\theta$, so that $Z_{qj} = (2kT/\mathfrak{m}\mathfrak{B})\sinh(\mathfrak{m}\mathfrak{B}/kT)$. From Eq. (13-21) the mean potential energy is

$$\langle -\mathfrak{m}\mathfrak{B}\cos\theta \rangle = kT - \mathfrak{m}\mathfrak{B}\coth(\mathfrak{m}\mathfrak{B}/kT)$$

$$\rightarrow \begin{cases} -\frac{1}{3}(\mathfrak{m}^2\mathfrak{B}^2/kT), & kT \gg \mathfrak{m}\mathfrak{B} \\ -\mathfrak{m}\mathfrak{B} + kT, & kT \ll \mathfrak{m}\mathfrak{B} \end{cases}$$

which is not equal to $\frac{1}{2}kT$.

Fluctuations of Simple Systems

A mass M, on the end of a spring of stiffness constant K, is constantly undergoing small, forced oscillations because of thermal fluctuations of the pressure of the gas surrounding it and also because of thermal fluctuations of the spring itself. In the absence of these fluctuations the mass will describe simple harmonic motion of amplitude A and frequency $(1/2\pi)(K/M)^{1/2}$, so its displacement, and its mean kinetic and potential energy would be

$$x = A \cos\left[t(K/M)^{1/2} + \alpha\right]; \qquad \langle \text{K.E.} \rangle = \tfrac{1}{4}KA^2 = \langle \text{P.E.} \rangle$$

where A and α are determined by the way the mass is started into motion. In the presence of the thermal fluctuations an irregular motion is superposed on this steady-state oscillation. Even if there is no steady-state oscillation the mass will never be completely at rest but will exhibit a residual motion having total energy, potential plus kinetic, of kT, having a mean-square amplitude such that $\tfrac{1}{2}KA_T^2 = kT$, or

$$A_T^2 = 2kT/K$$

With a mass of a few grams and a natural frequency of a few cycles per second ($K \simeq 1$), this mean-square amplitude is very small, of the order of 10^{-20} m^2, a root-mean-square amplitude of about 10^{-8} cm. This is usually negligible, but in some cases it is of practical importance. The human eardrum, plus the bony structure coupling it to the inner ear, acts like a mass-spring system. Even when there is no noise present the system fluctuates with thermal motion having a mean amplitude of about 10^{-8} cm. Sounds so faint that they drive the eardrum with less amplitude than this are "drowned out" by the thermal noise. In actual fact this thermal-noise motion of the eardrum sets the lower limit of audibility of sounds in the frequency range of greatest sensitivity of the ear (1000 to 3000 cps); if the incoming noise level is less than this "threshold of audibility," we "hear" the thermal fluctuations of our eardrums rather than the outside noises.

We notice that the root-mean-square amplitude of thermal motion of a mass on a spring, $(2kT/K)^{1/2}$, is independent of the density of the ambient air and thus independent of the number of molecular blows impinging on the mass per second. If the density is high the motion will be quite irregular because of the large number of blows per second; if the density is low the motion will be "smoother," but the mean-square amplitude will be the same if the temperature is the same, as illustrated in Figure 15-2. Even if the mass-spring system is in a vacuum the motion will still be present, caused by the thermal fluctuations of the atoms in the spring.

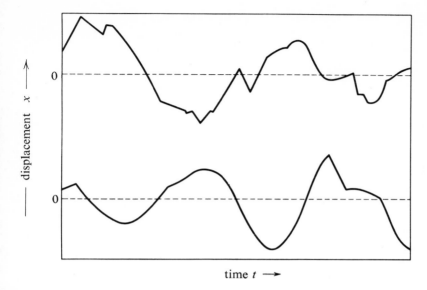

FIGURE 15-2

Brownian motion of a simple oscillator for two different mean times between collisions.

The same effect is present in more complex systems, each degree of freedom has mean kinetic energy $\frac{1}{2}kT$ and similarly for the potential energy if it depends quadratically on the displacement, as in a simple oscillator. A string of mass ρ per unit length and length L under tension Q can oscillate in any one of its standing waves; the displacement from equilibrium and the total energy of vibration of the nth wave is

$$y_n = A_n \sin(\pi n x/L) \cos[(\pi n t/L)(Q/\rho)^{1/2} + \alpha_n]$$

$$\langle \text{K.E.} \rangle + \langle \text{P.E.} \rangle = \tfrac{1}{2}QA_n^2(\pi n/L)^2 \int_0^L \sin^2(\pi n x/L)\, dx$$

$$= \tfrac{1}{4}(\pi^2 n^2 Q/L)A_n^2$$

When the string is at rest except for its thermal vibrations, each of the standing waves has a mean energy of kT, so the mean-square amplitude of motion A_n of the nth wave is $(4LkT/\pi^2 n^2 Q)$

and the mean-square amplitude of deflection of some central portion of the string is the sum of all the standing waves (because of the incoherence of the motion, we sum the squares),

$$\langle y^2 \rangle \simeq \tfrac{1}{4} \sum_{n=1}^{3N} \langle A_n^2 \rangle = (LkT/Q) \sum_{n=1}^{3N} (1/\pi^2 n^2) \simeq LkT/6Q$$

which is related to the result for the simple oscillator, $12Q/L$ being equivalent to the stiffness constant K of the simple spring. If the string is part of a string galvanometer these thermal fluctuations will mask the detection of any current that deflects the string by an amount less than $(LkT/6Q)^{1/2}$.

Density Fluctuations in a Gas

The thermal motion of the constituent molecules produces fluctuations of density, and thus of pressure, in a gas. We could analyze the fluctuation in terms of pressure waves in the gas, as we analyzed the motion of a string under tension in terms of standing waves. Instead of this, however, we shall work out the problem in terms of the potential energy of compression of the gas. Suppose we consider that part of the gas which, in equilibrium, would occupy a volume V_s and would contain N_s molecules. At temperature T the gas, at equilibrium, would have a pressure $P = N_s kT/V_s$ throughout. If the portion of the gas originally occupying volume V_s were compressed into a somewhat smaller volume $V_p - \Delta V (\Delta V \ll V_s)$, an additional pressure equal to $\Delta P = [N_s kT/(V_s - \Delta V)] - P \simeq P(\Delta V/V_s)$ would be needed and an amount of work

$$\int_0^{\Delta V} \Delta P \, d(\Delta V) = \tfrac{1}{2}(P/V_s)(\Delta V)^2 = \tfrac{1}{2}N_s kT(\Delta V/V_s)^2$$

would be required to produce this compression. When thus compressed, this portion of the gas would have a density greater than the equilibrium density ρ by an amount $\Delta \rho = \rho(\Delta V/V_s)$ and its pressure would be greater than P by an amount $\Delta P = P(\Delta V/V_s)$.

Therefore the potential energy corresponding to an increase of density of the part of the gas originally in volume V_s, from its equilibrium density ρ to a nonequilibrium density $\rho + \Delta \rho$, is $\tfrac{1}{2}N_s kT \times (\Delta \rho/\rho)^2$. For thermal fluctuations the mean potential energy is $\tfrac{1}{2}kT$, if the potential energy is a quadratic function of the variable $\Delta \rho$ (as it is here). Consequently the mean-square fractional fluctuation of density of a portion of the gas containing N_s molecules (and occupying volume V_s at equilibrium) is

$$\langle(\Delta\rho/\rho)^2\rangle = 1/N_s \qquad (15\text{-}6)$$

which is also equal to the mean-square fractional fluctuation of pressure, $\langle(\Delta P/P)^2\rangle$. Another derivation of this formula is given in Chapter 23.

We see that the smaller the fraction of the gas we look at (the smaller N_s is) the greater the fractional fluctuation of density and pressure is caused by thermal motion. If we watch a small group of molecules, their thermal motion will produce relatively large changes in their density. On the other hand if we include a large number of molecules in our sample, the large fluctuations in each small part of the sample will to a great extent cancel out, leaving a mean-square fractional fluctuation of the whole which is smaller the larger the number N_s of molecules in the sample. The root-mean-square fractional fluctuation of density or pressure of a portion of the gas is inversely proportional to the square root of the number of molecules sampled.

These fluctuations of density tend to scatter acoustical and electromagnetic waves as they travel through the gas. Indeed it is the scattering of light by the thermal fluctuations of the atmosphere which produces the blue of the sky. The fluctuations are independent of temperature, although at lower temperatures the N_s molecules occupy a smaller volume and thus the fluctuations are more "fine-grained" at lower temperatures.

Incidentally, we could attack the problem of density fluctuations by asking how many molecules happen to be in a given volume V_s at some instant, instead of asking what volume N_s molecules happen to occupy at a given instant, as we did here. The results will of course turn out the same, as will be shown in Chapter 23.

Brownian Motion

The fluctuating motion of a small particle in a fluid, caused by the thermal fluctuations of pressure on the particle, is called Brownian motion. We have already seen [in Eq. (15-3)] that the mean square of each of the velocity components of such motion is proportional to T and inversely proportional to the mass of the particle. The mean square of each of the position coordinates of the particle is not as simple to work out for the unbound particle as it was for the displacement of the mass on a spring, discussed in the preceding section.

In the case of the mass on the end of the spring, the displacement x from equilibrium is confined by the restoring force, the maximum displacement is determined by the energy possessed by the oscillator, and we can measure a mean-square displacement from equilibrium, $\langle x^2 \rangle$, by averaging the value of x^2 over any relatively long interval of time (the longer the interval, the more accurate the result, of course). But the x component of displacement of a free particle in a fluid is not so limited; the only forces acting on the particle (if we can neglect the force of gravity) are the fluctuations of pressure of the fluid, causing the Brownian motion, and the viscous force of the fluid, which tends to retard the particle's motion. If we measure the x component of the particle's position (setting the initial position at the origin) we shall find that, although the direction of motion often reverses, the particle tends to drift away from the origin as time goes on, and eventually it will traverse the whole volume of fluid, just as any molecule of the fluid does. If we allow enough time, the particle is eventually likely to be anywhere in the volume. This is in correspondence with the Maxwell–Boltzmann distribution; the potential energy does not depend on x (neglecting gravity) so the probability density f is independent of x; any value of x is equally likely in the end.

But this was not the problem at present. We assumed that the particle under observation started at $x = 0$; it certainly isn't likely to drift far from the origin right away. Of course the average value of x is zero, since the particle is as likely to drift to the left as to the right. But the expected value of x^2 must increase somehow with time. At $t = 0$ the particle is certainly at the origin; as time goes on the particle may drift farther and farther away from $x = 0$, in either direction. We wish to compute the expected value of x^2 as a function of time or, better still, to find the probability density of finding the particle at x after time t. Note that this probability is a conditional probability density; it is the probability of finding the x component of the position of the particle to be x at time t *if* the particle was at the origin at $t = 0$.

Random Walk

We can obtain a better insight into this problem if we consider the random-walk problem of Eqs. (11-13) and (11-14). A crude model of one-dimensional Brownian motion can be constructed as follows. Suppose a particle moves along a line with a constant speed v. At the end of each successive time interval τ it may change its direction of motion or not, the two alternatives being equally likely ($p = \frac{1}{2}$) and distributed at random. After N intervals (i.e., after time $N\tau$) the chance that the particle is displaced an amount $x_n = (2n - N)v\tau$ from its initial position is then the chance that during n of the N intervals it went to the right and during the other $(N - n)$ intervals it went to the left; it covered a distance $v\tau$ in each interval. According to Eq. (11-8) this probability is

$$P_n(N) = N! \tfrac{1}{2}^N / n!(N - n)! \tag{15-7}$$

since $p = \frac{1}{2}$ for this case.

The expected value of x_n and its variance are then obtained from Eqs. (11-13) and (11-14) (for $p = \frac{1}{2}$):

$$\langle x \rangle = \sum_{n=0}^{N} (2n - N)v\tau P_n(N) = 0$$

$$\langle x^2 \rangle = \sum_{n=0}^{N} (2n - N)^2 (v\tau)^2 P_n(N) = N(v\tau)^2 \tag{15-8}$$

The expected value of the displacement is zero, since the particle is as likely to go in one direction as in the other. Its tendency to stray from the origin is measured by $\langle x^2 \rangle$, which increases linearly with the number of time intervals N, and thus increases linearly

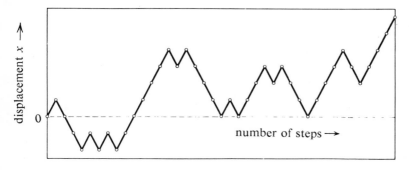

FIGURE 15-3

Displacement for random walk. At each dot the "walker" flipped a coin to decide whether to step forward or backward.

with time. If the particle, once started, continued always in the same direction ($p = 0$ or 1) the value of $\langle x^2 \rangle$ would be $(N v \tau)^2$, increasing quadratically with time. But with the irregular, to-and-fro motion of the random walk, $\langle x^2 \rangle$ increases only linearly with time. Figure 15-3 is a plot of x as a function of t for a random walk as described here. We note the irregular character of the motion and the tendency to drift away from $x = 0$. Compare it with the curves of Figure 15-1, and also with Figure 15-2, for a mass with restoring force.

The limiting case of N very large and τ very small is the case most nearly corresponding to Brownian motion. This limiting form was calculated at the end of Chapter 11. There we found that the probability distribution for displacement of the particle after N steps reduced, in the limit of N large, to the normal distribution of Eq. (11-17). The variance, in this case, as we just saw, is $\sigma^2 = (v\tau)^2 N$ (for $p = \frac{1}{2}$) and the mean value of x is $X = 0$. Since the time required for the N steps is $t = \tau N$ we can write the conditional probability density (that the particle will be at x at time t if it was at $x = 0$ at $t = 0$) as

$$F(x) = [1/(4\pi Dt)^{1/2}]e^{-x^2/4Dt}; \qquad D = \tfrac{1}{2}v^2\tau \tag{15-9}$$

so that the probability that the particle is between x and $x + dx$ at time t is $F(x)\,dx$. We see that the "spread" of the distribution increases as t increases, as the particle drifts away from its initial position. The mean-square value of x is

$$\langle x^2 \rangle = \sigma^2 = (v\tau)^2 N = 2Dt$$

which increases linearly with time. Thus the question of the dependence of $\langle x^2 \rangle$ on time, raised earlier in this section, is answered to the effect that $\langle x^2 \rangle$ is proportional to t. The value of the proportionality constant $2D$ for the actual Brownian motion of a particle in a fluid must now be determined.

In the case of the two curves of Figure 15-1, the root-mean-square velocity [equivalent to the v of Eq. (15-9)] is the same for the upper and lower curves. Since the mean free time τ is larger for the lower curves, the mean-square displacement $\langle x^2 \rangle = (v^2\tau)t$ is greater for the lower set of curves; the particle drifts away from the origin the more rapidly, the longer is the mean free time.

The Langevin Equation

To determine the value of constant D for a particle in a fluid we must study its equation of motion. As before, we study it in a single dimension first. The x component of the force on the particle can be separated into two parts, an average effect of the surrounding fluid plus a fluctuating part, caused by the pressure fluctuations of thermal motion of the fluid. The average effect of the fluid on the particle is a frictional force, caused by the fluid's viscosity. If the velocity of the particle in the x direction is $\dot{x}$, this average frictional force has an x component equal to $M\beta\dot{x}$, opposing the particle's motion, where the mechanical resistance to motion, $M\beta$, in a fluid of viscosity η, on a spherical particle of radius a is $M\beta = 6\pi a\eta$ (Stoke's law). The fluctuating component of the force on the particle can simply be written as $MA(t)$ (we write these functions with a factor M, the mass of the particle, so that M can be divided out in the resulting equation).

The equation of motion for the x component of the particle's position can thus be written as

$$M\ddot{x} = -M\beta\dot{x} + MA(t) \tag{15-10}$$

which is known as *Langevin's equation*. We note that β has the dimensions of reciprocal time. Multiplying the equation by x/M and using the identities

$$x\dot{x} = \frac{1}{2}\frac{d}{dt}(x^2) \qquad \text{and} \qquad x\ddot{x} = \frac{1}{2}\frac{d^2}{dt^2}(x^2) - (\dot{x})^2$$

we have

$$\frac{1}{2}\frac{d^2}{dt^2}(x^2) = -\tfrac{1}{2}\beta\frac{d}{dt}(x^2) + (\dot{x})^2 + xA(t)$$

This is an equation for one particular particle. If we had many identical particles in the fluid (or if we performed a sequence of similar observations on one particle) each particle would have different values of x and $\dot{x}$ at the end of a given time t, because of the effects of the random force $A(t)$.

Suppose we average the effects of the fluctuations by averaging the terms of Eq. (15-10) over all similar particles. The term $xA(t)$ will average out because both $\langle x \rangle$ and $\langle A \rangle$ are zero and the fluctuations of x and A are independent; the average value of $\dot{x}^2$, however, carries with it the mean effects of the fluctuating force $A(t)$. We showed in Eq. (15-3) that for a particle in thermal equilibrium at temperature T, its mean-square velocity component $\langle \dot{x}^2 \rangle$ is equal to kT/M.

If we now symbolize the mean-square displacement $\langle x^2 \rangle$ as $s(t)$, the average of the equation of motion written above turns out to be

$$\tfrac{1}{2}\ddot{s} = (kT/M) - \tfrac{1}{2}\beta\dot{s}$$

The solution of this equation is $\dot{s} = (2kT/M\beta) - Ce^{-\beta t}$. The transient exponential soon drops out, leaving for the steady-state solution $\dot{s} = (2kT/M\beta)$ and thus

$$s = \langle x^2 \rangle = (2kT/M\beta)t \tag{15-11}$$

This result answers the question raised at the end of the previous section; the constant $D = \tfrac{1}{2}\langle v^2 \tau \rangle$ used there now turns out to be $kT/M\beta$ and, for a spherical particle of radius a in a fluid of viscosity η, constant D is equal to $kT/6\pi a\eta$ from Stoke's law.

The innocuous looking result shown in Eq. (15-11) enabled Perrin and others first to measure Avogadro's number N_0 and thus, in a sense, first to make contact with atomic dimensions. They were able to measure N_0 in terms of purely macroscopic constants, plus observations of Brownian motion. A spherical particle was used, of known radius a, so that Stoke's law applied. The viscosity of the fluid in which the particle was immersed was measured as well as the temperature T of the fluid. The value of the gas constant R was known but at the time neither the value of the Boltzmann constant k nor the value of $N_0 = R/k$ was known. The x coordinates of the particle in the fluid were measured at the ends of successive intervals of time of length t; x_0 at $t = 0$, x_1 at t, x_2 at $2t$, and so on, and the average of the set of values $(x_{n+1} - x_n)^2$ was computed, which is equivalent to the $\langle x^2 \rangle$ of Eq. (15-11).

By making the measurements for several different values of the time interval t, it was verified that $\langle x^2 \rangle$ does indeed equal $2Dt$, and the value of D was determined. The value of Avogadro's number,

$$N_0 = R/k = RT/6\pi a\eta D$$

can thus be computed. By this method a value of N_0 was obtained which checks within about 5 parts in a thousand with values later obtained by more direct methods. Of course very small spheres had to be used, to make $\langle x^2 \rangle$ as large as possible, and careful observations with a microscope were made to determine the successive x_n's. Perrin used spheres of radius 2×10^{-5} cm and time intervals from a few seconds to a minute or more.

The Fokker–Planck Equations

Brownian motion is simply the fine details of the process of diffusion. If there were initially a concentration of particles in one region of the fluid, as time went on these particles would diffuse, by Brownian motion, to all parts of the fluid. The diffusion constant for a particle in a fluid is $D = kT/M\beta$, which is to be compared with Eq. (14-16) for the D for a molecule; in the molecular case β is evidently equal to $1/t_c$, whereas for a larger, spherical particle $\beta = 6\pi a\eta/M$. The mean concentration of the diffusing particles must satisfy a diffusion equation of the type given in Eq. (14-18).

This means that there is a close connection between the results of this chapter and those of the section on diffusion. Whether the diffusing entity undergoing Brownian motion is a molecule of the gas or a dust particle in the gas, the probability density of its presence at the point in space given by the vector $\mathbf{r}$ at time t, if it starts at $\mathbf{r}_0$ at $t = 0$, is given by the three-dimensional generalization of Eq. (15-9),

$$f_r(\mathbf{r},\, t) = (4\pi Dt)^{-3/2} \exp\left(- \frac{|\mathbf{r} - \mathbf{r}_0|^2}{4Dt}\right);$$
$$D = kT/M\beta \qquad\qquad\qquad (15\text{-}12)$$

which is a solution of the diffusion Eq. (14-18). The value of β appropriate for the particle under study must be used, of course.

The distribution function for diffusion by Brownian motion of Eq. (15-12) and the diffusion Eq. (14-18) that it satisfies can thus be derived by the methods of the previous chapter or else by those of this chapter. For example, it is possible to generalize the Langevin Eq. (15-10) and manipulate it to obtain the diffusion equation. Also, by either method, it can be shown that when an external force $\mathbf{F}$ acts on the diffusing particle (such as the force of gravity), the diffusion equation has the more general form

$$\partial f_r/\partial t = \text{div}[D \,\text{grad}\, f_r - (\mathbf{F}/M\beta)f_r] \qquad\qquad (15\text{-}13)$$

When f_r is the density of diffusing substance (molecules or heat, for example), Eq. (15-13), or its simple version (14-18) for $\mathbf{F} = 0$, is called the diffusion equation. When f_r is the distribution function for a particle undergoing Brownian motion and the equation is considered to be a first approximation to a generalized Langevin equation, then Eq. (15-13) is called a *Fokker–Planck equation*. The solutions behave the same in either case, of course.

The solution of (15-13) for $\mathbf{F} = 0$ and for the particle starting at $\mathbf{r} = \mathbf{r}_0$ when $t = 0$ is Eq. (15-12). From it can be derived all the characteristics of Brownian motion in regard to the possible position of the particle at time t.

A Fokker–Planck equation can also be obtained for the distribution in momentum, $f_p(\mathbf{p}, t)$ of the particle. It is

$$\partial f_p / \partial t = \beta \operatorname{div}_p (MkT \operatorname{grad}_p f_p + f_p \mathbf{p}) \tag{15-14}$$

For a particle that is started at $t = 0$ with a momentum $\mathbf{p} = \mathbf{p}_0$, the solution of this equation, which is the probability density of the particle in momentum space, is

$$f_p(\mathbf{p}, t) = [2\pi MkT(1 - e^{-2\beta t})]^{-3/2}$$
$$\times \exp\left[\frac{-|\mathbf{p} - \mathbf{p}_0 e^{-\beta t}|^2}{2MkT(1 - e^{-2\beta t})}\right] \tag{15-15}$$

This interesting solution shows that the expected momentum of the particle at time t is $\mathbf{p}_0 e^{-\beta t}$ [compare with the discussion of Eq. (11-17)], which is the momentum of a particle started with a momentum $\mathbf{p}_0$ and subjected to a frictional retarding force $-\beta \mathbf{p}$. As time goes on, the effect of the fluctuations "spreads out" the distribution in momentum; the variance of the momentum (i.e., its mean-square deviation from $\mathbf{p}_0 e^{-\beta t}$) is $kT(1 - e^{-2\beta t})$, starting as zero when $t = 0$, when we are certain that the particle's momentum is $\mathbf{p}_0$, and approaching asymptotically the value kT, which is typical of the Maxwell distribution. Thus Eq. (15-15) shows how an originally nonequilibrium distribution for a particle (or a molecule) in a fluid can change with time into the Maxwell distribution typical of an equilibrium state. Constant β, which equals $6\pi a\eta / M$ for a spherical particle or $1/t_c$ for a molecule in a gas, is thus equal to the reciprocal of the relaxation time for the distribution, which relates directly to the discussion of Eq. (14-8).

Of course the most general distribution function would be $f(\mathbf{r}, \mathbf{p}, t)$, giving the particle's distribution in both position and momentum at time t after initial observation. The equation for this f is, not surprisingly, closely related to the Boltzmann Eq. (13-6). It can be shown to be

$$\frac{\partial f}{\partial t} + \frac{\mathbf{p}}{M} \cdot \operatorname{grad}_r f + \mathbf{F} \cdot \operatorname{grad}_p f$$
$$= \beta \operatorname{div}_p [MkT \operatorname{grad}_p f + \mathbf{p} f] \tag{15-16}$$

The derivation of this equation, particularly of the right-hand side of it, involves a generalization of the arguments used in deriving Eq. (15-11). This right-hand side is another approximation to the collision term Q of Eq. (13-6).

3 STATISTICAL MECHANICS

CHAPTER

16

ENSEMBLES AND DISTRIBUTION FUNCTIONS

It is now time to introduce the final generalization, to present a theoretical model that includes all the special cases we have been considering heretofore. If we had been expounding our subject with mathematical logic we would have started at this point, presenting first the most general assumptions and definitions as postulates, working out the special cases as theorems following from the postulates, and only at the end demonstrating that the predictions, implicit in the theorems, actually do correspond to the "real world," as measured by experiment. We have not followed this procedure, for several reasons.

In the first place, most people find it easier to understand a new subject, particularly one as complex as statistical physics, by progressing from the particular to the general, from the familiar to the abstract.

A more important reason, however, is that physics itself has developed in a nonlogical way. Experiments first provide us with data on many particular cases, at first logically unconnected with each other, which have to be learned as a set of disparate facts. Then it is found that a group of these facts can be considered to be special cases of a "theory," an assumed relationship between defined quantities (energy, entropy, and the like) which will reproduce the experimental facts when the theory is appropriately specialized. This theory suggests more experiments, which may force modifications of the theory and may suggest further generalizations until, finally, someone shows that the whole subject can be "understood" as the logical consequences of a few basic postulates.

At this point the subject comes to resemble a branch of mathematics, with its postulates and its theorems logically deduced therefrom. But the similarity is superficial, for in physics the experimental facts are basic and the theoretical structure is erected to make it easier to "understand" the facts and to suggest ways of obtaining new facts. A logically connected theory turns out to be more convenient to remember than are vast arrays of unconnected data. This convenience, however, should not persuade us to accord the theory more validity than should inhere in a mnemonic device. We must not expect, for example, that the postulates and definitions should somehow be "the most reasonable" or "the logically inevitable" ones. They have been chosen for the simple, utilitarian reason that a logical structure reared on them can be made to correspond to the experimental facts. Thus the presentation of a branch of physics in "logical" form tends to exaggerate the importance and inevitability of the theoretical assumptions, and to make us forget that the experimental data are the only truly stable parts of the whole.

This danger, of ascribing a false sense of inevitability to the theory, is somewhat greater with statistical physics than with other branches of classical physics, because the connection between experiment and basic theory is here more indirect than usual. In classical mechanics the experimental verification of Newton's laws can be fairly direct; and the relationship between Faraday's and Ampere's experiments and Maxwell's equations of electromagnetic theory is clearcut. In thermodynamics, the experiments of Rumford, relating work and heat, bear a direct relationship to the first law, but the experimental verification of the second law is indirect and negative. Furthermore, the more accurate "proofs" that the Maxwell-Boltzmann distribution is valid for molecules in a gas, are experimentally circuitous. And finally, as we shall see later, there is no experiment, analogous to those of Faraday or Ampere, which directly verifies any of the basic assumptions of statistical mechanics; their validity must be proved piecemeal and inferentially. In the end, of course, the proofs are convincing from their very number and breadth of application.

However we have now reached a point in our exposition where the basic theory must be presented, and it is necessary to follow the pattern of mathematical logic for a time. Our definitions and postulates are bound to sound arbitrary until we have completed the demonstration that the theory does indeed correspond to a wide variety of observed facts. But we must keep in mind that they have been chosen solely to obtain this correspondence with observation, not because they "sound reasonable" or satisfy some philosophical "first principles."

Distribution Functions in Phase Space

In Chapters 13 and 15 we discussed distribution functions for molecules, and also for multimolecular particles in a gas. In statistical mechanics we carry this generalization to its logical conclusion, and deal with distribution functions for complete thermodynamic systems. A particular microstate of such a system (a gas of N particles, for example) can be specified by choosing values for the $6N$ position and momentum coordinates; the distribution function is the probability density that the system has these coordinate values. Geometrically speaking, an elementary region in this $6N$-dimensional phase space represents a microstate of the system; the point representing the system passes through all the microstates allowed by its thermodynamic state; the fraction of time spent by the system point in a particular region of phase space is proportional to the distribution function corresponding to the thermodynamic state. In other words, a choice of a particular thermodynamic state (a macrostate) is equivalent to a choice of a particular distribution function, and vice versa. The task of statistical mechanics is to devise methods for finding distribution functions which correspond to specific macrostates.

According to classical mechanics, the distribution function for a system with $\phi = 3N$ degrees of freedom is

$$f(q, p) = f(q_1, q_2,..., q_\phi, p_1, p_2,..., p_\phi)$$

where the q's are the coordinates and the p's the conjugate momenta [see page 188] which specify the configuration of the system as a whole. Then $f(q, p) \, dV_q \, dV_p$ [where $dV_q = h_1 \, dq_1 h_2 \, dq_2 \, \cdots \, h_\phi \, dq_\phi$ and $dV_p = (dp_1/h_1)(dp_2/h_2) \, \cdots \, (dp_\phi/h_\phi)$] is the probability that the system point is within the element of phase space $dV_q \, dV_p$ at position $q_1,..., p_\phi$, at any arbitrarily chosen instant of time.

More generally, the distribution function represents the probability density, not for one system, but for a collection of similar systems. Imagine a large number of identical systems, all in the same thermodynamic state but, of course, each of them in different possible microstates. This collection of sample systems is called an *ensemble* of systems, the ensemble corresponding to the specified macrostate. Different ensembles, representing different thermodynamic states, have different populations of microstates. The distribution function for the macrostate measures the relative number of systems in the ensemble which are in a given microstate at any instant. Thus it is a generalization of the distribution function of Eq. (13-10), which was for an ensemble of molecules

Each system in the ensemble has ϕ coordinates q_i and ϕ momenta p_i. The Hamiltonian function $H(p, q)$ for the system is the sum of the molecular Hamiltonians of Eq. (13-14), for every molecule in the system. It is the total energy of the system, kinetic plus potential, expressed as a function of the 2ϕ coordinates and momenta. Since each system in the ensemble is isolated from all others the magnitude of each system's H will be independent of time, although the various q's and p's for each system will change with time, as the atoms constituting the system move around and interact among themselves. The values of the q's and p's at a given instant determine the position in phase space of the point representing the system. Putting it the other way around, the position of the system point in the 2ϕ-dimensional phase space specifies the configuration and velocity of every part of the system, the position and momentum of every atom, at that instant.

The motion of the system point in phase space is determined by the equations of motion of the system. These can be expressed in many forms; the one most useful to us here is the set of Hamilton's equations, the generalization of Eqs. (13-3),

$$\dot{q}_i = (\partial H/\partial p_i); \qquad \dot{p}_i = -(\partial H/\partial q_i); \qquad i = 1, 2, ..., \phi \qquad (16\text{-}1)$$

where the dot over the letter indicates the time derivative. This gives the 2ϕ components of the velocity vector $\mathbf{U}$, of the system point in phase space, in terms of the 2ϕ components of the gradient of H in phase space. As noted in Chapter 13, the q-components of $\mathbf{U}$ are related to the p components of the gradient of H, and vice versa.

The ensemble of systems can thus be represented as a swarm of system points in phase space, each point moving in accordance with Eqs. (16-1); the velocity of the point in phase space is proportional to the gradient of H. The density of points in any region of phase space is proportional to the value of the distribution function $f(q, p)$ in that region.

We should emphasize yet again that the phase space we are

now discussing has 2ϕ dimensions, and that ϕ is usually a very large number. Our space intuition, accustomed to a mere three dimensions, may lead us astray. For example, if the system is a perfect gas of N point particles, the number of degrees of freedom ϕ will equal three times N and if the amount of gas is a mole, N will equal Avogadro's number, $N_0 = 6 \times 10^{26}$. The volume element

$$dV_q\,dV_p = dx_1\,dy_1\,dz_1\,dx_2 \cdots dz_N\,dp_{x1}\,dp_{y1}$$
$$\times\,dp_{z1}\,dp_{x2}\cdots dp_{yN}\,dp_{zN}$$

is thus a product of 36×10^{26} differentials, if $N = N_0$.

If the gas is enclosed in a vessel of volume V, the integral of dV_q over all the allowed ranges of all the q's (in this case the q's are the displacements x, y, z for each point particle) is

$$\int dx_1\,dy_1\,dz_1 \int dx_2\,dy_2\,dz_2 \int \cdots \int dx_N\,dy_N\,dz_N = V^N$$

the volume V, occupied by each particle, raised to an exceedingly large power. The position-space volume involved in this simple example is thus strongly dependent on N, as well as on the volume V occupied by the system in 3-space.

Since the system is isolated, the total kinetic energy of all the point particles,

$$U = \sum_{n=1}^{N}\left(\frac{1}{2m}\right)(p_{xn}^2 + p_{yn}^2 + p_{zn}^2) = \sum_{i=1}^{\phi}(p_i^2/2m) \qquad (\phi = 3N)$$

is constant and the system point consequently moves on the "surface" of a hypersphere of radius $\sqrt{2mU}$ in momentum space, such that the distance, $(\Sigma p_i^2)^{1/2} = \sqrt{2mU}$, from the origin in the ϕ-dimensional space, is constant. As will be demonstrated shortly, the volume of momentum space within such a spherical surface is

$$[(2\pi mU)^{\phi/2}/(\phi/2)!] = [(2\pi mU)^{3N/2}/(3N/2)!]$$

The volume of phase space enclosed within the hypersphere in momentum space and the "hypercube" of position space is thus $[V^N(2\pi mU)^{3N/2}/(3N/2)!]$, which is strongly dependent on N, the number of particles in the system. If it should happen that the system point is equally likely to be anywhere inside this volume, then the probability density $f(q, p)$ would be equal to

$$[(3N/2)!/(2\pi mU)^{3N/2}V^N]$$

Liouville's Theorem

If the thermodynamic state is not an equilibrium state, f will be a function of time. If the state is an equilibrium state, the density of system points in any specified region of phase space will be constant; as many system points will enter the region per unit of time as will leave it. The swarm of system points has some similarity to the swarm of particles in a gas. The differences are important, however. The system points are moving in a 2ϕ-dimensional phase space, not real space; also the system points do not collide. In fact the system points do not interact at all, for each system point represents a different system of the ensemble, and the separate systems cannot interact since they are but samples in an imaginary array of systems, assembled to represent a particular macrostate. Each individual system point, for example, may represent a whole gas of N atoms, or a crystal lattice, depending on the situation the ensemble has been chosen to represent. There can be no physical interaction between the individual sample systems.

This means that the equation for the change of f with time, the generalization of Eq. (13-4) to the ensemble, has no collision term Q. The equation,

$$\frac{\partial f}{\partial t} + \sum_{i=1}^{\phi} \frac{\partial}{\partial q_i}(\dot{q}_i f) + \sum_{i=1}^{\phi} \frac{\partial}{\partial p_i}(\dot{p}_i f) = 0 \qquad (16\text{-}2)$$

is simply the equation of continuity in phase space, and represents the fact that, as the swarm of system points moves about in phase space, no system point either appears or disappears.

Since each system in the ensemble obeys Hamilton's equations (16-1), this equation of continuity becomes

$$\frac{\partial f}{\partial t} + \sum_{i=1}^{\phi} \left[\frac{\partial}{\partial q_i}\left(f \frac{\partial H}{\partial p_i} \right) - \frac{\partial}{\partial p_i}\left(f \frac{\partial H}{\partial q_i} \right) \right] = 0$$

and since

$$\frac{\partial H}{\partial p_i}\frac{\partial f}{\partial q_i} = \frac{\partial}{\partial q_i}\left(f \frac{\partial H}{\partial p_i} \right) - f \frac{\partial^2 H}{\partial q_i \partial p_i}$$

etc., we have [see discussion following Eq. (13-4)]

$$\frac{\partial f}{\partial t} + \sum_{i=1}^{\phi} \left[\left(\frac{\partial H}{\partial p_i}\frac{\partial f}{\partial q_i} \right) - \left(\frac{\partial H}{\partial q_i}\frac{\partial f}{\partial p_i} \right) \right] = 0$$

or

$$\frac{\partial f}{\partial t} + \mathbf{U} \cdot \text{grad} \ f \equiv \frac{\partial f}{\partial t} + \sum_{i=1}^{\phi} \dot{q}_i \frac{\partial f}{\partial q_i} + \sum_{i=1}^{\phi} \dot{p}_i \frac{\partial f}{\partial p_i} \equiv \frac{df}{dt} = 0$$

$$(16\text{-}3)$$

where df/dt is the change in time of the distribution function f in a coordinate system which moves with the system points. Because of the relationship between $\dot{q}$ and p, $\dot{p}$ and q, inherent in Hamilton's equations, the density of system points near a given system point of the ensemble remains constant as the swarm moves about. If, at $t = 0$, the swarm has a high density in a localized region of phase space, this concentration of system points moves about as time goes on but it does not disperse; it keeps its original high density. This result is known as *Liouville's theorem*.

We can use Liouville's theorem to devise distribution functions which are independent of time, i.e., which represent equilibrium macrostates. For example, if f had the same value everywhere in phase space, it would be independent of time; as a part of the swarm moved away from a given region of phase space a different part of the swarm would move in and, since all parts of the swarm have (and keep) the same density, the density in a given region would not change. We can also devise less trivial stationary distributions, for the path traversed by any one system point does not cover all of phase space; it confines itself to the hypersurface on which the Hamiltonian function $H(q, p)$ is constant; an isolated system cannot change its total energy. Therefore if the distribution function is the same for all regions of phase space for which H is the same (i.e., if f is a function of H alone) the density of system points in a given region cannot change as the points move along their constant-H paths.

We shall deal with several different types of distribution functions, corresponding to different specifications regarding the thermodynamic state of the system. The simplest one is for f to be zero everywhere in phase space except on the hypersurface corresponding to $H(q, p) = E$, a constant; the ensemble corresponding to this is called a *microcanonical ensemble*. A more useful case is for f to be proportional to $\exp[-H(q, p)/kT]$, corresponding to what is called the *canonical ensemble*. Other ensembles, with f's which are more complicated functions of H, will also prove to be useful. But, in order for any of them to represent actual thermodynamic macrostates, we must assume a relationship between the distribution function f for an ensemble and the corresponding thermodynamic properties of the macrostate which the ensemble is supposed to represent. The appropriate relationship turns out to be between the entropy of the macrostate and the distribution function of the corresponding ensemble.

Quantum States and Phase Space

Before we state the basic postulate of statistical mechanics, relating entropy and ensembles, we should discuss the modifications which quantum theory makes in our definitions. In some respects the change is in the direction of simplification, the summation over denumerable quantum numbers being substituted for integration over continuous coordinates in phase space. Instead of Hamilton's equations (16-1), there is a Schrödinger equation for a wave function $\Psi(q_1, q_2, ..., q_\phi)$ and an allowed value E of energy of the system, both of which depend on the ϕ quantum numbers $v_1, v_2, ..., v_\phi$, which designate the quantum state of the system.

For example, if the system consists of N particles in the simple crystal model of Eq. (13-2), the classical Hamiltonian for the whole system can be written as

$$H(q, p) = \sum_{j=1}^{\phi} [(1/2m)p_j^2 + (m\omega^2/2)q_j^2]; \qquad \phi = 3N \qquad (16\text{-}4)$$

where $q_{3i-2} = x_i, q_{3i-1} = y_i, q_{3i} = z_i, p_{3i-2} = p_{xi}$, etc. Hamilton's equations become

$$\dot{q}_i = (1/m)p_i; \qquad -\dot{p}_i = m\omega^2 q_i \qquad (16\text{-}5)$$

and Schrödinger's equation for the system is $H\Psi = E\Psi$, where each p_i in the H is changed to $(\hbar/i)(\partial/\partial q_i)$. For (16-4) it is

$$\sum_{i=1}^{\phi} \left[\tfrac{1}{2}m\omega^2 q_i^2 - \frac{\hbar^2}{2m}\frac{\partial^2}{\partial q_i^2} \right]\Psi = E\Psi \qquad (16\text{-}6)$$

where Planck's constant h is equal to $2\pi\hbar$. Solution of this equation, subject to the requirement that Ψ is finite everywhere, results in the following allowed values of the energy E,

$$E_{v_1, v_2, ..., v_\phi} = \sum_{i=1}^{\phi} \hbar\omega(v_i + \tfrac{1}{2}) \qquad (16\text{-}7)$$

The distribution function for an equilibrium state is a function of $E_{v_1, ..., v_\phi}$ and thus is a function of the ϕ quantum numbers

$v_1, v_2,...., v_\phi$, instead of being a function of $H(q, p)$ and thus a function of the 2ϕ continuous variables $q_1, q_2,..., q_\phi, p_1,..., p_\phi$, as it was in classical mechanics. Function $f(v_1,..., v_\phi)$ is the probability that the system is in the quantum state characterized by the quantum numbers $v_1,..., v_\phi$; as contrasted with the probability $f(q_1,..., p_\phi)$ $dV_q \, dV_p$ for phase space. These statements apply to any system, not just to the simple crystal model. The quantum state for any system with ϕ degrees of freedom, no matter what conservative forces its particles are subjected to, is characterized by ϕ quantum numbers $v_1,..., v_\phi$. To simplify notation we shall often write the single symbol v instead of the ϕ individual numbers $v_1,..., v_\phi$, just as we write q for $q_1,..., q_\phi$, etc. Thus f_v is the probability that the system is in the quantum state $v \equiv v_1,..., v_\phi$. The sum $\Sigma_v f_v$ over all allowed states of the system must be unity, of course.

The probability f_v also determines the probability that the system point is in the element of position space dV_q at $q_1, q_2,..., q_\phi$. For the square of the magnitude of the wave function $\Psi_v(q_1,..., q_\phi)$ is the probability density of the system point when the system is in the quantum state v. Therefore for a macrostate characterized by the distribution function f_v, the probability density in position space is the weighted sum

$$\sum_v f_v |\Psi_v(q_1, q_2,..., q_\phi)|^2 \tag{16-8}$$

over all occupied quantum states v. We shall have more to say about wave functions in Chapter 27.

We consequently have two alternative ways of expressing the microstates of the system, and thus of writing the distribution function. The quantum way, saying that each quantum state of the system is a separate and distinct microstate, is the correct way, but it sometimes leads to computational difficulties. The classical way, of representing a microstate as a region of phase space, is only an approximate representation, good for large energies; but when valid it is often easier to handle mathematically. The quantitative relationship between these two ways is obtained by working out the volume of phase space "occupied" by one quantum state of the system.

The connection between the classical coordinates q_i and momenta p_i and the quantum state is provided by the Heisenberg uncertainty principle, $\Delta q_i \cdot \Delta p_i \geq h$. A restatement of this is that, in the phase space of one degree of freedom, one quantum state occupies a "volume" $\Delta q_i \, \Delta p_i$ equal to h. For example, the one-dimensional harmonic oscillator has a Hamiltonian $H_i = (1/2m)p_i^2 + (m\omega^2/2)q_i^2$. When in the quantum state v_i, with energy $\hbar\omega(v_i + \frac{1}{2}) = (h\omega/2\pi)(v_i + \frac{1}{2})$, its phase-space orbit is an ellipse in phase space, with semiminor axis $q_m = [(h/\pi m\omega)(v_i + \frac{1}{2})]^{1/2}$ along q_i and semimajor axis $p_m = [(hm\omega/\pi)(v_i + \frac{1}{2})]^{1/2}$, which encloses an area

$$A(v_i) = \pi p_m q_m = h(v_i + \frac{1}{2})$$

The area between successive ellipses, for successive values of v_i, is the area "occupied" by one quantum state. We see that $A(v_i + 1) - A(v_i) = h$, as stated above.

Thus a volume element $dq_i\, dp_i$ corresponds, on the average, to $(dq_i\, dp_i/h)$ quantum states. Similarly, for the whole system, with ϕ degrees of freedom, the volume element $dV_q\, dV_p = dq_1 \cdots dp_\phi$ will correspond, on the average, to $(dV_q\, dV_p/h^\phi)$ quantum states. Thus the correspondence between volume of phase space and number of quantum states is

$$\text{No. of microstates} = (1/h^\phi)(\text{vol. of phase space}) \qquad (16\text{-}9)$$

when the system has ϕ degrees of freedom.

When the volume of phase space occupied by the swarm of system points, representing a particular ensemble, is very large compared to h^ϕ, the classical representation, in terms of the continuous variables $q_1,...,p_\phi$ can be safely used. But when the volume occupied by the swarm is not large compared to h^ϕ, the classical representation is not likely to be valid and the quantum representation is needed [see Eq. (19-8) et seq.].

CHAPTER

17

ENTROPY AND ENSEMBLES

As pointed out in the preceding chapter, we are presenting statistical mechanics in "logical" order, with definitions and basic postulates first, theorems and connections with experiment later. The last chapter was devoted to definitions. Each thermodynamic macrostate of a system may be visualized as an ensemble of systems in a variety of microstates, or may be represented quantitatively in terms of a distribution function, which is the probability f_v that a system chosen from the ensemble is in the quantum state $v \equiv v_1, v_2, ..., v_\phi$ or is the probability density $f(q, p)$ that the system point has the coordinates $q_1, q_2, ..., p_\phi$ in phase space, if the macrostate is such that classical mechanics is valid. In this chapter we shall introduce the essential postulates.

Entropy and Information

The basic postulate, relating the distribution function f_v to the thermodynamic properties of the macrostate which the ensemble represents, was first stated by Boltzmann and restated in more general form by Planck. In the form appropriate for our present discussion it relates the entropy S of the system to the distribution function f_v by the equation

$$S = - k \sum_v f_v \ln(f_v); \qquad \sum_v f_v = 1 \tag{17-1}$$

where k is the Boltzmann constant and where the summation is over all the quantum states present in the ensemble (i.e., for which f_v differs from zero). We note that this definition is somehow related to the entropy density of Eq. (14-13).

The formula satisfies our earlier statements that S is a measure of the degree of disorder of the system [see discussion preceding Eq. (6-14)]. A system that is certainly in its single, lowest quantum state is one in perfect order, so its entropy should be zero. Such a system would have the f_v for the lowest quantum state equal to unity and all the other f's would be zero. Since $\ln(1) = 0$ and $x \ln(x) \to 0$ as $x \to 0$, the sum on the right-hand side of Eq. (17-1) is zero for this case. On the other hand, a disorderly system would be likely to be in any of a number of different quantum states; the larger the number of states it might occupy the greater the disorder. If $f_v = 1/N$ for N different microstates (label them $v = 1, 2,..., N$) and f_v is zero for all other states then

$$S = - k \sum_{v=1}^{N} (1/N) \ln(1/N) = k \ln N$$

which increases as N increases. Thus Eq. (17-1) satisfies our preconceptions of the way entropy should behave. It also provides an opportunity to be more exact in regard to the measurement of disorder.

Disorder, in the sense we have been using it, implies a lack of information regarding the exact state of the system. A disordered system is one about which we lack complete information. Equation (17-1) is the starting point for Shannon's development of information theory. It will be useful to sketch a part of this development, for it will cast further light on the meaning of entropy, as postulated in Eq. (17-1).

Information Theory

A gasoline gauge, with a pointer and scale, gives us more information about the state of the gasoline tank of an automobile than does a red light, which lights if the tank is nearly empty. How much more? Information comes to us in messages and to convey information each message must tell us something new, i.e., something not completely expected. Quantitatively, if there are N possible messages that could be received, and if the chance that the ith one will be sent is f_i, then the information I that would be gained *if* message i were received must be a function $I(f_i)$, which increases as

$1/f_i$ increases. The less likely the message, the greater the information conveyed *if* the message *is* sent.

We can soon determine what function $I(f_j)$ must be, for we require that information be additive; if two messages are received and if the messages are independent, then the information gained should be the sum of the I's for each individual message. If the probability of message i be f_i and that for j be f_j then, if the two are independent, Eq. (11-3) requires that the probability that both messages happen to be sent is $f_i f_j$. The additive requirement for information then requires that

$$I(f_i f_j) = I(f_i) + I(f_j)$$

and this, in turn, requires that function I be a logarithmic function of f,

$$I(f_i) = -C \ln f_i$$

where C is a constant. This is the basic definition of information theory. Since $0 \le f_i \le 1$, I is positive and increases as $1/f_i$ increases, as required.

The definition satisfies our preconceptions of how information behaves. For example, if we receive a message that is completely expected (i.e., its a priori probability is unity) we receive no information and I is zero. The less likely is the message (the smaller is f_i) the greater the amount of information conveyed if it does get sent. The chance that the warning light of the gasoline gauge is off (showing that the tank is not nearly empty) is 0.9, say, so the information conveyed by the fact that the light is not lit is a small amount, equal to $-C \ln 0.9 \simeq 0.1C$. On the other hand if the gauge has a pointer and five marks, each of which represents an equally likely state of the tank, then the information conveyed by a glance at the gauge is $C \ln 5 \simeq 1.6C$, roughly 16 times the information conveyed by the unlit warning light (the information conveyed by a lit warning light, however, is $C \ln 10 \simeq 2.3C$, a still larger amount).

To see how these definitions relate to our discussion of information and disordered systems, let us return to an ensemble, corresponding to some thermodynamic state, with its distribution function f_ν. If we wish to find out exactly what microstate the system happens to be in at any instant, we would subject it to detailed measurement designed to tell us. The results of the measurement would be a kind of message to us, giving us information. If the measurements happened to show that the system is in microstate v, the information gained by the measurement would be $-C \ln f_v$, for f_v is the probability that the system would happen to be in state v. We of course cannot make a detailed enough examination to determine exactly what microstate a complicated system should happen to be in, nor would we wish to do so even if we could. But we can use the expected amount of information we *would* obtain, if we made the examination, as a measure of our present *lack* of knowledge of the system, i.e., of the system's disorder.

The expected amount of information we would obtain if we did examine the system in detail is the weighted mean of $-C \ln f_v$ over all quantum states v in the ensemble, the weighting factor being the probability f_v of receiving the message that the system is in state v. This is the sum

$$-C \sum_v f_v \ln f_v = (C/k)S$$

according to Eq. (17-1). Thus the entropy S is proportional to our *lack of detailed information* regarding the system when it is in the thermodynamic state corresponding to the distribution function f_v. The less we know about the system, the more microstates there are in which the system may be, and the greater the entropy. Here again Eq. (17-1) corresponds to our preconceptions regarding the entropy S of a system.

Entropy for Equilibrium States

But we do not wish to use postulate (17-1) to compute the entropy of a thermodynamic state when we know its distribution function; we wish to use Eq. (17-1) to *find* the distribution function for specified thermodynamic states, particularly those for equilibrium states. In order to do this we utilize a form of the second law. We noted in our initial discussion of entropy [see Eq. (6-5)] that in an isolated system S tends to increase until, at equilibrium, it is as large as it can be, subject to the restrictions on the system. If the sum of Eq. (17-1) is to correspond to the entropy, defined by the second law, it too must be a maximum, subject to restrictions, for an equilibrium state. These requirements should serve to determine the form of the distribution function, just as the wave equation, plus boundary conditions, determines the form of a vibrating string.

To show how this works, suppose we at first impose no restrictions on f_v, except that $\Sigma_v f_v = 1$ and that the number of microstates in the ensemble represented by f_v is finite (so that the quantum number v can take on the values 1, 2,..., W, where W is a finite integer). Then our problem is to determine the value of each f_v so that

$$S = -k \sum_{v=1}^{W} f_v \ln f_v \text{ is maximum, subject to } \sum_{v=1}^{W} f_v = 1$$

$$(17\text{-}2)$$

This is a fairly simple problem in the calculus of variations, which can be solved by the use of Lagrange multipliers. But to show how Lagrange multipliers work, we shall first solve the problem by straightforward calculus.

The requirement that $\Sigma f_v = 1$ means that only $W - 1$ of the f's can be varied independently. One of the f's, for example f_W, depends on the others through the equation

$$f_W = 1 - \sum_{v=1}^{W-1} f_v$$

Now S is a symmetric function of all the f's, so we can write it $S(f_1, f_2,..., f_W)$, where we can substitute for f_W in terms of the others. In order that S be maximum we should have the partial derivative of S with respect to each independent f be zero. Taking into account the fact that $f_{W,}$ depends on all the other f's, these equations become

$$(\partial S/\partial f_1) + (\partial S/\partial f_W)(\partial f_W/\partial f_1) \equiv (\partial S/\partial f_1) - (\partial S/\partial f_W) = 0$$
$$(\partial S/\partial f_2) - (\partial S/\partial f_W) = 0 \qquad\qquad (17\text{-}3)$$

. . . .

. . . .

$$(\partial S/\partial f_{W-1}) - (\partial S/\partial f_W) = 0$$

The values of the f's which satisfy these equations, plus the equation $\Sigma f_v = 1$, are those for which Eq. (17-2) is satisfied. For these values of the f's, the partial derivative $\partial S/\partial f_W$ will have some value; call it $-\alpha_0$. Then we can write Eqs. (17-3) in the more symmetric form

$$\left(\frac{\partial S}{\partial f_1}\right) + \alpha_0 = 0 \quad \left(\frac{\partial S}{\partial f_2}\right) + \alpha_0 = 0 \quad \cdots \quad \left(\frac{\partial S}{\partial f_W}\right) + \alpha_0 = 0$$

However this is just the set of equations we would have obtained if, instead of the requirement that $S(f_1,..., f_W)$ be maximum, subject to $\Sigma f_v = 1$ of Eq. (17-2), we had instead used the requirement that

$$S(f_1,..., f_W) + \alpha_0 \sum_{v=1}^{W} f_v \text{ be maximum,}$$

$$\alpha_0 \text{ determined so that } \sum_{v=1}^{W} f_v = 1$$

(17-4)

Constant α_0 is a *Lagrange multiplier*.

Let us now solve the set of Eqs. (17-4), inserting Eq. (17-1) for S. We set each of the partials of $S + \alpha_0 \Sigma f_v$ equal to zero. For example, the partial with respect to f_κ is

$$0 = \frac{\partial}{\partial f_\kappa} \left[\alpha_0 \sum_{v=1}^{W} f_v - k \sum_{v=1}^{W} f_v \ln f_v \right] = \alpha_0 - k \ln f_\kappa - k$$

or

$$f_\kappa = \exp[(\alpha_0/k) - 1]$$

The solution indicates that all the f's are equal, since neither α_0 nor k depend on κ. The determination of the value of α_0, and thus of the magnitude of f_v, comes from the requirement $\Sigma f_v = 1$: $f_v = (1/W)$, so that

$$S = - k \sum_{v=1}^{W} \left(\frac{1}{W}\right) \ln\left(\frac{1}{W}\right) = k \ln W$$

(17-5)

For a system restricted to a finite number W of microstates, and with no other restrictions, the state of maximum entropy is that for which the system is equally likely to be in any of the W microstates, and the corresponding maximal value of the entropy is k times the natural logarithm of the number W (which is sometimes called the *statistical weight* of the equilibrium macrostate). Again we see that the greater the number of microstates that are included in a macrostate, the greater is the entropy of the macrostate.

Application to a Perfect Gas

To show how much is inherent in these abstract-sounding results, we apply them to a gas of N point particles, confined in a container of volume V. To say that the system is confined to a finite number of quantum states is equivalent, classically, to saying that the system point is confined to a finite volume in phase space. In fact, from Eq. (16-9), the volume of phase space Ω, within which the system point is to be found, is $\Omega = h^{3N} W$, where $\phi = 3N$ is the number of degrees of freedom of the gas of N particles. And, from Eq. (17-5) we see that the system point is *equally likely* to be anywhere within this volume Ω. Thus, classically the analogue of Eq. (17-5) is

$$f(q, p) = 1/\Omega; \qquad S = k \ln(\Omega/h^{3N}) \tag{17-6}$$

As long as the volume of the container V is considerably larger than atomic dimensions, Ω is likely to be considerably larger than h^{3N}, so the classical description, in terms of phase space, is valid [see the discussion following Eq. (16-9)].

The volume Ω can be computed by integrating $dV_q \, dV_p$ over the region allowed to the system. Since each particle is confined to the volume V, the integration over the position coordinates is

$$\int \cdots \int dV_q = \int \cdots \int dx_1 \, dy_1 \, dz_1 \cdots dx_N \, dy_N \, dz_N = V^N \tag{17-7}$$

as we showed in Chapter 16. The integration of dV_p will be discussed in the next chapter; here we shall simply write it as Ω_p. Therefore,

$$\Omega = V^N \Omega_p \qquad \text{and} \qquad S = Nk \ln V + k \ln(\Omega_p/h^{3N}) \tag{17-8}$$

Comparison with Eq. (6-10) shows that the entropy of a perfect gas does indeed have a term $Nk \ln V$ ($Nk = nR$).

Moreover in this case of uniform distribution within Ω, the mean energy of the gas, $U = \Sigma f_v E_v$, will be given by the integral

$$U = \frac{1}{\Omega} \int \cdots \int dV_q \int \cdots \int H \, dV_p = \frac{1}{\Omega_p} \int \cdots \int H \, dV_p$$

where

$$H = (1/2m) \sum_{j=1}^{N} (p_{xj}^2 + p_{yj}^2 + p_{zj}^2)$$

is the total energy of the perfect gas. Thus U is a function of Ω_p, as well as of m and N and the shape of the volume in phase space, within which the ensemble is confined; we can emphasize this by writing it as $U(\Omega_p)$. Note that this is so only when H is independent of the q's. However, from Eq. (17-8) we have

$$\Omega_p = (h^3/V)^N e^{S/k} \qquad \text{so} \qquad U = U(h^{3N} V^{-N} e^{S/k})$$

Thus the formalism of Eq. (17-4) has enabled us to determine something about the dependence of the internal energy on V and S

for a system with H independent of the q's. We do not know the exact form of the dependence, but we do know that it is via the product $V^{-N}e^{S/k}$. From this one fact we can derive the equation of state for the perfect gas. We first refer to Eqs. (8-1). If the postulates (17-1) and (17-4) are to correspond to experiment, the partials of the function $U(h^{3N}V^{-N}e^{S/k})$ with respect to S and V must equal T and $-P$, respectively. But

$$\left(\frac{\partial U}{\partial S}\right)_v = \left(\frac{h^{3N}}{kV^N}\right)e^{S/k}\left(\frac{\partial U}{\partial \Omega_p}\right)$$

and

$$-\left(\frac{\partial U}{\partial V}\right)_s = N\left(\frac{h^{3N}}{V^{N+1}}\right)e^{S/k}\frac{\partial U}{\partial \Omega_p} = \frac{kN}{V}\left(\frac{\partial U}{\partial S}\right)_v \qquad (17\text{-}9)$$

Thus, if the first partial is to equal the thermodynamic temperature T and minus the second partial is the pressure P for any system with H dependent only on the momenta, the relationship between T, P, and V must be $P = (kN/V)T = nRT/V$, which is the equation of state of a perfect gas. Postulate (17-1) does indeed correspond satisfactorily with the thermodynamic properties of entropy (at least for the simple examples we have used). We should investigate in more detail, however.

CHAPTER

18

THE MICROCANONICAL ENSEMBLE

We now must utilize the definitions of Chapter 16 and the basic postulate of Eq. (17-1), which related the entropy S to the distribution function f_v, in order to find the dependence of S and f_v on the thermodynamic variables of the system. This can be carried out in several ways, each of them having their own special advantages. The simplest way is to consider an ensemble of systems, all having the same energy, and then adjust this energy, and the related entropy, to satisfy the requirements of the second law of thermodynamics. If all the systems in the ensemble have the same energy then, by Eq. (17-5), every quantum state having this energy must be equally represented or, classically, the probability density $f(q, p)$ must have the same value everywhere on the constant-energy surface, must be zero elsewhere in phase space.

Such an ensemble is called a *microcanonical ensemble*. To use it to determine S, U, etc., we first find the number $W(U)$ of quantum states of the system which have the energy U or, classically, we find the "area" Ω of the hypersurface of energy U in the 2ϕ-dimensional phase space of the Liouville theorem. Then, according to Eq. (17-5), the ensemble which is to represent the system being in equilibrium with internal energy U has entropy $S = k \ln W$ or $k \ln(\Omega/h^\phi)$, where W (or Ω) is a function of U.

At this point neither U nor S is a function of T, the value of U has been arbitrarily chosen. But the second law specifies the relationships among U and S and T in order that the system be at equilibrium *at a given temperature*. If the volume V of the system is kept constant, then the Helmholtz function $F = U - TS$ must have a minimal value; if pressure is constant then $G = U + PV - TS$ must be minimal. As we saw in Chapter 17, W (or Ω) is a function of V as well as of U.

For example, for a system kept at constant volume V, the microcanonical ensemble satisfying the second law at temperature T has energy U and entropy S obtained by the following procedure:

$$f_v = [1/W(U)] \qquad \text{or} \qquad [1/\Omega(U)]$$

when $E = U$ or when $H(q, p) = U$, zero otherwise; then

$$S = k \ln[W(U)] \qquad \text{or} \qquad k \ln[\Omega(U)/h^\phi]$$

finally U must be adjusted so that $F = U - TS$ is a minimum, holding V constant. In other words

$$T = \frac{1}{(\partial S/\partial U)_v}$$

(18-1)

[see Eq. (6-4)]. The minimization of F determines the U which is appropriate for temperature T and thus determines the functional dependence of U, S, and f_v on T. Here we can see directly the balance between energy and entropy which is required by the second law of thermodynamics. With T small, so that S has little influence on F, the tendency is for the energy U of the system to be quite small. But as T is increased the entropy has more effect. As we shall soon see, S increases quite rapidly as U increases, so that for a time $F = U - TS$ decreases as U increases, reaching a minimum when $(\partial S/\partial U)_v = (1/T)$ or $(\partial W/\partial U)_v = (W/kT)$. Thus U increases as T increases, in order that S correspondingly increase and $(\partial F/\partial U)_v$ remain zero.

The procedures outlined here will be more easily understood when illustrated by several examples.

A Simple Crystal Model

The first example is the simple crystal model of Eqs. (13-7) and (16-4), with each of the N atoms in the crystal lattice being a three-dimensional harmonic oscillator of frequency $\omega/2\pi$. Here we shall use quantum theory, since the formula for the allowed energies of a quantized oscillator is simple. The allowed energy for the ith degree of freedom is $\hbar\omega(v_i + \frac{1}{2})$, where $\hbar = h/2\pi$ and v_i is the quantum number for the ith degree of freedom. Therefore the allowed energy of vibration of this crystal is the sum of Eq. (16-7)

$$E_v = \hbar\omega \sum_{i=1}^{\phi} v_i + \tfrac{1}{2}\phi\hbar\omega; \qquad \phi = 3N \qquad (18\text{-}2)$$

and the total internal energy, including the potential energy of static compression [see Eqs. (3-6) and (13-20)] is

$$\begin{aligned} U &= E_v + [(V - V_0)^2/2\kappa V_0] \\ &= \hbar\omega M + \tfrac{3}{2}N\hbar\omega + [(V - V_0)^2/2\kappa V_0] \end{aligned} \qquad (18\text{-}3)$$

where $M = (v_1 + v_2 + \cdots + v_\phi)$ and $\phi = 3N$.

A microcanonical ensemble would consist of equal proportions of all the states for which M is a constant integer, and W is the number of different permutations of the quantum numbers v_i whose sum is M. This number can be obtained by induction. When $\phi = 1$, there is only one state for which $v_1 = M$, so $W = 1$. When $\phi = 2$, there are $M + 1$ different states for which $v_1 + v_2 = M$, one for $v_1 = 0$, $v_2 = M$, another for $v_1 = 1$, $v_2 = M - 1$, and so on to $v_1 = M$, $v_2 = 0$. When $\phi = 3$, there are $M + 1$ different combinations of v_2 and v_3 when v_1 is 0, M different ones when $v_1 = 1$, and so on, so that

$$\begin{aligned} W &= (M + 1) + M + (M - 1) + \cdots + 2 + 1 \\ &= \tfrac{1}{2}(M + 1)(M + 2) \qquad \text{for } \phi = 3 \end{aligned}$$

Continuing as before, we soon see that for ϕ different v's (i.e., ϕ degrees of freedom),

$$W = \frac{(M + \phi - 1)!}{M!(\phi - 1)!} = \frac{(M + 3N - 1)!}{M!(3N - 1)!} \qquad (18\text{-}4)$$

From this and from Eq. (18-3) we can obtain W as a function of U; then from Eq. (18-1) or Eq. (17-5) we can obtain S as a function of U. Probability f_v that the system is in one of the combinations of v_i's which add up to M is then $(1/W)$.

Since M and ϕ are large numbers we can use the asymptotic formula for the factorial function [see Eq. (11-18)]

$$n! \simeq (2\pi n)^{1/2} n^n e^{-n} \qquad n \gg 1 \qquad (18\text{-}5)$$

which is called Stirling's formula. By using it, we can obtain a simple approximation for the number W of different quantum states which have the same value of M and thus of U;

$$W \simeq \left[\frac{M + \phi}{2\pi M\phi}\right]^{1\,2} (M + \phi)^{M + \phi - 1} M^{-M}\phi^{-\phi + 1}$$

$$\simeq \left[\frac{\phi}{2\pi M(M + \phi)}\right]^{1\,2}\left(1 + \frac{\phi}{M}\right)^{M}\left(1 + \frac{M}{\phi}\right)^{\phi} \tag{18-6}$$

where, since $\phi = 3N$ is large, we have substituted ϕ for $\phi - 1$ in the parentheses. Therefore the entropy of the simple crystal is

$$S = k \ln W \simeq k(3N + M) \ln\left(1 + \frac{M}{3N}\right)$$

$$- kM \ln\left(\frac{M}{3N}\right) \tag{18-7}$$

where we have neglected the logarithm of the square root, since it is so much smaller than the other terms.

We have thus obtained both U and S as functions of M, the sum of the quantum numbers. As indicated in Eqs. (18-1), for a crystal at constant V and T, the function $F = U - TS$ reaches a minimum at thermodynamic equilibrium. In Figure 18-1 we have plotted the part of U which depends on M (divided by $3NkT$ so it will be dimensionless) for different values of $(\hbar\omega/kT)$ and the corresponding curve for $-(TS/3NkT)$. The sum of these curves is $(F_v/3NkT)$ and the value M_0 of M for equilibrium is that for which F_v is minimal. We see that M_0 is smaller the larger $(\hbar\omega/kT)$ is. A little estimation of slopes indicates that the value of $(U_v/3NkT)$ for $M = M_0$ is never larger than unity (i.e., that U_v is never larger than $3NkT$).

Quantitatively, the minimal value of the Helmholtz function is for that value M_0 of M for which $(\partial F/\partial M)_{VT} = 0$;

$$F = \hbar\omega M + \tfrac{3}{2}N\hbar\omega + \frac{(V - V_0)^2}{2\kappa V_0}$$

$$- kT\left[(M + 3N)\ln\left(1 + \frac{M}{3N}\right) - M\ln\left(\frac{M}{3N}\right)\right]$$

$$(\partial F/\partial M)_{VT} = 0 \text{ corresponds to } (\hbar\omega/kT) = \ln\left(1 + \frac{3N}{M_0}\right)$$

or

$$M_0 = 3N(e^{\hbar\omega/kT} - 1)^{-1} \tag{18-8}$$

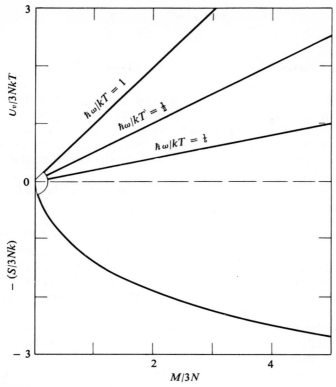

FIGURE 18-1

Curves for $U_v = \hbar\omega M$ and $-ST$, divided by $3NkT$, the two parts of $(F_v/3NkT)$, as functions of the quantum number parameter $(M/3N)$, for different values of the temperature parameter $(\hbar\omega/kT)$, for the Einstein model of a crystal lattice.

Substituting this in the expressions for internal energy and entropy results in the formulas

$$U = U_0 + \frac{3N\hbar\omega}{e^x - 1} \rightarrow \begin{cases} U_0 + 3N\hbar\omega e^{-x}; & kT \ll \hbar\omega \\ U_0 + 3NkT; & kT \gg \hbar\omega \end{cases}$$

$$S = -3Nk \ln(1 - e^{-x}) \tag{18-9}$$

$$+ \frac{3Nkx}{e^x - 1} \rightarrow \begin{cases} 3Nkx^{-x}; & kT \ll \hbar\omega \\ 3Nk \ln(e/x); & kT \gg \hbar\omega \end{cases}$$

$$C_v = \frac{3Nkx^2 e^x}{(e^x - 1)^2} \rightarrow \begin{cases} (3Nkx^2)e^{-x}; & kT \ll \hbar\omega \\ 3Nk; & kT \gg \hbar\omega \end{cases} \qquad \begin{array}{c} (18\text{-}9) \\ (cont.) \end{array}$$

where $x = (\hbar\omega/kT)$ and $U_0 = (3N\hbar\omega/2) + (1/2\kappa V_0)(V - V_0)^2$

This set of formulas was first obtained by Einstein, early in the development of quantum theory. It and the Planck formula, which will be discussed in Chapter 25, were the first triumphs of the quantum theory. Here, for the first time, was a formula for C_v which predicted a drop to zero at $T = 0$, as the experimental curves of Figure 3-1 require.

We see that, at temperatures high enough so that kT is large compared to the spacing $\hbar\omega$ between the vibrational levels, the crystal behaves like the classical model of Eqs. (9-2) and (9-3), with $U \simeq 3NkT + U_0$, $S \simeq 3Nk \ln(T/T_0)$, and $C_v \simeq 3Nk$. In this temperature range the sum of quantum numbers M is considerably larger than $3N$, the number of degrees of freedom, so the average quantum number $(M/3N)$ per oscillator is much larger than unity. Each atom is vibrating vigorously, so it is not surprising that the classical model is adequate.

On the other hand when kT is small compared to $\hbar\omega$, the quantum number sum $M \simeq 3Ne^{-\hbar\omega/kT}$ is much smaller than $3N$ and most of the atoms are "motionless" ($v_i = 0$). With quantum theory limiting the allowed energies of vibration, each oscillator cannot gradually reduce its amplitude of vibration as T is reduced. At low temperatures it must either be quiescent or else be vibrating more energetically than many atoms can afford to do, so most of them give up. Few atoms are moving; the entropy drops to zero as more and more atoms are quiescent. Also the heat capacity drops to zero as $T \rightarrow 0$, as is required by the experimental curve of Figure 3-1. Many fewer than $3N$ degrees of freedom carry any heat energy at all.

The heat capacity $(C_v/3Nk)$, as predicted by the Einstein formula of Eqs. (18-9), is plotted as the dashed curve of Figure 20-2. We see that it drops to zero too rapidly as $T \rightarrow 0$; the solid curve, which will be derived in Chapter 20, fits the experimental data better. The simple model of Eq. (18-6) overdoes the quantum effect; the model of Chapter 20, which takes into account interactions between neighboring atoms, is a better fit at low temperatures.

Microcanonical Ensemble for a Perfect Gas

In the case of a perfect gas of N point particles in a volume V of "normal" size, the energy levels are so closely spaced that we can use classical physics for temperatures greater than a fraction of a degree Kelvin. Thus a microcanonical ensemble for such a system is represented by a distribution function $f(q, p)$ which is zero everywhere in phase space except on the "surface,"

$$H \equiv \sum_{i=1}^{3N} (1/2m)p_i^2 = U \text{ (a constant)} \tag{18-10}$$

where it is $1/\Omega$, Ω being the integral of $dV_q \, dV_p$ over this surface (i.e., Ω is the "area" of the surface). The entropy of the microcanonical ensemble is then given by Eq. (18-1). This classical approximation should be valid for $T \gg 0.01\,°\mathrm{K}$, as will be shown later [see the discussion following Eq. (21-5)].

Since the energy of a perfect gas is independent of the positions of the particles, the integral of dV_q, as shown in Eq. (17-7), is simply the nth power of the volume V of the container. The integral of dV_p, however, is the "area" of the surface in momentum space defined by Eq. (18-10). This surface is the $3N$-dimensional generalization of a spherical surface; the coordinates are the p's and the radius is $R = (2mU)^{1/2}$,

$$p_1^2 + p_2^2 + \cdots + p_\phi^2 = R^2; \qquad \phi = 3N; \qquad R^2 = 2mU \tag{18-11}$$

Once the area Ω_p of this hyperspherical surface is computed, the rest of the calculation is easy, for the volume of phase space occupied is $\Omega = V^N \Omega_p$ and f and S are given by Eq. (17-8).

To find the area we need to define some hyperspherical coordinates. Working by induction:

For two dimensions:
$x_1 = R \cos \theta_1, \ x_2 = R \sin \theta_1, \ x_1^2 + x_2^2 = R^2$
Element of length of circle: $ds = R \, d\theta_1$

For three dimensions:
$x_1 = R \cos \theta_1, \ x_2 = R \sin \theta_1 \cos \theta_2, \ x_3 = R \sin \theta_1 \sin \theta_2$
Element of area of sphere: $dA = R^2 \sin \theta_1 \, d\theta_1 \, d\theta_2$

For four dimensions:
$x_1 = R \cos \theta_1, \ x_2 = R \sin \theta_1 \cos \theta_2,$

$x_3 = R \sin \theta_1 \sin \theta_2 \cos \theta_3, \ x_4 = R \sin \theta_1 \sin \theta_2 \sin \theta_3$
Element of surface: $dA = R^3 \sin^2 \theta_1 \sin \theta_2 \, d\theta_1 \, d\theta_2 \, d\theta_3$

For ϕ dimensions:

$$x_1 = R \cos \theta_1, \, x_2 = R \sin \theta_1 \cos \theta_2,... \tag{18-12}$$

$$x_{\phi-1} = R \sin \theta_1 \cdots \sin \theta_{\phi-2} \cos \theta_{\phi-1}, \, x_\phi = R \sin \theta_1 \cdots \sin \theta_{\phi-1}$$

$$dA = R^{\phi-1} \sin^{\phi-2}\theta_1 \sin^{\phi-3}\theta_2 \cdots \sin \theta_{\phi-2} \, d\theta_1 \, d\theta_2 \cdots d\theta_{\phi-1}$$

where angle $\theta_{\phi-1}$ goes from 0 to 2π and angles $\theta_1,...,\theta_{\phi-2}$ go from 0 to π. To integrate this area element we need the formula

$$\int_0^\pi \sin^n \theta \, d\theta = \sqrt{\pi}\left[(\tfrac{1}{2}n - \tfrac{1}{2})!/(\tfrac{1}{2}n)!\right] \tag{18-13}$$

where $m!$ is the factorial function

$$m! = \int_0^\infty x^m e^{-x} \, dx = m \cdot (m-1)!; \qquad 0! = 1! = 1$$

$$(-\tfrac{1}{2})! = \sqrt{\pi} = 2 \cdot (\tfrac{1}{2})! \tag{18-14}$$

with asymptotic values given by Stirling's formula (18-5). Thus the total area of the hypersphere is {neglecting such factors as $\sqrt{2}$ and $\sqrt{[1 - (1/\phi)]}$}

$$A = \Omega_p = R^{\phi-1} \int_0^\pi \sin^{\phi-2}\theta_1 \, d\theta_1 \cdots$$

$$\times \int_0^\pi \sin \theta_{\phi-2} \, d\theta_{\phi-2} \int_0^{2\pi} d\theta_{\phi-1}$$

$$= 2\pi R^{\phi-1} \pi^{(\phi-2)/2} \frac{(\tfrac{1}{2}\phi - \tfrac{3}{2})!}{(\tfrac{1}{2}\phi - 1)!} \frac{(\tfrac{1}{2}\phi - 2)!}{(\tfrac{1}{2}\phi - \tfrac{3}{2})!} \cdots \frac{(0)!}{(\tfrac{1}{2})!}$$

$$= \frac{2\pi^{(\phi/2)} R^{\phi-1}}{(\tfrac{1}{2}\phi - 1)!}$$

$$\simeq \left(\frac{4\pi m U}{\phi}\right)^{(\phi-1)/2} \left(1 - \frac{2}{\phi}\right)^{-(\phi/2)} e^{(\phi-2)/2}$$

$$\simeq \left(\frac{4\pi m U e}{\phi}\right)^{(\phi/2)}; \qquad \phi \gg 1 \tag{18-15}$$

and the final expression for the volume of phase space occupied is

$$\Omega \simeq V^N (4\pi m U e/3N)^{(3/2)N} \tag{18-16}$$

where we have used Eq. (18-5), have replaced $[(\phi - 1)/2]$ by $\frac{1}{2}\phi = \frac{3}{2}N$ and have used the limiting formula for the exponential function

$$\left(1 + \frac{x}{n}\right)^n \to e^x; \qquad n \to \infty \qquad (18\text{-}17)$$

The e in the formula for Ω is, of course, the base of the natural logarithm, $e = 2.71828$.

Consequently the entropy of the gas is, from Eq. (17-6),

$$S = Nk \ln[V(4\pi m U e/3Nh^2)^{3/2}]$$

or

$$U = (3Nh^2/4\pi me)V^{-2/3}e^{2S/3Nk} \qquad (18\text{-}18)$$

which is to be compared with Eq. (17-9) as well as with the discussion following Eq. (6-6). As with the discussion of Eq. (17-9), we can now obtain the thermodynamic temperature and pressure,

$$T = (\partial U/\partial S)_v = (h^2/2\pi mek)V^{-2/3}e^{2S/3Nk}$$

$$P = -(\partial U/\partial V)_s = (Nh^2/2\pi me)V^{-5/3}e^{2S/3Nk} = NkT/V \qquad (18\text{-}19)$$

Also

$$U = \tfrac{3}{2}NkT; \qquad C_v = (\partial U/\partial T)_v = \tfrac{3}{2}Nk$$

So the microcanonical ensemble does reproduce the thermodynamic behavior of a perfect gas in complete detail. With Eq. (17-9) we were able to obtain the equation of state, but now that we have computed the dependence of Ω_p on U and N, the theoretical model also correctly predicts the dependence of U on T and hence the heat capacity of the gas. (However, see page 306.)

The Maxwell Distribution

The microcanonical ensemble can also predict the velocity distribution of molecules in the gas; if we have the distribution function for the whole gas we can obtain from it the distribution function for a constituent particle. Utilizing Eq. (18-19) we can restate our results thus far. Since $R = (2mU)^{1/2}$ and $U = \tfrac{3}{2}NkT$, we can say that the system points in the microcanonical ensemble for a perfect gas of point particles are uniformly distributed on the surface of a hypersphere in $3N$-dimensional momentum space of radius $(3NmkT)^{1/2}$. The probability that the point representing a particular system is within any given region dA on this surface is equal to the ratio of the area of the region dA to the total area $A = \Omega_p$ of the surface. The region near where the p_1 axis cuts the surface, for example, corresponds to the microstates in which the x component of momentum of particle 1 carries practically all the energy $\frac{1}{2}\phi kT$ of the whole system. It is an interesting property of hyperspheres of large dimensionality (as we shall show) that the areas close to any axis are negligibly small compared to the areas well away from any axis (where the energy is relatively evenly divided between all the degrees of freedom). Therefore the chance that one degree of freedom will turn out to have most of the energy of the whole gas, $\frac{1}{2}\phi kT$, and that the other components of momentum are zero is negligibly small.

To show this, and incidentally to provide yet another "derivation" of the Maxwell distribution, we note that the probability that the momentum coordinate, which we happen to have labeled by the subscript 1, has a value between p_1 and $p_1 + dp_1$ can be obtained easily, since we have chosen our angle coordinates such that $p_1 = R \cos \theta_1$. Thus the probability

$$dA = \frac{[\frac{1}{2}\phi - 1]!}{2\pi^{(1/2)\phi}} \sin^{\phi-2}\theta_1 \, d\theta_1 \sin^{\phi-3}\theta_2 \, d\theta_3 \cdots$$
$$\times \sin \theta_{\phi-2} \, d\theta_{\phi-2} \, d\theta_{\phi-1} \qquad (18\text{-}20)$$

as a function of θ_1, is only large near $\theta_1 = \frac{1}{2}\pi$ [i.e., where

$$p_1 = (\phi mkT)^{1/2} \cos \theta_1$$

is very small compared to $(\phi mkT)^{1/2}$] and drops off very rapidly, because of the large power of $\sin \theta_1$, whenever the magnitude of p_1 increases. The factor $\sin^{\phi-2}\theta_1$ ensures that the probability is very small that the degree of freedom labeled 1 carries most of the total kinetic energy $\frac{1}{2}\phi kT$ of the gas. This would be true for each degree of freedom. It is much more likely that each degree of freedom carries an approximately equal share, each having an amount near $\frac{1}{2}kT$.

The formula for the probability that degree of freedom 1 have momentum between p_1 and $p_1 + dp_1$, irrespective of the other momenta, is obtained by integrating dA/A over $\theta_2, \theta_3,..., \theta_{\phi-1}$. Using the results of Eqs. (18-13) to (18-17) produces

$$f(p_1) \, dp_1 = \frac{-(\frac{1}{2}\phi - 1)!}{\sqrt{\pi}(\frac{1}{2}\phi - \frac{3}{2})!} \sin^{\phi-2}\theta_1 \, d\theta_1$$
$$= \frac{1}{(\pi\phi mkT)^{1/2}} \frac{(\frac{1}{2}\phi - 1)!}{(\frac{1}{2}\phi - \frac{3}{2})!} \left(1 - \frac{p_1^2}{\phi mkT}\right)^{(\phi-3)/2} dp_1$$

since

$$-d\theta_1 = (1/\phi mkT)^{1/2}(dp_1/\sin\theta_1)$$

and

$$\sin^2\theta_1 = 1 - (p_1^2/\phi mkT)$$

To obtain the Maxwell distribution in its usual form we utilize Eqs. (18-5) and (18-17) and consider factors like $[1 - (2/\phi)]^{1/2}$ and $[1 - (p_1^2/\phi mkT)]^{-3/2}$ to equal unity [but not such factors to the $\frac{1}{2}\phi$ power, of course]. The calculations go as follows:

$$f(p_1)\,dp_1 \simeq \frac{1}{(2\pi mkT)^{1/2}}\left[\frac{1 - (2/\phi)}{1 - (3/\phi)}\right]^{(\phi/2)} e^{-1/2}$$
$$\times\, [1 - (p_1^2/\phi mkT)]^{(\phi-3)/2}\,dp_1$$
$$\simeq \frac{1}{(2\pi mkT)^{1/2}}\exp(-p_1^2/2mkT)\,dp_1 \qquad (18\text{-}21)$$

which is the familiar Maxwell distribution for one degree of freedom [see Eq. (12-7)].

This time we arrived at the Maxwell distribution as a consequence of requiring that the total kinetic energy of the gas be $\frac{1}{2}\phi kT$ and that all possible distributions of this energy between the ϕ degrees of freedom be equally likely. For $\phi = 3N$ large, by far the majority of these configurations represent the energy being divided more or less equally between all ϕ degrees of freedom, with a variance for each p equal to $2m$ times the mean kinetic energy per degree of freedom, $\frac{1}{2}kT$.

Paramagnetic Crystals

Although it will require the full machinery of the canonical ensemble to clarify fully the magnetic behavior of paramagnetic salts at very low temperatures, a discussion using the microcanonical ensemble will more clearly delineate the phenomena involved and their relative magnitudes. The phenomenon was discussed earlier, in Chapters 8 and 13, where we computed the magnetic equation of state but could not obtain the entropy. The orientation of the atomic magnets in a paramagnetic salt is not appreciably affected by the lattice vibrations which we have treated earlier in this chapter, so we can work out the entropy S_m and energy U_m of the magnets as separate terms, which can be added to the S and U of Eqs. (18-9) for the vibration and compression effects. In the present discussion we shall assume that the magnetic units are electrons, with spin quantum number $J = \frac{1}{2}$, rather than carrying out the general case for $J > \frac{1}{2}$, as was done in Chapter 13 and will be done again in Chapter 20. Therefore each magnet has only two possible orientations, one parallel to the magnetic field, with moment component $\frac{1}{2}\mathfrak{m}_B$, where $\mathfrak{m}_B$ is the Bohr magneton $(eh/4\pi m)$ of Eq. (13-24), and the other antiparallel to the field, with component $-\frac{1}{2}\mathfrak{m}_B$. The energy of interaction with the magnetic field $\mathfrak{B}$ is $-\frac{1}{2}\mathfrak{m}_B\mathfrak{B}$ for the parallel case and $+\frac{1}{2}\mathfrak{m}_B\mathfrak{B}$ for the antiparallel case.

If there are N atomic magnets in the crystal and if n of them are oriented parallel to $\mathfrak{B}$ (and thus $N - n$ are antiparallel) the magnetic energy U_m and the magnetization $\mathfrak{M} = \mu_0 V \mathfrak{P}$ are

$$U_m = \tfrac{1}{2}(N - 2n)\mathfrak{m}_B\mathfrak{B}; \qquad \mathfrak{M} = \tfrac{1}{2}\mu_0(2n - N)\mathfrak{m}_B \qquad (18\text{-}22)$$

The quantum number n is independent of the quantum number M of Eq. (18-3) and thus it can be adjusted, to make $U_m - TS_m$ a minimum, separately from the vibrational terms dealt with earlier.

To find S_m we must count the number of different arrangements of the N magnets, of which n are parallel to $\mathfrak{B}$ and $N - n$ are antiparallel. The discussion preceding Eq. (11-8) indicates that this number is

$$W_m = \frac{N!}{n!(N - n)!} = \frac{N!}{(\frac{1}{2}N - \frac{1}{2}j)!(\frac{1}{2}N + \frac{1}{2}j)!}; \qquad j = 2n - N$$

$$\simeq \sqrt{\frac{2\pi N}{\pi^2(N^2 - j^2)}} \frac{(2N)^N}{(N - j)^{N-j}(N + j)^{N+j}} \simeq 2^N e^{-j^2/2N}$$

$$(18\text{-}23)$$

using Stirling's formula, as with the discussion following Eq. (11-24). We have omitted the square root factor in the second form, since its logarithm is much smaller than the logarithm of 2^N, and we shall only use the logarithm of W_m.

Thus the number W_m of possible atomic-magnet configurations is largest when half the magnets are parallel and half are antiparallel ($j = 0$ or $n = \frac{1}{2}N$). Here, when U_m and $\mathfrak{B}$ are zero, W_m is approximately equal to 2^N, corresponding to the two possible orientations of each magnet. We note that if the magnets have angular momentum J greater than $\frac{1}{2}$, as in Eq. (13-6), the maximum value of W_m is $(2J + 1)^N$. As the "unbalance parameter" j deviates from zero in either positive or negative direction, W_m diminishes, the dependence being that of the normal distribution of Figure 11-1.

For $j = \pm\frac{1}{2}N$ ($n = 0$ or N), $W_m = 1$; there is only one way of having all the magnets parallel or all antiparallel.

Using the crudest approximation of Eqs. (18-23), the entropy can be written $S_m \simeq Nk \ln 2 - k(j^2/2N)$ and the internal energy as $U_m = -\frac{1}{2}jm_B\mathfrak{B}$. For $F_m = U_m - TS_m$ to be a minimum, j must adjust itself so that

$$(\partial F_m/\partial j) = -\tfrac{1}{2}m_B\mathfrak{B} + kT(j/N) = 0$$

or

$$j_0 = (Nm_B\mathfrak{B}/2kT)$$

or

$$\mathfrak{M} = \tfrac{1}{2}\mu_0 j_0 m_B = (N\mu_0 m_B^2\mathfrak{B}/4kT) \qquad (18\text{-}24)$$

which is the Curie formula (13-22), with

$$\mathfrak{m} = \sqrt{3/4}\,m_B = \sqrt{J(J+1)}\,m_B$$

We can improve this calculation in two ways. First, we can use the more accurate expression for W_m, to obtain the Langevin formula (13-26) for $J = \frac{1}{2}$. Second, we can include the small forces of coupling between the magnets and the crystal lattice, which must be included if the formulas are to be valid at very low temperatures, when the magnets are held fixed in direction. Equations (18-23) do not include this and as a result the resulting S_m, for $\mathfrak{B} = 0$, does not go to zero when kT becomes smaller than the energy of binding of the magnets to the lattice. As a result all the formulas, including those for the specific heat, are incorrect in the limit of very low temperatures.

At very low temperatures and zero $\mathfrak{B}$ the magnets cease being randomly oriented and align themselves in a unique pattern along the crystal axes. The pattern involves equal numbers of parallel and antiparallel magnets in any chosen direction, as does the random orientation. But the pattern in the bound state is a regular pattern, not a random one, although the polarization is still zero. Any dissarranging of the pattern will require a small increase of internal energy, the increase being proportional to the amount of disarray. We can express this approximately, by saying that only one of the 2^N possible configurations for $j = 0$ is the minimal-energy bound state. Therefore, as T is dropped below T_0 (the energy of binding of one magnet to the lattice, divided by k) the entropy of the magnets drops suddenly from $Nk \ln 2$ to zero, as the magnets settle down into their unique, bound pattern of alignment. The internal energy U_m also drops suddenly, which means that the heat capacity $C_v = (\partial U_m/\partial T)_v$, for $\mathfrak{B} = 0$, has a narrow peak there, dropping to zero as $|T - T_0|$ increases.

But the microcanonical ensemble is unsuitable for working out these effects quantitatively. It is not easy to devise a form for which W_m which will bring in the change caused by binding. Therefore we postpone further discussion of the problem until Chapter 20.

CHAPTER

19

THE CANONICAL ENSEMBLE

The microcanonical ensemble has sufficed to demonstrate that the basic postulates of statistical mechanics correspond to the facts of thermodynamics as well as of kinetic theory. But it has several drawbacks, hindering its general use. In the first place, the computation of the number of microstates that have a given energy is not always easy. It actually would be easier to calculate average values with a distribution function that included a range of energies, rather than one that differs from zero only when the energy has a specific value.

In the second place (and perhaps more importantly) the microcanonical ensemble corresponds to a system with energy U, completely isolated from the rest of the universe, which is not the way a thermodynamic system is usually prepared. We usually do not know the exact value of the system's energy; we much more often know its temperature, which means that we know its *average* energy. In other words, we do not usually deal with completely isolated systems, but we do often deal with systems kept in contact with a heat reservoir at a given temperature, so that its energy varies somewhat from instant to instant, but its time average is known. This changes the boundary conditions of Eq. (17-2) and the resulting distribution function will differ from that of Eq. (18-1).

If we can work out an ensemble that has the temperature already in it we shall not have the additional task of minimizing

$U - TS$ to obtain U and S as functions of T. An ensemble with T in it will have done this already.

Solving for the Distribution Function

Suppose we prepare an ensemble as follows: Each system has the same number of particles N and has the same forces acting on the particles. Each system is placed in a furnace and brought to equilibrium at a specified temperature, with each system enclosed in a volume V. Thus, although we do not know the exact energy of any single system, we do require that the mean energy, averaged over the ensemble, has the relationships between S and T expressed in Eqs. (6-3) and (8-8), for example. The distribution function for such an ensemble, corresponding to a system in contact with a heat reservoir, should satisfy the following requirements:

$$S = - k \sum_v f_v \ln f_v \text{ is maximum,}$$

subject to $\sum_v f_v = 1$ and $\sum_v f_v E_v = U$, the internal energy.

We solve for f_v by using Lagrange multipliers. We require

$$S + \alpha_0 \sum_v f_v + \alpha_e \sum_v f_v E_v \text{ be maximum,}$$

with α_0 and α_e adjusted so that

$$\sum_v f_v = 1 \qquad \text{and} \qquad \sum_v f_v E_v = U \tag{19-1}$$

where U, for example, satisfies the equation $U = F + TS$. Setting the partials of this function, with respect to the f's, equal to zero we obtain

$$- k \ln f_v - k + \alpha_0 + \alpha_e E_v = 0 \tag{19-2}$$

or

$$f_v = \exp[(\alpha_0 - k + \alpha_e E_v)/k]$$

The value of the Lagrange multiplier α_0 is adjusted to satisfy the first subsidiary condition,

$$e^{(\alpha_0/k) - 1} \sum_v e^{\alpha_e E_v/k} = 1$$

or

$$\sum_v e^{\alpha_e E_v/k} = Z = e^{1 - (\alpha_0/k)} \tag{19-3}$$

where we define the quantity Z as being the sum of the terms $e^{\alpha_e E_v / k}$ over all possible states v of the system. Thus α_0 is determined in terms of Z.

Next we must adjust the Lagrange multiplier α_e so that U and S are related to T, as in the formula $F = U - TS$ which we used in the microcanonical ensemble. The results of the maximization are embodied in Eq. (19-2). If we relate this to the equation $F = U - TS$ we shall have made the connection. Taking the average value of Eq. (19-2), by multiplying it by f_v and summing over v, we obtain

$$-k \sum_v f_v \ln f_v - (k - \alpha_0) \sum_v f_v + \alpha_e \sum_v f_v E_v = 0$$

or

$$S + \alpha_e U = (k - \alpha_0) = k \ln Z$$

where we have used Eqs. (17-1), (19-1), and then (19-3).

This equation can be given the desired form $S - (U/T) = -(F/T)$ by setting α_e equal to $-(1/T)$ and $k \ln Z$ equal to $-(F/T)$. In this manner we shall have introduced the temperature T into our ensemble and shall have defined F as the quantity that is minimized, on the average, for equilibrium at constant T and V. Thus the solution of the requirements (19-1), consistent with the thermodynamic definitions of entropy, temperature, and Helmholtz function, is

$$f_v = (1/Z)e^{-E_v/kT}; \qquad Z = \sum_v e^{-E_v/kT} = e^{-F/kT}$$

$$S = k \sum_v f_v [\ln Z + (E_v/kT)] = \frac{U - F}{T} = -\left(\frac{\partial F}{\partial T}\right)_v \tag{19-4}$$

The ensemble corresponding to this distribution is called the *canonical ensemble*. The normalizing constant Z, considered as a function of T and V, is called the *partition function*. Part of the computational advantage of the canonical ensemble is the fact that all the thermodynamic functions can be computed from the partition function. For example,

$$F = -kT \ln Z; \qquad S = -(\partial F/\partial T)_v;$$
$$P = -(\partial F/\partial V)_T \tag{19-5}$$

When the separations between successive allowed energies E_v are considerably less than kT, classical mechanics can be used and instead of sums over the quantum states v of the system we can use integrals over phase space. The distribution function is the probability density $f(q, p)$, and, for a system with ϕ degrees of freedom,

$$f(q, p) = (1/h^\phi Z)e^{-H(q,p)/kT}$$

$$Z = (1/h^\phi) \int \cdots \int e^{-H/kT} \, dV_q \, dV_p \qquad (19\text{-}6)$$

where $H(q, p)$ is the Hamiltonian function of the system, the kinetic plus potential energy, expressed in terms of the q's and p's [see Eqs. (13-9) and (16-1)]. From Z one can then obtain F, S, etc., as per Eq. (19-5). The H of Eq. (19-6) is the total energy of the system, whereas the H of Eq. (13-10) is the energy of a single molecule. One might say that the canonical distribution function is the Maxwell-Boltzmann distribution for a whole system. It is an exact solution, whereas the f of Eq. (13-10) for a molecule is only valid in the limit of vanishing interactions between molecules.

General Properties of the Canonical Ensemble

The canonical ensemble includes sample systems of more than a single energy; its distribution function f_v is not equal to the constant $(1/W)$, it depends on the energy E_v as well as on the temperature T. Thus the ensemble averages are not functions of U, as they were with the microcanonical ensemble (with the relationship to T having to be worked out separately) but are explicit functions of T and V, as was desired. We can consider the canonical ensemble as being a collection of microcanonical ensembles, the one for energy E_v entering with relative weight proportional to the number of states $W(E_v)$ having energy E_v, times the factor $e^{-E_v/kT}$, which adjusts the weight appropriately for the temperature T.

The balance between the "multiplicity factor" $W(E_v)$ and the temperature factor $e^{-E_v/kT}$ determines the relative probability of finding a system with energy E_v in the canonical ensemble. This balance is the analogue of the balance between $-TS$ and U, which we had to impose on the microcanonical ensemble; here it is automatically done by the ensemble. The number of states $W(E_v)$ with energy E_v increases rapidly as E_v increases; W is approximately $(2\pi e E_v/3N\omega)^{3N}$ for the simple crystal of Eq. (18-6), when $M \gg \phi = 3N$, and is $(\Omega/h^{3N}) = V^N(4\pi m E_v e/3Nh^2)^{3N/2}$ for the perfect gas of Eq. (18-16). The product

$$W(E_v)e^{-E_v/kT} = \exp\left[\frac{1}{kT}(kT \ln W - E_v)\right] = \exp\left[\frac{TS_v - E_v}{kT}\right]$$

at first increases as E_v increases and then, for E_v large enough, the exponential "takes over" and the product drops back down to virtually zero. The maximum value obviously comes at the minimal value of $E_v - TS_v = F_v$.

The value of E_v that has the most representatives in the canonical ensemble is thus the value for which $We^{-E_v/kT}$ is maximum. For the gas this value is $E_v = U = \frac{3}{2}NkT$ and for the crystal it is $3NkT$; in each case it is equal to the average value U of energy of the ensemble of systems. The number of systems in the ensemble with energy larger or smaller than this mean value U diminishes quite sharply as $|E_v - U|$ increases. Although some systems with energy $E_v \neq U$ do occur, there are not many of them, for the fractional deviation from the mean ($\Delta E/U$) of the canonical distribution turns out to be inversely proportional to $\sqrt{\phi}$ and thus is quite small when ϕ is large.

The argument is the same for those cases where classical mechanics can be used. Instead of E_v we have the Hamiltonian $H(q, p)$ and the number of states with energy H is $[\Omega(H)/h^\phi]$, where $\Omega(H)$ is the "area" of the surface of constant H in the 2ϕ-dimensional phase space representing the system's configuration and momentum. The fraction of systems in the ensemble that have energy H are then proportional to $\Omega(H)e^{-H/kT}$, which has its maximum at $H = U$, where U satisfies the requirement that $F = U - TS$ is minimal for T and V constant.

The canonical ensemble has other advantages, particularly in respect to the demonstration of general thermodynamic properties of various systems. As was shown in Eq. (19-5), these properties can all be obtained from the partition function Z, the sum of terms $e^{-E_v/kT}$ over all possible quantum states of the system. The various properties can be most clearly demonstrated by various rearrangements of the terms in the sum for Z. As just indicated the relation between U and S and T is best demonstrated by grouping together all the $W(E_v)$ terms in Z having the same value E_v of energy,

$$Z = \sum_{v_1 v_2 \cdots v_\phi} exp[-E_{v_1 \cdots v_\phi}/kT] = \sum_{E_v} W(E_v)e^{-E_v/kT} \qquad (19\text{-}7)$$

where the first sum is over all values of all the quantum numbers $v_1, v_2,..., v_\phi$, without regard to order of the terms, but where, in the second sum, we have gathered together all the terms for a given value E_v of the energy and then summed over E_v.

Other arrangements of the terms in Z may show that Z can be split into a product of independent factors, $Z = z_1 z_2 \cdots z_N$. When this is possible then $F = -kT \ln Z$ can be split into a sum of terms, $F = F_1 + F_2 + \cdots + F_N$, where $F_j = -kT \ln z_j$ and all the other thermodynamic functions are likewise sums of terms, each of which can be discussed and computed separately. In order that Z be a product of several factors the energy $E_{v_1 \cdots v_\phi}$ must be a sum of terms, each dependent on a discrete set of quantum numbers. For example if the system contains a subsystem, with δ degrees of freedom, with energy $\varepsilon_{v_1 \cdots v_\delta}$ dependent only on its own δ quantum numbers (we then order the v's so these are $v_1 \cdots v_\delta$) and the energy $E_{v_{\delta+1} \cdots v_\phi}$ of the rest of the system does not depend on $v_1 \cdots v_\delta$, then the total sum for Z can be split into two factors,

$$
\begin{aligned}
Z &= \sum_{v_1 \cdots v_\phi} \exp[-(\varepsilon_{v_1 \cdots v_\delta} + E_{v_{\delta+1} \cdots v_\phi})/kT] \\
&= \sum_{v_1 \cdots v_\delta} \exp(-\varepsilon_{v_1 \cdots v_\delta}/kT) \cdot \sum_{v_{\delta+1} \cdots v_\phi} \exp(-E_{v_{\delta+1} \cdots v_\phi}/kT)
\end{aligned}
$$

each of which can be calculated separately.

In many cases the system consists of N separate, independent subsystems, the ith having δ_i degrees of freedom

$$
\text{(so that} \quad \phi = \sum_{i=1}^{N} \delta_i)
$$

each subsystem having negligible interaction with any other, although there may be strong forces holding each subsystem together. For a perfect gas of N molecules, the molecules are the subsystems, the number of degrees of freedom of each molecule being three times the number of particles per molecule. For a tightly bound crystal lattice the "subsystems" are the different normal modes of vibration of the crystal, and so on. Whenever such a separation is possible, the partition function turns out to be a product of N factors, one for each subsystem,

$$
Z = \sum_{\text{all } v} \exp(-E_{v_1 v_2 \cdots v_\phi}/kT) = z_1 \cdot z_2 \cdots z_j \cdots z_N
$$

where

$$
z_j = \sum \exp[-\varepsilon_j(v_j, v_{j+1}, ..., v_{j+\delta_j})/kT] \tag{19-8}
$$

Other examples can be found. For instance the energy of interaction between the magnetic field and the orientation of the atomic magnets in a paramagnetic solid is, to a good approximation, independent of the motion of translation or vibration of these and other atoms in the crystal. Consequently the magnetic term in the Hamiltonian, the corresponding factor in the particular function, and the resulting additive terms in F and S can be discussed and calculated separately from all the other factors and terms required to describe the thermodynamic properties of the paramagnetic material. This of course is what was done in Chapters 13 and 18, and will be done in Chapter 20, when we take up the problem of paramagnetism once more.

The Effects of Quantization

The canonical ensemble enables us to demonstrate the general effect on a system's thermodynamic behavior of the fact that a system's energy is quantized, that not all energy values are allowed. If the system can be separated into subsystems, the energy levels of the jth system can be arranged in order, with the lowest level being ε_{j1}, the next lowest ε_{j2}, and so on. The lowest level may be multiple, of course; there may be g_{j1} different quantum states, all with this same lowest energy. The next lowest energy can be labeled $\varepsilon_{j,2}$; it may have multiplicity g_{j2}; and so on. Thus we have replaced the set of δ_j quantum numbers for the jth subsystem by the single index number v, which runs from 1 to ∞, and for which $\varepsilon_{j,v+1} > \varepsilon_{j,v}$, the vth level having multiplicity g_{jv}.

Thus the jth factor in the partition function can be written

$$z_j = \sum_{v=1}^{\infty} g_{jv}e^{-\varepsilon_{jv}/kT} \tag{19-9}$$

Note the parallel between this and the general sum of (19-7) for the whole system. Weighting factor $W(E_v)$ is the multiplicity of the level E_v of the whole system, g_{jv} the multiplicity of level ε_{jv} of the subsystem. If the energy differences $\varepsilon_{j2} - \varepsilon_{j1}$ and $\varepsilon_{j3} - \varepsilon_{j2}$, between the lowest three allowed energy levels of the jth subsystem are quite large compared to kT, then the second term in the sum for z_j is small compared to the first and the third term is appreciably smaller yet, so that

$$z_j \simeq g_{j1}e^{-\varepsilon_{j1}/kT}[1 + (g_{j2}/g_{j1})e^{-(\varepsilon_{j2} - \varepsilon_{j1})/kT}] \tag{19-10}$$

for kT small compared to $\varepsilon_{j2} - \varepsilon_{j1}$. The factor in brackets becomes practically independent of T when kT is small enough.

The Helmholtz function for the system is a sum of terms, one for each subsystem,

$$F = -kT \ln Z = \sum_{j=1}^{N} F_j; \qquad F_j = -kT \ln z_j \tag{19-11}$$

and the entropy, pressure, and the other thermodynamic potentials are then also sums of terms, one for each subsystem. Whenever any one of the subsystems has energy levels separated farther apart than kT, the corresponding terms in F, S, and U have the limiting forms, obtained from Eq. (19-10),

$$F_j \simeq -kT \ln g_{j1} + \varepsilon_{j1} - (g_{j2}/g_{j1})kTe^{-(\varepsilon_{j2}-\varepsilon_{j1})/kT}$$

$$S_j \simeq k \ln g_{j1} + (g_{j2}/g_{j1})\left[k + \frac{\varepsilon_{j2} - \varepsilon_{j1}}{T} \right]e^{-(\varepsilon_{j2}-\varepsilon_{j1})/kT} \quad (19\text{-}12)$$

$$U_j = F_j + S_j T \simeq \varepsilon_{j1} + (g_{j2}/g_{j1})(\varepsilon_{j2} - \varepsilon_{j1})e^{-(\varepsilon_{j2}-\varepsilon_{j1})/kT}$$

Thus whenever the jth subsystem has a single lowest state ($g_{j1} = 1$, i.e., when the subsystem is a simple one) its entropy goes to zero when T is reduced so that kT is much smaller than the energy separation between the two lowest quantum states of the subsystem. On the other hand, if the lowest state is multiple, S goes to $k \ln g_{j1}$ as $T \to 0$. In either case, however, the heat capacity $C_{jv} = (\partial U_j/\partial T)_v$ of the subsystem vanishes at $T = 0$. Since all the subsystems have nonzero separations between their energy levels, these results apply to all the subsystems, and thus to the whole system, when T is made small enough. We have thus "explained" the shape of the curve of Figure 3-1 and the statements made at the beginning of Chapter 9. They are ways of stating the third law of thermodynamics, discussed in Chapter 6.

The High-Temperature Limit

When T is large enough so that many allowed levels of a subsystem are contained in a range of energy equal to kT, the exponentials in the partition function sum of Eq. (19-9) vary slowly enough with v so that the sum can be changed to a classical integral over phase space, of the form given in Eq. (19-6). In this case, of course, the dependence of Z on T and V is determined by the dependence of the Hamiltonian H on p and q. For example, if the subsystem is a particle in a perfect gas occupying a volume V, $H_j = (p_{jx}^2 + p_{jy}^2 + p_{jz}^2)/2m$ depends only on the momentum, and the factor in the partition function for the jth particle is

$$z_j = (1/h^3) \int\int\int dV_q \int\int\int \exp[-(p_x^2 + p_y^2 + p_z^2)/2mkT]\, dV_p$$

$$= (V/h^3)(2\pi mkT)^{3/2} \quad (19\text{-}13)$$

and if there are N particles,

$$F = -NkT \ln V - \tfrac{3}{2}NkT \ln(2\pi mkT/h^2); \qquad U = \tfrac{3}{2}NkT$$

[but see Eq. (21-13)].

On the other hand, if the "subsystem" is one of the normal modes of vibration of a crystal, $H_j = (p_j^2/2m) + (m\omega_j^2 q_j^2/2)$ depends on both q and p, so that

$$z_j = (1/h) \int e^{-m\omega_j^2 q_j^2/2kT} \, dq_j \int e^{-p_j^2/2mkT} \, dp_j$$

$$= 2\pi kT/h\omega_j = kT/\hbar\omega_j \qquad\qquad (19\text{-}14)$$

and, if there are $3N$ modes

$$F = kT \sum_{j=1}^{3N} \ln(h\omega_j) - 3NkT \ln(kT); \qquad U = 3NkT$$

the difference between $U = \tfrac{3}{2}NkT$ and $U = 3NkT$ being caused by 'the presence of the q's in the expression for H in the latter case.

For intermediate temperatures we may have to use equations like (19-12) for those subsystems with widely spaced levels and classical equations like (19-13) or (19-14) for those with closely packed energy levels. The mean energy of the former subsystems is practically independent of T, whereas the mean energy of the latter depends linearly on T: thus only the latter contribute appreciably to the heat capacity of the whole. In a gas of diatomic molecules, for example, the energy levels of translational motion of the molecules are very closely packed, so that for T larger than $1°K$, the classical integrals are valid for the translational motions, but the rotational, vibrational, and electronic motions only contribute to C_v at higher temperatures.

CHAPTER

20

STATISTICAL
MECHANICS OF
A CRYSTAL

Two examples of the use of the canonical ensemble will be discussed here; the thermal properties of a crystal lattice and those of a diatomic gas. Both of these systems have been discussed before, but we now have developed the techniques to enable us to work out their properties in detail and to answer various questions and paradoxes that have been raised earlier.

We have already used the microcanonical ensemble to work out the statistical mechanics of a very simple crystal model, one which neglects all coupling between the vibration of one atom and that of its neighbors. Equations (18-9), the Einstein formulas, predict a behavior which is distinctly closer to actuality than is the classical model. They predict a rapid drop in entropy and in heat capacity when kT falls below $\hbar\omega$, the spacing between the energy levels of the vibrating atoms. As a matter of fact the predicted drop is too rapid; measurements indicate that C_v is more nearly proportional to T^3 at very low temperatures. What is needed is a model for which the allowed frequencies of vibration are not all equal to one value of ω. A distribution of ω's from zero up to some maximum value would "spread out" the drop in C_v as $T \to 0$. Such a spread can be obtained by adopting a model that includes coupling between neighboring atomic vibrations. The model was first worked out by Debye.

Normal Modes of Crystal Vibration

The crystal lattice is held together by interatomic forces which are partly electric, partly quantum mechanical. These forces are quite large, so that static pressure P produces relatively small changes in volume, the compressibility $\kappa = -(1/V)(\partial V/\partial P)$ being of the order of magnitude of 10^{-11} m^2/newton for many crystals. The potential energy of static compression,

$$\varphi_c = [(V - V_0)^2/2\kappa V_0] \tag{20-1}$$

therefore becomes quite large if the crystal volume V should depart much from its equilibrium value V_0.

In addition the binding forces resist the displacement of any individual atom from its equilibrium position. We can label these displacements as $x_1, y_1, z_1, ..., x_N, y_N, z_N$ or $q_1, q_2, q_3, ..., q_{\phi-2}, q_{\phi-1}, q_\phi$, where $\phi = 3N$. The restoring forces depend in a complicated way on the displacements of whole groups of atoms. If the displacements are small, the forces depend linearly on the relative displacements and thus the potential energy is a combination of quadratic terms such as $\frac{1}{2}K_i q_i^2$, depending on the displacements from equilibrium of one of the atoms [which were included in the simplified model of Eq. (20-1)] but also terms such as $\frac{1}{2}K_{ij}(q_i - q_j)^2$, corresponding to a force of interaction between one atom and another. Although many of the K_{ij}'s are small or zero, some are not. The total potential energy is thus

$$\varphi_v = \frac{1}{2} \sum_{i=1}^{3N} \left[K_i q_i^2 + \sum_{j>i}^{3N} K_{ij}(q_i - q_j)^2 \right] = \frac{1}{2} \sum_{ij=1}^{3N} A_{ij} q_i q_j$$

$$A_{ii} = K_i + \sum_{j=1}^{i-1} K_{ji} + \sum_{j=i+1}^{3N} K_{ij}; \qquad A_{ij} = K_{ij} \quad \text{or} \quad K_{ji}$$

Therefore the Hamiltonian for the crystal is

$$H = \frac{1}{2m} \sum_{j=1}^{3N} p_j^2 + \frac{1}{2} \sum_{i,j=1}^{3N} A_{ij} q_i q_j + \varphi_c \tag{20-2}$$

Actually there are six coordinates not represented in the sum over the q's, those for the motion of the crystal as a rigid body; so the total number of coordinates in the second sum is $3N - 6$ rather than $3N$. However, 6 is so much smaller than $3N$ that we can ignore this discrepancy between the sums, by leaving out the kinetic energy of rigid motion and calling $3N - 6$ the same as $3N$.

The solution of a dynamical problem of this sort is discussed in all texts of dynamics. The matrix of coefficients A_{ij} determines a set of normal coordinates, Q_m, with conjugate momenta P_m, in terms of which the Hamiltonian becomes a sum of separated terms, each of which is dependent on just one coordinate pair,

$$H = \tfrac{1}{2} \sum_{n=1}^{3N} [(1/m_n)P_n^2 + m_n\omega_n^2 Q_n^2] + [(V - V_0)^2/2\kappa V_0] \quad (20\text{-}3)$$

Application of Hamilton's equations (16-1), $(\partial H/\partial P_m) = \dot{Q}_m$ and $(\partial H/\partial Q_m) = -\dot{P}_m$, results in a set of equations

$$P_n = m_n\dot{Q}_n; \qquad \ddot{Q}_n + \omega_n^2 Q_n = 0 \tag{20-4}$$

which may be solved to obtain the classical solution $Q_n = Q_{0n}e^{i\omega_n t}$. Thus $\omega_n/2\pi$ is the frequency of oscillation of the nth normal mode of oscillation of the crystal.

These normal modes of the crystal are its various standing waves of free vibration. The lowest frequencies are in the sonic range, corresponding to wavelengths a half or a third or a tenth of the dimensions of the crystal. The highest frequencies are in the infrared and correspond to wavelengths of the size of the interatomic distances. Because there are $3N$ degrees of freedom there are $3N$ different standing waves (or rather $3N - 6$ of them, to be pedantically accurate); some of them are compressional waves and some are shear waves.

Quantum States for the Normal Modes

According to Eq. (16-7), the allowed energies of a single normal mode, with Hamiltonian $(1/2m_j)P_j^2 + \tfrac{1}{2}m_j\omega_j^2 Q_j^2$ are given by the formula $\hbar\omega_j[v_j + \tfrac{1}{2}]$, where v_j is an integer, the quantum number of the jth normal mode. Sometimes the quantized standing waves are called *phonons*; v_j is the number of phonons in the jth wave. Microstate v of the crystal corresponds to a particular choice of value for each of the v_j's. The energy of the phonons in microstate v is then

$$E_v = E_0(V) + \hbar \sum_{j=1}^{3N} \omega_j v_j;$$

$$E_0 = [(V - V_0)^2/2\kappa V_0] + \tfrac{1}{2}\hbar \sum_{j=1}^{3N} \omega_j \tag{20-5}$$

each term in the sum being the energy of a different standing wave.

The difference between this and the less accurate Einstein formulas of Eqs. (18-2) is that in the previous case the ω's were the same for all the oscillators, whereas inclusion of atomic interaction in the present model has spread out the resonant frequencies, so that each standing wave has a different value of ω.

According to Eq. (19-4) the partition function is

$$Z = \sum_{\text{all } v_j\text{'s}} \exp\left(\frac{-E_0 - \hbar \sum_j \omega_j v_j}{kT}\right) = e^{-E_0/kT} z_1 z_2 \cdots z_{3N}$$

where

$$z_j = \sum_{v_j} e^{-\hbar\omega_j v_j/kT} = (1 - e^{-\hbar\omega_j/kT})^{-1} \qquad (20\text{-}6)$$

where we have used the formula

$$(1 - x)^{-1} = \sum_{n=0}^{\infty} x^n \qquad |x| < 1 \qquad (20\text{-}7)$$

Thus, from Eq. (19-5), the Helmholtz function for the crystal is

$$F = -kT \ln Z = E_0(V) + kT \sum_{j=1}^{3N} \ln(1 - e^{-\hbar\omega_j/kT}) \qquad (20\text{-}8)$$

We can then compute the probability f_v that the system is in the microstate specified by the quantum numbers $v \equiv v_1, v_2,..., v_{3N}$. It is the product [see Eq. (19-4)],

$$f_v = (1/Z)e^{-E_v/kT} = f_1 f_2 f_3 \cdots f_{3N}$$

where

$$f_j = (1/z_j)e^{-\hbar\omega_j v_j/kT} = e^{-\hbar\omega_j v_j/kT} - e^{-\hbar\omega_j(v_j + 1)/kT} \qquad (20\text{-}9)$$

is the probability that the jth standing wave of thermal vibration is in the v_jth quantum state. The probability that the crystal is in the microstate v is of course the product of the probabilities that the various normal modes are in their corresponding states.

When kT is small compared to $\hbar\omega_j$ for all the standing waves of crystal vibration, all the z_j's are practically unity, F is approximately equal to $E_0(V)$, independent of T, and the entropy is very small. When kT is large compared to any $\hbar\omega_j$, each of the terms in parentheses in Eq. (20-8) will be approximately equal to $\hbar\omega_j/kT$ and consequently the Helmholtz function will contain a term $-3NkT \ln(kT)$, the temperature-dependent term in the entropy will be $3Nk \ln kT$, and the heat capacity will be $3Nk = 3nR$, as expected. To find values for the intermediate temperatures we must carry out the summation over j in Eq. (20-8) or, what is satisfactory here, we must approximate the summation by an integral and then carry out the integration.

Summing over the Normal Modes

The crucial question in changing from sum to integral is: How many standing waves are there with frequencies (times 2π) between ω and $\omega + d\omega$? There are three kinds of waves in a crystal, a set of compressional waves and two sets of mutually perpendicular shear waves. If the crystal is a rectangular parallelopiped of dimensions l_x, l_y, l_z, the pressure distribution of one of the compressional waves would be

$$p = \alpha Q_j \sin(\pi k_j x/l_x) \sin(\pi m_j y/l_y) \sin(\pi n_j z/l_z)$$

where $Q_j(t)$ is the amplitude of the normal mode j, with equations of motion (20-4), α is the proportionality constant relating Q_j and the pressure amplitude of the compressional wave, and k_j, m_j, n_j are integers giving the number of standing-wave nodes along the x, y, and z axes, respectively, for the jth wave.

The value of ω_j, 2π times the frequency of the jth mode, is given by the familiar formula

$$\omega_j^2 = (\pi c k_j/l_x)^2 + (\pi c m_j/l_y)^2 + (\pi c n_j/l_z)^2 \qquad (20\text{-}10)$$

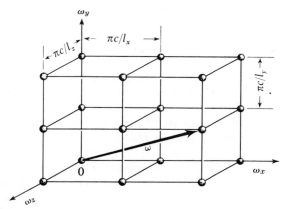

FIGURE 20-I

Representation of allowed values of ω in ω space.

where c is the velocity of the compressional wave. Each different j corresponds to a different trio of integers k_j, m_j, n_j. A similar discussion will arrive at a similar formula for each of the shear-wave sets, except that the value of c is that appropriate for shear waves. The problem is to determine how many allowed ω_j's have values between ω and $\omega + d\omega$.

To visualize the problem, imagine the allowed ω_j's to be plotted as points in "ω space," as shown in Figure 20-1. They form a lattice of points in the first octant of the space, with a spacing in the "ω_x" direction of $\pi c/l_x$, a spacing in the "ω_y" direction of $\pi c/l_y$, and a spacing in the "ω_z" direction of $\pi c/l_z$, with the allowed value of ω given by the distance from the origin to the point in question, as shown by the form of Eq. (20-10). The point closest to the origin can be labeled $j = 1$, the next $j = 2$, etc. The spacing between the allowed points is therefore such that there are, on the average, $l_x l_y l_z / \pi^3 c^3 = V/\pi^3 c^3$ points in a unit volume of "ω space," where $V = l_x l_y l_z$ is the volume occupied by the crystal.

Therefore all the allowed ω_j's having value less than ω are represented by those points inside a sphere of radius ω (with center at the origin). The volume of the part of the sphere in the first octant is $\frac{1}{8}(4\pi\omega^3/3)$ and, because there are $V/\pi^3 c^3$ allowed points per unit volume, there must be $(V/\pi^3 c^3)(\pi\omega^3/6)$ standing waves with values of ω_j less than ω. Differentiating this with respect to ω, we see that the average number of ω_j's with value between ω and $\omega + d\omega$ is

$$dj = (V/2\pi^2 c^3)\omega^2 \, d\omega \tag{20-11}$$

Several comments must be made about this formula. In the first place, the formula is for just one of the three sets of standing waves, and thus the dj for all the normal modes is the sum of three such formulas, each with its appropriate value of c, the wave velocity. But we can combine the three by using an average value of c, and say that, approximately, the total number of standing waves with ω_j's between ω and $\omega + d\omega$ is

$$dj = (3V/2\pi^2 c^3)\omega^2 \, d\omega \tag{20-12}$$

where c is an appropriate average of the wave velocities of the compressional and shear waves. Next we should note that Eq. (20-11) was derived for a crystal of rectangular shape. However, a more-detailed analysis of standing waves in crystals of more-general shapes shows that these equations still hold for the other shapes as long as V is the crystal volume. For a differently shaped crystal, the lattice of allowed points in ω space is not that shown in Fig.

20-1, but in spite of this the *density* of allowed points in ω space is the same, $V/\pi^3 c^3$.

Next we remind ourselves that there is an upper limit to the allowed values of the ω_j's; in fact there can only be $3N$ different normal modes in a crystal with N atoms ($3N - 6$, to be pedantically exact). Therefore our integrations should go to an upper limit ω_m, where

$$3N = \sum_{j=1}^{3N} 1 = \int_0^{\omega_m} dj = (3V/2\pi^2 c^3) \int_0^{\omega_m} \omega^2 \, d\omega = (V\omega_m^3/2\pi^2 c^3)$$

or

$$\omega_m = (6\pi^2 N c^3/V)^{1/3} \tag{20-13}$$

Finally we note that both Eqs. (20-12) and (20-13) are approximations of the true state of things, first because we have tacitly assumed that c is independent of ω, which is not exactly true at the higher frequencies, and second because we have assumed that the highest compressional frequency is the same as the highest shear frequency, namely, $\omega_m/2\pi$, and this is not correct either. All we can do is to hope our approximations tend to average out and that our final result will correspond reasonably well to the measured facts.

The Debye Formulas

Returning to Eq. (20-8), we change from a sum over j to an integral over dj, using Eq. (20-12) and integrating by parts; we obtain

$$F = [(V - V_0)^2/2\kappa V_0] + \int_0^{\omega_m} [\tfrac{1}{2}\hbar\omega_j + kT \ln(1 - e^{-\hbar\omega_j/kT})] \, dj$$

$$= E_0(V) + (3kTV/2\pi^2 c^3) \int_0^{\omega_m} \ln(1 - e^{-\hbar\omega/kT})\omega^2 \, d\omega$$

$$= \bar{E}_0 + \frac{kTV\omega_m^3}{2\pi^2 c^3} \ln(1 - e^{-\hbar\omega_m/kT})$$

$$\qquad\qquad\qquad - \frac{kTV\omega_m^3}{6\pi^2 c^3} D\left(\frac{\hbar\omega_m}{kT}\right) \tag{20-14}$$

where

$$\bar{E}_0 - [(V - V_0)^2/2\kappa V_0] = \tfrac{1}{2}\int_0^{\omega_m} \hbar\omega_j \, dj = (3V\hbar\omega_m^4/16\pi^2 c^3)$$

The function D, defined by the integral

$$D(x) = (3/x^3) \int_0^x [z^3 \, dz/(e^z - 1)] \to \begin{cases} \pi^4/5x^3 & x \gg 1 \\ 1 - (3/8)x & x \ll 1 \end{cases} \quad (20\text{-}15)$$

is called the *Debye function*, after the originator of the formula.

We now can express the temperature scale in terms of the *Debye temperature* $\theta = \hbar\omega_m/k$ (which is a function of V) and then write down the thermodynamic functions of interest,

$$F = [(V - V_0)^2/2\kappa V_0] + \tfrac{9}{8}Nk\theta$$
$$+ NkT[3\ln(1 - e^{-\theta/T}) - D(\theta/T)]$$
$$\to \begin{cases} [(V - V_0)^2/2\kappa V_0] + \tfrac{9}{8}Nk\theta - (\pi^4 NkT^4/5\theta^3) & T \ll \theta \\ [(V - V_0)^2/2\kappa V_0] + \tfrac{3}{2}Nk\theta \\ \qquad\qquad + 3NkT\ln(\theta/T) - NkT & T \gg \theta \end{cases}$$

$$S = Nk\left[-3\ln(1 - e^{-\theta/T}) + 4D\left(\frac{\theta}{T}\right) \right]$$
$$\to \begin{cases} (4\pi^4 NkT^3/5\theta^3) & T \ll \theta \\ 3Nk\ln(Te^{4/3}/\theta) & T \gg \theta \end{cases}$$

$$U = [(V - V_0)^2/2\kappa V_0] + \tfrac{9}{8}Nk\theta + U_v(T);$$
$$U_v = 3NkTD(\theta/T) \qquad\qquad (20\text{-}16)$$

$$C_v = 3Nk\left[4D\left(\frac{\theta}{T}\right) - \frac{(3\theta/T)}{e^{\theta/T} - 1} \right]$$
$$\to \begin{cases} (12\pi^4 NkT^3/5\theta^3) & T \ll \theta \\ 3Nk & T \gg \theta \end{cases}$$

$$P = [(V_0 - V)/\kappa V_0] - \tfrac{9}{8}Nk\theta' - 3NkT(\theta'/\theta)D(\theta/T)$$
$$\to \begin{cases} [(V_0 - V)/\kappa V_0] - \tfrac{9}{8}Nk\theta' - \tfrac{3}{5}\pi^4 Nk\theta'(T/\theta)^4 & T \ll \theta \\ [(V_0 - V)/\kappa V_0] - 3NkT(\theta'/\theta) & T \gg \theta \end{cases}$$

where $\theta' = d\theta/dV = (h/k)(d\omega_m/dV)$ is a negative quantity. Referring to Eq. (3-6) we see that the empirical equation of state is approximately the same as the last line of Eqs. (20-16) if $-(3Nk\theta'/\theta)$ is equal to β/κ of the empirical formula. This relationship can be used to predict values of β if θ' can be computed, or it can be used to determine θ' from measurements of β and κ.

The functions $D(x) = [xU_v(\theta/x)/3Nk\theta]$ and $C_v/3Nk$ are given in Table 20-1 as functions of $x = \theta/T$. A curve of $C_v/3Nk$ versus T/θ is given in Figure 20-2 (solid curve).

TABLE 20-I
Debye Functions for a Crystal Lattice

x	$D(x)$	$C_v/3Nk$	x	$D(x)$	$C_v/3Nk$
0.0	1.0000	1.0000	4.0	0.1817	0.5031
0.1	0.9627	0.9995	5.0	0.1177	0.3689
0.2	0.9270	0.9980	6.0	0.0776	0.2657
0.5	0.8250	0.9882	8.0	0.0369	0.1382
1.0	0.6745	0.9518	10	0.0193	0.0759
1.5	0.5473	0.8960	12	0.0113	0.0448
2.0	0.4411	0.8259	15	0.0056	0.0230
2.5	0.3540	0.7466	20	0.0024	0.0098
3.0	0.2833	0.6630	25	0.0012	0.0050

Comparison with Experiment

Several checks with experiment are possible. By adjusting the value of θ we can fit the curve for C_v, predicted by Eq. (20-16) and drawn in Figure 20-2, to the experimental curve. That the fit is excellent can be seen from the check between the circles and triangles and the solid line. We see, for example, that the Debye formula, which takes into account (approximately) the coupling between atomic vibrations, fits better than the Einstein formula, which neglects interaction, the discrepancy being greatest at low temperatures.

From the fit one, of course, obtains an empirical value of $\theta = \hbar\omega_m/k$ for each crystal measured, and thus a value of ω_m for each crystal. However, by actually measuring the standing-wave frequencies of the crystal and by summing as per Eq. (20-13), we can find out what ω_m (and thus θ) ought to be, and then check it against the θ that gives the best fit for C_v. These checks are also quite good, as can be seen from Table 20-2.

TABLE 20-2
Comparison of Debye Temperatures

Substance	θ, °K from C_v fitting	θ, °K from elastic data
NaCl	308	320
KCl	230	246
Ag	237	216
Zn	308	305

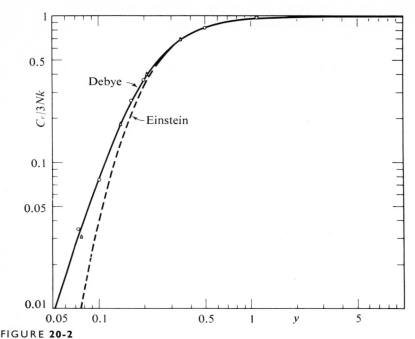

FIGURE 20-2

Specific-heat curves for a crystal. Ordinate y for the Debye curve is T/θ = $kT/\hbar\omega_m$; ordinate for the Einstein curve is $3kT/4\hbar\omega$. Circles are experimental points for graphite, triangles for KCl.

Thus formulas (20-16) represent a very good check with experiment for many crystals. A few differences do occur, however, some of which can be explained by using a somewhat more complicated model. In a few cases, lithium for example, the normal modes are so distributed that the approximation of Eq. (20-12) for the number of normal modes with ω_j's between ω and $\omega + d\omega$ is not good enough, and a better approximation must be used [which modifies Eqs. (20-13) and (20-14)]. In the case of most metals the C_v does not fit the Debye curve at *very* low temperatures (below about $2\,°K$); in this region the C_v for metals turns out to be more nearly linearly dependent on T than proportional to T^3, as the Debye formula predicts. The discrepancy is caused by the free electrons present in metals, as will be shown in Chapter 26.

An Ensemble Appropriate for Paramagnetism

The presence of permanent atomic dipoles in the crystal also requires modification of the Debye model. Ferromagnetic and antiferromagnetic behavior (see page 197) involves quantum and indistinguishability effects which are too specialized for us to consider in this volume. On the other hand the thermal behavior of the paramagnetic salts, such as iron ammonium alum, which are neither ferro- nor antiferromagnetic, is simple enough to warrant further discussion here. It will provide an opportunity to demonstrate the application of the canonical ensemble to situations where the Gibbs function, rather than the Helmholtz function, is minimal. In this case three independent variables, T, V, and $\mathfrak{H}$ are important, and if they are all to be thermodynamic variables, they must all be included in the ensemble.

At the end of Chapter 18 we pointed out that to portray the behavior of these paramagnetic salts at very low temperatures we had to include the small interaction between the magnets and the lattice. At extremely low temperatures and zero applied field, the magnets are oriented in an orderly pattern with respect to the crystal lattice, with correspondingly small entropy. A small increase in temperature is required for the thermal vibrations to shake the magnets loose from this bound pattern and to raise the magnetic entropy to the value $Nk \ln 2$ of Eq. (18-24) for $\mathfrak{B} = 0$. This interaction with the lattice was difficult to include in our microcanonical ensemble, so we postponed its quantitative treatment.

The machinery to cope with the problem is provided by the canonical ensemble. The interaction of the magnets with the lattice can be assumed to be independent of the lattice vibrations, at least at the low temperatures where the interaction effects are perceptible. Other models, including interaction with the vibrations, can be worked out but the calculations are more tedious and the results are not much better. Neglecting interaction with the vibration allows us to deal with the N_m magnets in the crystal as a separate subsystem in the sense of Eq. (19-8). Their partition function is simply another factor, to be multiplied to the Z of Eq. (20-6) and their entropy, internal energy, and other thermodynamic functions are added to the S, U, etc., of Eqs. (20-16).

In the present case, however, we have added another pair of thermodynamic variables, $\mathfrak{H}$ and $\mathfrak{M}$, one of which is an additional independent variable for the system and thus should enter explicitly into the distribution function. The most useful trio of variables is T, V, and $\mathfrak{H}$, so the appropriate thermodynamic potential is the magnetic Gibbs function $G_m = U - TS - \mathfrak{H}\mathfrak{M}$. The appropriate modification of the canonical ensemble goes as follows:

$$S = -k \sum f_v \ln f_v \qquad \text{is to be maximum}$$

subject to

$$\sum f_v = 1; \quad \sum E_v f_v = U \quad \text{and also} \quad \sum \mathfrak{M}_v f_v = \mathfrak{M} \qquad (20\text{-}17)$$

where E_v is the nonmagnetic part of the magnet-system's energy, the part dependent on V, not on $\mathfrak{H}$, and $\mathfrak{M}_v$ is its magnetization when it is in the vth quantum state. If we are to consider $\mathfrak{M}$ and $\mathfrak{H}$ as separate variables, we should not "bury" the magnetic energy in U, but should carry it as an extra term.

Using Lagrange multipliers, we must maximize

$$\sum f_v [-k \ln f_v + \alpha_0 + \alpha_e E_v + \alpha_m \mathfrak{M}_v]$$

which results in the equations

$$k \ln f_v = (\alpha_0 - k) + \alpha_e E_v + \alpha_m \mathfrak{M}_v \qquad (20\text{-}18)$$

Multiplying this by f_v and summing, we see our solution corresponds to

$$-S = (\alpha_0 - k) + \alpha_e U + \alpha_m \mathfrak{M}$$

and for this to be equivalent to $-S = (G_m - U + \mathfrak{H}\mathfrak{M})/T$ we must have $(\alpha_0 - k) = G_m/T$, $\alpha_e = -1/T$, and $\alpha_m = \mathfrak{H}/T$. Finally the normalizing condition for f_v requires that

$$1 = \sum f_v = \sum \exp \left\{ \frac{1}{k} [(\alpha_0 - k) + \alpha_e E_v + \alpha_m \mathfrak{M}_v] \right\}$$
$$= e^{G_m/kT} \sum e^{(\mathfrak{H}\mathfrak{M}_v - E_v)/kT} = e^{G_m/kT} Z$$

or

$$Z = \sum_v e^{(\mathfrak{H}\mathfrak{M}_v - E_v)/kT} = e^{-G_m/kT} \qquad (20\text{-}19)$$

where Z is the partition function for the crystal-magnet system. From the magnetic Gibbs function G_m we can then obtain the other thermodynamic functions of interest,

$$G_m = -kT \ln Z; \qquad (\partial G_m/\partial T)_{V\mathfrak{H}} = -S;$$

$$(\partial G_m/\partial \mathfrak{H})_{TV} = -\mathfrak{M}; \qquad (\partial G_m/\partial V)_{T\mathfrak{H}} = -P \qquad (20\text{-}20)$$

The Coupling between Magnets and Lattice

To utilize this ensemble we must be able to write down the nonmagnetic energy E_v and the magnetization $\mathfrak{M}_v$ of the system in state v, and be able to count how many different states there are with the same values of E_v and $\mathfrak{M}_v$, so that the partition function can be written out. In addition to the energy of vibration and compression of Eq. (20-5) there is the energy of coupling between the atomic magnets and the crystal lattice, which is essentially nonmagnetic. At absolute zero the magnets are "frozen in" to the lattice in groups of λ magnets, each group so oriented with respect to the lattice and to each other that the net magnetic moment of the group is zero. When in this bound state the λ magnets are in their lowest quantum state, of zero energy. As the temperature is increased, one group after another comes uncoupled from the lattice and its magnets take up a random, rather than a specified, orientation. It requires an energy $\lambda\gamma$ to uncouple each group of λ magnets from its bound state; the energy γ is very small, of the order of $(k/100)$.

Therefore there are two possible quantum levels of each group of λ magnets, in regard to coupling with the lattice; a bound state with zero energy and zero magnetization and an uncoupled state with energy γ per magnet and a possible magnetization depending on the applied field $\mathfrak{H}$. The energy E_v thus consists of two independent parts; the vibrational-compressional part given in Eq. (20-5) and the magnet-lattice coupling part

$$E_{m1} = \sum_{i=1}^{N_m/\lambda} \lambda\gamma\delta_i; \qquad \delta_i = 0 \text{ or } 1$$

N_m is the number of atomic magnets in the system, (N_m/λ) is the number of groups (presumably an integer), and δ_i is the quantum number which determines the state of coupling of the ith group to the lattice; $\delta_i = 0$ for the group being bound and $\delta_i = 1$ for it being unbound.

If the atomic magnets have angular momentum quantum number J, then in their uncoupled state they can each take on $2J + 1$ orientations with respect to the magnetic field [see Eq. (13-24)]. The magnetic moment of the n_ith magnet in the ith group is $m_{n_i}\mathfrak{m}_B\delta_i$, being zero if the group is bound ($\delta_i = 0$) or having possible quantized values ranging from $-J\mathfrak{m}_B$ to $+J\mathfrak{m}_B$ if the group is unbound ($\delta_i = 1$). Thus the quantity $E_v - \mathfrak{H}\mathfrak{M}_v$ entering in the partition function Z of Eq. (20-20) is

$$E_0 + \hbar \sum_{j=1}^{3N} \omega_j v_j + \sum_{i=1}^{N_m/\lambda} \left[\lambda\gamma - \mu_0\mathfrak{H} \sum_{n_i=1}^{\lambda} m_{n_i}\mathfrak{m}_B \right]\delta_i$$

where the magnetic quantum number m_{n_i} for the n_ith magnet can have the values $(-J, -J + 1,..., J - 1, J)$ and where the factor μ_0 has been inserted to change from polarization to magnetization. The quantity $\mathfrak{m}_B = (eh/4\pi m)$ is the Bohr magneton.

The partition function thus factors into two parts, one being the partition function of Eq. (20-6), which can be manipulated separately to obtain the energy, entropy, etc., of vibration-compression, treated earlier in this chapter. The other factor is the magnetic part,

$$Z_m = \sum_{\text{all } \delta,m} \exp\left\{ \frac{-1}{kT} \sum_{i=1}^{N_m/\lambda} \left[\lambda\gamma - \mu_0\mathfrak{H} \sum_{n_i=1}^{\lambda} m_{n_i}\mathfrak{m}_B \right]\delta_i \right\} = \prod_{i=1}^{N_m/\lambda} z_{mi}$$

(20-21)

a sum over all possible combinations of the quantum numbers δ_i and m_{n_i}. This also factors into the product of (N_m/λ) identical factors, one for each group of magnets. The one for the ith group is

$$z_{mi} = \left\{ 1 + e^{-\lambda\gamma/kT} \sum_{\text{all } m} \exp\left[\frac{\mathfrak{m}_B\mathfrak{B}}{kT}(m_1 + m_2 + \cdots + m_\lambda) \right] \right\}$$

$$= 1 + e^{-\lambda\gamma/kT} \left[\sum_{m_1=-J}^{J} e^{m_1\mathfrak{m}_B\mathfrak{B}/kT} \right] \cdots \left[\sum_{m_\lambda=-J}^{J} e^{m_\lambda\mathfrak{m}_B\mathfrak{B}/kT} \right]$$

$$= 1 + e^{-\lambda\gamma/kT}\left\{ \frac{\sinh[(J + \tfrac{1}{2})\mathfrak{m}_B\mathfrak{B}/kT]}{\sinh(\mathfrak{m}_B\mathfrak{B}/2kT)} \right\}^\lambda$$

(20-22)

where $\mathfrak{B} = \mu_0\mathfrak{H}$. The first term in this sum corresponds to the bound state of the group, the other terms correspond to the unbound states, for all the different orientations of the λ different magnets. This is to be compared with the normalizing constant of Eq. (13-25).

To develop facility in working out partition functions, and also because the example will be pertinent in some of the discussion in the next chapter, we shall derive this partition function Z_m by a different route. The energy of decoupling of the magnet groups from the lattice is equal to the number n of groups which are decoupled times the energy $\lambda\gamma$ required to decouple one group. But there are (N_m/λ) different groups and the number of ways in which n of them are decoupled and $(N_m/\lambda) - n$ are bound is, of course [see Eq. (11-8)], $\{(N_m/\lambda)!/n![(N_m/\lambda) - n]!\}$; thus there will be this number of terms in the partition function that have the factor $e^{-n\lambda\gamma/kT}$. Also the magnetization of the crystal is equal to

$-J\mu_0\mathfrak{m}_B$ times the number n_{-J} of decoupled magnets in the crystal which have magnetic quantum number m equal to $-J$, plus $-(J-1)\mu_0\mathfrak{m}_B$ times the number n_{-J+1} with m equal to $-J+1$ plus, and so on, up to $J\mu_0\mathfrak{m}_B$ times the number n_J with m equal to J. But, according to Eq. (11-12), the number of ways of arranging the $n\lambda$ decoupled magnets so that n_{-J} of them have magnetic quantum number $m = -J$, etc., up to n_J of them having $m = J$ is

$$\frac{(n\lambda)!}{(n_{-J})!(n_{-J+1})!\cdots(n_{J-1})!(n_J)!}$$

where

$$\sum_{m=-J}^{J} n_m = n\lambda$$

and this is the number of terms in Z_m which have the factor

$$\exp(\mathfrak{H}\mathfrak{M}_v/kT) = \exp\left\{\frac{\mathfrak{H}}{kT}\left[n_{-J}[-J\mathfrak{m}_B\mathfrak{B}/kT]\right.\right.$$
$$+ n_{-J+1}[(-J+1)\mathfrak{m}_B\mathfrak{B}/kT]$$
$$\left.\left.+ \cdots + n_{J-1}[(J-1)\mathfrak{m}_B\mathfrak{B}/kT] + n_J[J\mathfrak{m}_B\mathfrak{B}/kT]\right]\right\}$$

Therefore the partition function Z_m can be written as a sum over the number n of magnet groups which are decoupled from the lattice and also over the numbers n_m, of the $n\lambda$ decoupled magnets, which have magnetic quantum numbers m, for $m = -J, -J+1,..., J-1, J$:

$$Z_m = \sum_{n=0}^{N_m/\lambda} \frac{(N_m/\lambda)!}{n![(N_m/\lambda)-n]!} \sum_{\text{all }n_m} \frac{(n\lambda)!}{(n_{-J})!\cdots(n_J)!} \cdot$$

$$\times \exp\left\{\frac{-n\lambda\gamma}{kT} + \mathfrak{H}\sum_{m=-J}^{J} n_m\left(\frac{m\mathfrak{m}_B\mathfrak{B}}{kT}\right)\right\}; \qquad \sum_{m=-J}^{J} n_m = n\lambda$$

The multinomial sum over the n_m's is

$$\sum_{\text{all }n_m} \frac{(n\lambda)!}{(n_{-J})!\cdots(n_J)!} [Y(-J)]^{n_{-J}}[Y(-J+1)]^{n_{-J+1}}\cdots[Y(J)]^{n_J}$$

$$= \left\{\sum_{m=-J}^{J} Y(m)\right\}^{n\lambda} = \left\{\frac{\sinh[(J+\frac{1}{2})\mathfrak{m}_B\mathfrak{B}/kT]}{\sinh(\mathfrak{m}_B\mathfrak{B}/2kT)}\right\}^{n\lambda};$$

$$Y(m) = \exp\left(\frac{m\mathfrak{m}_B\mathfrak{B}}{kT}\right)$$

and the magnetic partition function is

$$Z_m = \sum_{n=0}^{N_m/\lambda} \frac{(N_m/\lambda)!}{n![(N_m/\lambda)-n]!} \left\{ e^{-\lambda\gamma/kT} \frac{\sinh^\lambda[(J+\frac{1}{2})m_B\mathfrak{B}/kT]}{\sinh^\lambda(m_B\mathfrak{B}/2kT)} \right\}^n$$

$$= \left\{ 1 + e^{-\lambda\gamma/kT} \frac{\sinh^\lambda[(J+\frac{1}{2})m_B\mathfrak{B}/kT]}{\sinh^\lambda(m_B\mathfrak{B}/2kT)} \right\}^{N_m/\lambda} \tag{20-23}$$

which is the same as that obtained from Eqs. (20-21) and (20-22).

We shall carry the rest of the analysis through for the simplest case, $J = \frac{1}{2}$, which was the case discussed in the last section of Chapter 18. The cases for $J > \frac{1}{2}$ can be worked out without much additional trouble. For $J = \frac{1}{2}$, the ratio of hyperbolic sines in (20-23) becomes $2^\lambda \cosh^\lambda(m_B\mathfrak{B}/2kT)$, so that

$$Z_m = (1 + X^\lambda)^{N_m/\lambda},$$

where

$$X = 2e^{-x}\cosh(y), \qquad x = (\gamma/kT); \qquad y = (m_B\mathfrak{B}/2kT)$$

so that

$$G_m = -\frac{1}{\lambda} N_m kT \ln(1 + X^\lambda)$$

$$S_m = -(\partial G_m/\partial T) = \frac{1}{\lambda} N_m k \ln(1 + X^\lambda)$$

$$+ \frac{N_m k X^\lambda}{1 + X^\lambda}(x - y\tanh y)$$

$$\mathfrak{M} = -(\partial G_m/\partial\mathfrak{H}) = -N_m\mu_0 m_B \tanh(y)\frac{X^\lambda}{1+X^\lambda} \tag{20-24}$$

$$U_m = G_m + TS_m + \mathfrak{H}\mathfrak{M} = N_m\gamma\frac{X^\lambda}{1+X^\lambda}$$

$$C_\mathfrak{H} = T\left(\frac{\partial S}{\partial T}\right)$$

$$= N_m k \frac{X^\lambda}{1+X^\lambda}\left\{\frac{\lambda}{1+X^\lambda}(x - y\tanh y)^2 + y^2 \operatorname{sech}^2 y\right\}$$

Parameter λ is the number of atomic magnets in the group which link together to cancel moments when they become bound to the lattice.

Magnetic Entropy at Very Low Temperatures

The results are in accord with those discussed qualitatively at the end of Chapter 18, and are in general agreement with measurement. At temperatures considerably larger than (γ/k) (i.e., larger than about 0.1°K), x is very small and $e^{-\lambda x}$ can be set equal to 1. The majority of the magnets are decoupled from the lattice and the entropy for $\mathfrak{H} = 0$ $(\gamma = 0)$ is approximately $(1/\lambda)N_m k \ln(1 + 2^\lambda)$ $\simeq N_m k \ln 2$, in agreement with Eq. (18-24). There is complete disorder in orientation of the N_m magnets, so W is 2^{N_m} and S is $N_m k \ln 2$. Application of a magnetic field reduces the entropy,

$$ S_m \simeq N_m k \left\{ \ln\left[2\cosh\left(\frac{\mathfrak{m}_B \mathfrak{B}}{2kT}\right)\right] - \left(\frac{\mathfrak{m}_B \mathfrak{B}}{2kT}\right)\tanh\left(\frac{\mathfrak{m}_B \mathfrak{B}}{2kT}\right)\right\} \quad \gamma \ll kT $$

which reduces to zero as $\mathfrak{B} \to \infty$ and the magnets are all lined up. Also the expression for the magnetization for $kT \gg \gamma$,

$$ \mathfrak{M} \simeq \tfrac{1}{2}N_m \mu_0 \mathfrak{m}_B \tanh(\mathfrak{m}_B \mathfrak{B}/2kT) $$

is that given in Eq. (13-26) for $J = \tfrac{1}{2}$.

As T becomes smaller than (γ/k) the entropy, for $\mathfrak{H} = 0$, drops from $N_m k \ln 2$ to zero; the larger λ, the bound-group size, the more abrupt is the drop. The internal energy U_m (nonmagnetic) also drops from $N\gamma$ to zero as the magnets settle down, by groups, into their bound state. The magnetization $\mathfrak{M}$, for $\tfrac{1}{2}\mathfrak{m}_B \mathfrak{B}$ smaller than γ, also drops to zero as T is made smaller than (γ/k). When $\tfrac{1}{2}\mathfrak{m}_B \mathfrak{B}$ is larger than γ, the torque of the magnetic field is large enough to pull the magnets free from their coupling with the lattice and $\mathfrak{M}$ again becomes nearly equal to $\tfrac{1}{2}N_m \mu_0 \mathfrak{m}_B \tanh(\mathfrak{m}_B \mathfrak{B}/2kT)$.

Plots of entropy S_m versus T for several values of $\mathfrak{B} = \mu_0 \mathfrak{H}$ are given in Figure 20-3. The range of temperature plotted is so low that the entropy of lattice vibration, of Eq. (20-16), is negligible compared to S_m. These curves enable us to complete our discussion of adiabatic demagnetization, begun in Chapters 4 and 8. Starting at temperature T_i and zero field, the entropy has value S_i. We then apply a field isothermally, keeping $T = T_i$. The entropy decreases, reaching a value S_b or S_d, depending on the field strength. Insulating the material thermally, we decrease the field again, this time adiabatically, keeping S constant. At zero field the temperature has reduced to T_c or T_e, depending on the strength of the initial field.

We should notice three facts about this process. First, the lower limit to the temperature that can be attained is determined

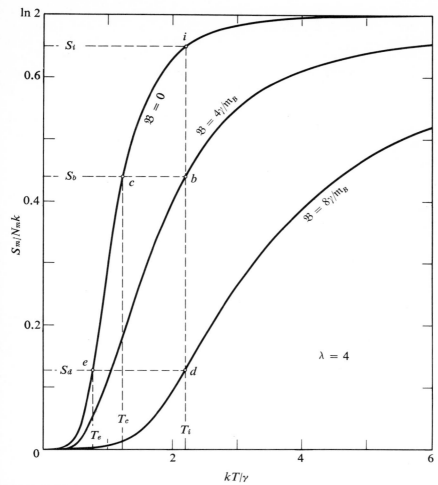

FIGURE 20-3

Entropy curves for paramagnetic salt, illustrating temperature reduction by adiabatic demagnetization.

by the value of the lattice coupling constant γ, for S drops rapidly to zero when T is reduced below (γ/k). The less coupled the magnets are to the lattice the lower the temperature which can be attained, though this temperature may have to be reached by a sequence of steps, not one step. The lowest temperatures reached in practice, less than $0.0001\,°K$, have been reached by a two-stage process—first by demagnetizing the electron magnets of a paramagnetic material, which is then used to cool a material having nuclei of large magnetic moment, and then by demagnetizing the nuclear magnets. These nuclear magnets are nearly completely shielded by the electron shells from any coupling with the neighboring atoms; the coupling is small enough so that (γ/k) is less than $0.0002\,°K$ for them.

The second point to note is the corroboration of the third law, discussed in Chapter 6. The curves of Figure 20-3 correspond to those of Figure 6-3, with the T and S axes reversed. All the curves, for different values of $\mathfrak{B}$, coalesce to the same value of S at $T = 0$, in this case $S_0 = 0$.

Finally we should notice the behavior of the magnetic heat capacity $C_\mathfrak{H}$. At zero field Eq. (20-24) indicates that

$$C_\mathfrak{H} = \lambda N_m k \left[\left(\frac{\gamma}{kT}\right)^2 2^\lambda e^{\lambda\gamma/kT}/(2^\lambda + e^{\lambda\gamma/kT})^2\right]; \qquad \mathfrak{H} = 0$$

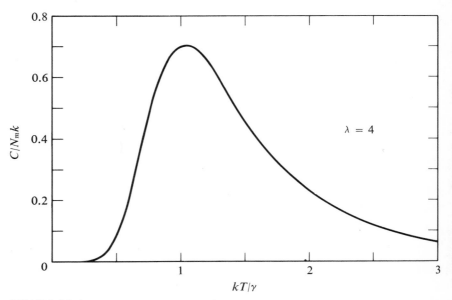

FIGURE **20-4**

Magnetic heat capacity $C_\mathfrak{H}$ for $\mathfrak{H} = 0$ for a paramagnetic salt, showing the quenching effect of lattice binding energy γ.

which is plotted in Figure 20-4 for $\lambda = 4$. We see that for $kT > \gamma$ this heat capacity increases as T decreases, approximately proportional to $(1/T)^2$, as was assumed in obtaining Eq. (8-29). As T is reduced, $C_{\mathfrak{H}}$ for $\mathfrak{H} = 0$ increases to a maximum at $T \simeq (\gamma/k \ln 2)$; as T is reduced still further $C_{\mathfrak{H}}$ drops sharply to zero. The heat capacity of the magnets therefore behaves as though the change from free to bound states for the magnet groups is analogous to a phase change of the second kind (see the penultimate section of Chapter 9). This is true in a sense; the disordered, unbound state of the magnet groups is analogous to a liquid and the regularly ordered bound state is analogous to a solid. There is no discontinuity in C_v, but the drop from its maximum value at $T \simeq (\gamma/k \ln 2)$ to zero, as T is diminished, is quite rapid if λ is large.

When kT is considerably smaller than the binding energy γ all quantities are quite sensitive to an applied magnetic field, changing abruptly as $\mathfrak{B}$ rises through the value $(2\gamma/k\mathfrak{m}_B)$, where the magnetic field just overpowers the lattice binding, and the magnets rearrange themselves, from arrangement with respect to the lattice to alignment along the field. The entropy has a sudden change here, and thus the heat capacity has a peak. In fact when both $|\frac{1}{2}\mathfrak{m}_B\mathfrak{B}|$ and γ are larger than kT the heat capacity is approximately

$$C_{\mathfrak{H}} \simeq \lambda N_m k \, \frac{e^{\lambda(|y|-x)}}{1 + e^{\lambda(|y|-x)}}\left[4y^2 e^{-2|y|} + \lambda \frac{(|y| - x)^2}{1 + e^{\lambda(|y|-x)}} \right]$$

with $x = (\gamma/kT)$ and $y = (\mathfrak{m}_B\mathfrak{B}/2kT)$ as before. This has a sharp peak at $|y| = x$, or $|\frac{1}{2}\mathfrak{m}_B\mathfrak{B}| = \gamma$, the field just sufficient to decouple a group of magnets from the lattice. At this field strength (when $kT \ll \gamma$) the system of magnets is particularly sensitive to change, as is evidenced by the rapid rise in entropy and the consequent peak in heat capacity.

We thus see that the canonical ensemble is capable of "explaining" many of the thermal properties of matter in bulk. In the next chapter we shall see that it is capable of exposing a major misconception which classical statistics made regarding the degree of identity possessed by the fundamental particles, composing all matter, and of indicating the ways by which quantum statistics corrects the misconception.

CHAPTER

21

STATISTICAL
MECHANICS OF A GAS

We turn now to the low-density gas phase. A gas, filling volume V, is composed of N similar molecules which are far enough apart so the forces between molecules are small compared to the forces within a molecule. At first we assume that the intermolecular forces are negligible. This does not mean that the forces are completely nonexistent; there must be occasional collisions between molecules so that the gas can come to equilibrium. We do assume, however, that the collisions are rare enough so that the mean potential energy of interaction between molecules is negligible compared to the mean kinetic energy of the molecules.

Factoring the Partition Function

The total energy of the system will therefore be just the sum of the separate energies $\varepsilon(v_{\text{mole}})$ of the individual molecules, each one depending only on their own quantum numbers (which we can symbolize by v_{mole}) and the partition function can be split into N molecular factors, as explained in Eq. (19-8):

$$Z = (z_{\text{mole}})^N; \qquad z_{\text{mole}} = \sum_{v_{\text{mole}}} \exp[-\varepsilon(v_{\text{mole}})/kT]$$

[but see Eq. (21-12)].

In this case the partition function can be still further factored, for the energy of each molecule can be split into an energy of translation H_{tr} of the molecule as a whole, an energy of rotation H_{rot}, as a rigid body, an energy of vibration H_{vib} of the constituent atoms with respect to the molecular center of mass, and finally an energy of electronic motion H_{el}:

$$H_{mole} = H_{tr} + H_{rot} + H_{vib} + H_{el} \qquad (21\text{-}1)$$

To the first approximation these energy terms are independent; the coordinates that describe H_{tr}, for example, do not enter the functions H_{rot}, H_{vib}, or H_{el} unless we include the effect of collisions, and we have assumed this effect to be negligible. This independence is not strictly true for the effects of rotation, of course; the rotation does affect the molecular vibration and its electronic states to some extent. But the effects are usually small and can be neglected to begin with.

Consequently, each molecular partition function can be, approximately, split into four separate factors,

$$z_{mole} = z_{tr} \cdot z_{rot} \cdot z_{vib} \cdot z_{el}$$

and the partition function for the system can be divided correspondingly,

$$Z = Z_{tr} \cdot Z_{rot} \cdot Z_{vib} \cdot Z_{el}$$

where

$$Z_{tr} = (z_{tr})^N, \qquad Z_{rot} = (z_{rot})^N, \qquad \text{etc.} \qquad (21\text{-}2)$$

The individual molecular factors are sums of exponential terms, each corresponding to a possible state of the individual molecule, with quantized energies,

$$z_{tr} = \sum_{k,m,n} \exp(-\varepsilon_{kmn}^{tr}/kT); \qquad z_{rot} = \sum_{\lambda,v} g_{\lambda v} \exp(-\varepsilon_{\lambda v}^{rot}/kT) \qquad (21\text{-}3)$$

and so on, where k, m, n are the quantum numbers for the state of translational motion of the molecule, λ, v those for rotation, etc., and where $g_{\lambda v}$ are the multiplicities of the rotational states (the g's for the translational states are all 1, so they are not written).

The energy separation between successive translational states is very much smaller than the separation between successive rotational states, and these are usually much smaller than the separations between successive vibrational states of the molecule; the separations between electronic states are still another order of magnitude larger. To standardize the formulas, we shall choose the energy origin so the ε for the first state is zero; thus the first term in each z sum is unity.

The Translational Factor

Therefore there is a range of temperature within which several terms in the sum for z_{tr} are nonnegligible, but only the first term is nonnegligible for z_{rot}, z_{vib}, and z_{el}. In this range of temperature the total partition function for the gas system has the simple form

$$Z \simeq Z_{tr} = \left[\sum_{k,m,n} \exp(-\varepsilon_{kmn}^{tr}/kT) \right]^N \qquad (21\text{-}4)$$

all the other factors being practically equal to unity. To compute Z for this range of temperature we first compute the energies ε_{kmn} and then carry out the summation. From it we can calculate F, S, etc., for a gas of low density at low temperatures.

The Schrödinger Eq. (16-6) for the translational motion of a molecule of mass M is

$$(\hbar^2/2M)[(\partial^2\psi/\partial x^2) + (\partial^2\psi/\partial y^2) + (\partial^2\psi/\partial z^2)] = -\varepsilon^{tr}\psi$$

If the gas is in a rectangular box of dimensions l_x, l_y, l_z and volume $V = l_x l_y l_z$, with perfectly reflecting walls, the allowed wave functions and energies turn out to be

$$\psi_{kmn} = A \sin(\pi kx/l_x) \cdot \sin(\pi my/l_y) \cdot \sin(\pi nz/l_z)$$
$$\varepsilon_{kmn}^{tr} = (\pi^2\hbar^2/2M)[(k/l_x)^2 + (m/l_y)^2 + (n/l_z)^2] = p^2/2M \quad (21\text{-}5)$$

where p is the momentum of the molecule in state k, m, n. For a molecule of molecular weight 30 and for a box 1 cm on a side, $\pi^2\hbar^2/2Ml^2 \simeq 10^{-38}$ joule. Since $k \simeq 10^{-23}$ joule/°K, the spacing of the translational levels is very much smaller than kT even when $T = 1$°K, and we can safely change the sum for z_{tr} into an integral over dk, dm, and dn,

$$z_{tr} = \int_0^\infty \!\! \int \!\! \int \exp\left\{ -\frac{h^2}{8MkT}\left[\left(\frac{k}{l_x}\right)^2 + \left(\frac{m}{l_y}\right)^2 + \left(\frac{n}{l_z}\right)^2 \right] \right\} dk\,dm\,dn$$

$$= (V/h^3)(2\pi MkT)^{3/2} \qquad (21\text{-}6)$$

[but see Eq. (21-13)] by using Eqs. (12-6) (note that we have changed from $\hbar$ back to $h = 2\pi\hbar$).

This result has been obtained by summing over the quantized states. But with the levels so closely spaced we should not have difficulty in obtaining the same result by integrating over phase space. The translational Hamiltonian is $p^2/2M$ and the integral is

$$z_{\text{tr}} = \int\limits_{-\infty}^{\infty} \int \int \exp\left[\frac{-1}{2MkT}(p_x^2 + p_y^2 + p_z^2)\right]$$

$$\times h^{-3}\, dx\, dy\, dz\, dp_x\, dp_y\, dp_z$$

$$= (V/h^3)(2\pi MkT)^{3/2} \tag{21-7}$$

[but see Eq. (21-13)] as before. Integration in (21-7) goes just the same as in (21-6), except that we integrate over p_x, p_y, p_z from $-\infty$ to $+\infty$, whereas we integrated over k, m, n from 0 to ∞; the result is the same.

The probability f_{kmn} that a molecule has translational quantum numbers k, m, n is thus $(1/z_{\text{tr}})\exp(-\varepsilon_{kmn}^{\text{tr}}/kT)$ and the probability density that a molecule has translational momentum $\mathbf{p}$ and is located at $\mathbf{r}$ in V is

$$f(q, p) = (1/V)(2\pi MkT)^{-3/2}\exp(-p^2/2MkT)$$

which is the Maxwell distribution again. Also, in the range of temperature where only Z_{tr} changes appreciably with temperature, the Helmholtz function and the entropy of the gas are

$$F = -kT\ln(Z_{\text{tr}}) = -NkT[\ln(V) + \tfrac{3}{2}\ln(2\pi MkT/h^2)]$$
$$S = Nk\ln(V) + \tfrac{3}{2}Nk\ln(2\pi MkT/h^2) + \tfrac{3}{2}Nk \tag{21-8}$$

[but see Eq. (21-14)].

There is a major defect in this pair of formulas. Neither F nor S satisfies the requirement that it be an extensive variable, as illustrated with regard to U in the discussion preceding Eq. (6-5) [see also Eq. (8-22)]. Keeping intensive variables constant, increasing the amount of material in the system by a factor λ should increase all extensive variables by the same factor λ. If we increase N to λN in formulas (21-8), the temperature term will increase by the factor λ but the volume term will become $\lambda Nk\ln(\lambda V)$, which is *not* λ times $Nk\ln(V)$. The corresponding terms in Eqs. (8-22), giving the thermodynamic properties of a perfect gas, are $Nk\ln(V/V_0)$, and when N changes to λN, V goes to λV and also V_0 goes to λV_0, so that the term becomes $\lambda Nk\ln(\lambda V/\lambda V_0)$, which *is* just λ times $Nk\ln(V/V_0)$. Evidently the term $Nk\ln(V)$ in (21-8) should be

$Nk \ln(V/N)$, or something like it, and thus the partition function of (21-7) should have had an extra factor N^{-1}, or the partition function for the gas should have had an extra factor N^{-N} (or something like it). The trouble with the canonical ensemble for a gas seems to be in the way we set up the partition function.

If we remember Stirling's formula (18-5) we might guess that somehow we should have divided the Z of Eq. (21-1) by $N!$ to obtain the correct partition function for the gas. The resolution of this dilemma, which is another aspect of Gibbs' paradox, mentioned at the end of Chapter 6, lies in the degree of distinguishability of individual molecules.

The Indistinguishability of Molecules

Before the advent of quantum mechanics we somehow imagined that, in principle, we could distinguish one molecule from another—that we could paint one blue, for example, so we could always tell which one was the blue one. This is reflected in our counting of translational states of the gas, for we talked as though we could distinguish between the state, where molecule 1 has energy ε_1 and molecule 2 has energy ε_2, from the state, where molecule 1 has energy ε_2, and molecule 2 has energy ε_1, for example. But quantum mechanics has taught us that we cannot so distinguish between molecules; a state where molecule 1 has quantum numbers k_1, m_1, n_1, molecule 2 has k_2, m_2, n_2, and so on, not only has the same energy as the one where we have reshuffled the quantum numbers among the molecules, it is *really the same state*, and should only be counted once, not $N!$ times. We have learned that physical reality is represented by the wave function, and that the square of a wave function gives us the probability of presence of *a* molecule but does not specify *which* molecule is present. Different states correspond to different wave functions, not to different permutations of molecules.

At first sight the answer to this whole set of problems would seem to be to divide Z by $N!$. If particles are distinguishable, there are $N!$ different ways in which we can assign N molecules to N different quantum states. If the molecules are indistinguishable there is only one state instead of $N!$ ones. This is a good-enough answer for our present purposes. But the correct answer is not so simple as this, as we shall indicate briefly here and investigate in detail later. The difficulty is that, for many states of some systems, the N particles are not distributed among N *different* quantum states; sometimes several molecules occupy the same state.

To illustrate the problem, let us consider a system with five particles, each of which can be in quantum state 0 with zero energy or else in quantum state 1 with energy ε. The possible energy levels E_v of the system of five particles are, therefore,

$E_0 = 0$ all five particles in lower state
$E_1 = \varepsilon$ one particle in upper state, four in lower
$E_2 = 2\varepsilon$ two particles in upper state, three in lower
$E_3 = 3\varepsilon$ three particles in upper state, two in lower
$E_4 = 4\varepsilon$ four particles in upper state, one in lower
$E_5 = 5\varepsilon$ all five particles in upper state

(Note that we must distinguish between system states, with energies E_v, and particle states, with energies 0 and ε.) There is only one system state with energy E_0, no matter how we count states. All particles are in the lower particle state and there is no question of which particle is in which state. In this respect, a particle state is like the mathematician's urn, from which he draws balls; ordering of particles inside a single urn has no meaning; they are either in the urn or not

Distinguishability does come into the counting of the system states having energy E_1, however. If we can distinguish between particles we shall have to say that five different system states have energy E_1; one with particle 1 in the upper state and the others all in the lower "urn," another with particle 2 excited and 1, 3, 4, and 5 in the ground state, and so on. In other words the multiplicity W_1 of Eq. (19-7) is 5 for the system state $v = 1$. On the other hand, if we cannot distinguish between particles, there is only one state with energy E_1, the one with one particle excited and four in the lower state (and it has no meaning to ask which particle is excited; they all are at one time or other, but only one is excited at a time).

For distinguishable particles, a count of the different ways we can put five particles into two urns, two in one urn, and three in the other, will show that the appropriate multiplicity for energy E_2 is $W_2 = 10$. And so on; $W_3 = 10$, $W_4 = 5$, $W_5 = 1$. Therefore, for *distinguishable particles*, the partition function for this simple system would be

$$Z = \sum_{v=0}^{5} W_v e^{-v\varepsilon/kT} = 1 + 5x + 10x^2 + 10x^3 + 5x^4 + x^5$$

$$= (1 + x)^5 \qquad \text{where } x = e^{-\varepsilon/kT}$$

and where we have used the binomial theorem to take the last step. Thus such a partition function factors into single-particle factors

$z = 1 + e^{-\varepsilon/kT}$, as was assumed in Eqs. (19-8). See also the derivations of Eqs. (20-22) and (20-23).

On the other hand, if the particles are *indistinguishable*, all the multiplicities W are unity and

$$Z = 1 + x + x^2 + x^3 + x^4 + x^5$$

which does *not* factor into five single-particle factors.

Counting the System States

Generalizing, we can say that if we have N distinguishable particles, distributed among M different quantum states, n_j of them in particle state j,

with energy ε_j (so that $\displaystyle\sum_{j=1}^{M} n_j = N$),

then the number of different ways we can distribute these N particles among the M particle states,

so that the system energy $E_\nu = \displaystyle\sum_{j=1}^{M} n_j \varepsilon_j$, is

$$W_\nu = \frac{N!}{n_1! n_2! \cdots n_M!} \tag{21-9}$$

$N!$, the number of different ways all N particles can be permuted, being reduced by the numbers $n_j!$ of different ways the particles could be permuted in each of the M "urns," since permutation inside an urn does not count. The Z for distinguishable particles then is

$$Z_{\text{dist}} = \sum_\nu W_\nu \exp - \left(\sum_{j=1}^{M} n_j \varepsilon_j / kT \right)$$

$$= \sum_\nu \frac{N!}{n_1! n_2! \cdots n_M!} x_1^{n_1} x_2^{n_2} \cdots x_M^{n_M} = z^N \tag{21-10}$$

where $x_j = \exp(-\varepsilon_j/kT)$, $z = x_1 + x_2 + \cdots x_M$, where the sum is over all values of the n_j's for which

$$\sum_{j=1}^{M} n_j = N,$$

and where we have used the multinomial theorem of Eq. (11-12) to make the last step. Again this partition function factors into single-particle factors.

Again, if the particles are indistinguishable, the partition function is

$$Z_{\text{ind}} = \sum_v x_1^{n_1} x_2^{n_2} \cdots x_M^{n_M} \tag{21-11}$$

with the sum again over all values of the n_j's for which $\Sigma n_j = N$. This sum does not factor into single-particle factors.

We thus have reached a basic difficulty with the canonical ensemble. As long as we could consider the particles in the gas as distinguishable, our partition functions came out in a form that could be factored into N z's, one for each separate particle. As we have seen, this makes the calculations relatively simple. If we now have to use the canonical ensemble for indistinguishable particles, this factorability is no longer possible, and the calculations become much more difficult. In later chapters we shall find that a more general ensemble enables us to deal with indistinguishable particles nearly as easily as with distinguishable ones. But in this chapter we are investigating whether, under some circumstances, the partition function for the canonical ensemble can be modified so that indistinguishability can approximately be taken into account, still retaining the factorability we have found so useful. Can we divide Z_{dist} of Eq. (21-10) by some single factor so it is, at least approximately, equal to the Z_{ind} of Eq. (21-11)?

There are many terms in the sum of (21-10) which have multiplicity $W_v = N!$. These are the ones for the system states v, for which all the n_j's are 0 or 1, for which no particle state is occupied by more than one particle. We shall call these system states the *sparse* states, since the possible particle states are sparsely occupied. On the other hand, there are other terms in (21-10) with multiplicity less than $N!$. These are the terms for which one or more of the n_j's are larger than 1; some particle states are occupied by more than one of the particles. Such system states can be called *dense* states, for some particle states are densely occupied. If the number and magnitude of the terms for the sparse states in (21-10) are much larger than the number and magnitude of the terms for the dense states, then it will not be a bad approximation to say that all the W_v's in (21-10) are equal to $N!$ and thus that $(Z_{\text{dist}}/N!)$ does not differ much from the correct Z_{ind}. And $(Z_{\text{dist}}/N!)$ can still be factored, although Z_{ind} cannot.

To see when this advantageous situation will occur, we should examine the relative sizes of the terms in the sum of Eq. (21-10).

The term for which the factor $x_{1_1}^n \cdots x_M^{n_M}$ is largest is the one for which $n_1 = N$, $n_j = 0$ $(j > 1)$ (i.e., for which all particles are in the lowest state). This term has the value $1 \cdot \exp(-N\varepsilon_1/kT)$. It is one of the "densest" states. The largest term for a sparse state is the one for which $n_1 = n_2 = \cdots = n_N = 1$, $n_j = 0$ $(j > N)$ (i.e., for which one particle is in the lowest state, one in the next, and so on up to the Nth state). Its value is

$$(N!)\exp[-(\varepsilon_1 + \varepsilon_2 + \cdots + \varepsilon_N)/kT]$$

$$\simeq \sqrt{2\pi N}\exp\{N[\ln(N/e) - (\bar{\varepsilon}_N/kT)]\}$$

where we have used Stirling's formula (18-5) for $N!$ and we have written $\bar{\varepsilon}_N$ for the average energy $[(\varepsilon_1 + \varepsilon_2 + \cdots + \varepsilon_N)/N]$ of the first N particle states. Consequently, whenever $\ln(N/e)$ is considerably larger than $(\bar{\varepsilon}_N - \varepsilon_1)/kT$, the sum of sparse-state terms in Z_{dist} is so much larger than the sum of dense-state terms that Z_{dist} is practically equal to a sum of the sparse-state terms only, and in this case $Z_{\text{dist}} \simeq N!Z_{\text{ind}}$. This situation is the case when kT is considerably larger than the spacing between particle-state energy levels, which is the case when classical mechanics holds.

The Classical Correction Factor

Therefore whenever the individual particles in the system have energy levels sufficiently closely packed, compared to kT, so that classical phase-space integrals can be used for at least part of the z factor, it will be a good approximation to correct for the lack of distinguishability of the molecules by dividing Z_{dist} by $N!$. In this case there are enough low-lying levels so that each particle can occupy a different quantum state and our initial impulse, to divide Z by $N!$, the number of different ways in which we can assign N molecules to N different states, was a good one. Instead of Eq. (21-1) we can use the approximate formula

$$Z \simeq (1/N!)(z_{\text{mole}})^N \simeq (ez_{\text{mole}}/N)^N \tag{21-12}$$

[omitting the factor $\sqrt{(2\pi N)}$ in the second form because its logarithm is negligible compared to the logarithm of the rest].

Since the translational energy levels of a gas are so closely spaced, this method of correcting for the indistinguishability of the molecules should be valid for $T > 1\,°K$.

The correction factor can be included in the translational factor, so that, instead of Eqs. (21-4) to (21-8), we should use

$$Z_{tr} = (1/N!)V^N(2\pi MkT/h^2)^{(3/2)N} \simeq (eV/N)^N(2\pi MkT/h^2)^{(3/2)N}$$
$$= (eV/nl_T^3)^N \tag{21-13}$$

where $n = N/N_0$ is the number of moles and where the "thermal length" $l_T = hN_0^{1/3}/\sqrt{(2\pi MkT)}$ is equal to 1.47×10^{-2} meter for protons at $T = 1°K$ (for other molecules or temperatures divide by the square root of the molecular weight or of T). The values of the translational parts of the various thermodynamic quantities for the gas, corrected for molecular indistinguishability, are then

$$F_{tr} = -NkT[\ln(eV/N) + \tfrac{3}{2}\ln(2\pi MkT/h^2)]$$
$$S_{tr} = Nk[\ln(V/N) + \tfrac{3}{2}\ln(2\pi MkT/h^2)] + \tfrac{5}{2}Nk$$
$$U = \tfrac{3}{2}NkT; \qquad C_v = \tfrac{3}{2}Nk; \qquad P = (NkT/V)$$
$$H = \tfrac{5}{2}NkT; \qquad C_p = \tfrac{5}{2}Nk \tag{21-14}$$

The equation for S is called the *Sackur-Tetrode formula*.

Comparison with Eqs. (8-21) shows that statistical mechanics has indeed predicted the thermodynamic properties of a perfect gas. It has done more, however; it has given the value of the constants of integration S_0, T_0, and V_0 in terms of the atomic constants h, M, and k, and it has indicated the conditions under which a collection of N molecules can behave like a perfect gas of point particles.

For a gas mixture of N_1 molecules of one kind and N_2 molecules of another kind, the molecules of one kind cannot be distinguished from their fellows, but can be distinguished from those of the other kind. Therefore, for this system

$$Z_{tr} = (eV/N_1)^{N_1}(eV/N_2)^{N_2}(2\pi M_1 kT/h^2)^{(3/2)N_1}(2\pi M_2 kT/h^2)^{(3/2)N_2}$$
$$F_{tr} = -N_1kT\ln(eV/N_1) - N_2kT\ln(eV/N_2)$$
$$\qquad - \tfrac{3}{2}NkT\ln(2\pi MkT/h^2) \tag{21-15}$$
$$S_{tr} = N_1k\ln(V/N_1) + N_2k\ln(V/N_2)$$
$$\qquad + \tfrac{3}{2}Nk\ln(2\pi MkT/h^2) + \tfrac{5}{2}Nk$$

where $N = N_1 + N_2$, and $M = (M_1^{N_1} \cdot M_2^{N_2})^{1/N}$. From these equations we can obtain Eq. (6-15) for the entropy of mixing of two *different* gases. When two quantities of the same gas are mixed there is no change in entropy. Thus we find that we now can solve Gibbs' paradox, started at the end of Chapter 6. Mixing two different gases does change the entropy by the amount given in Eq. (6-15). But mixing together two portions of the same gas produces no change in entropy. If the molecules on both sides of the diaphragm are identical, there is really no increase in disorder after the diaphragm is removed. One can never tell (in fact one must never even ask) from which side of the diaphragm a given molecule came, so one cannot say that the two collections of identical molecules "intermixed" after the diaphragm was removed.

We also note that division by $N!$ was not required for the crystal discussed in Chapter 20. In a manner of speaking, $N!$ was already divided out. We never tried to include, in our count, the number of ways the N atoms could be assigned to the different lattice points, and so we did not have to divide out the number again. More will be said about this in Chapter 27.

The Effects of Molecular Interaction

We have shown several times [see Eqs. (17-9), (18-19), and (21-14)] that, when the interaction between separate molecules in a gas is neglected completely, the resulting equation of state is that of a perfect gas. Before we finish discussing the translational partition function for a gas, we should show how the effects of molecular interaction can be taken into account. We shall confine our discussion to modifications of the translational terms, since these are the most affected. Molecular interactions do change the rotational, vibrational, and electronic motions of each molecule, but the effects are smaller.

The first effect of molecular interactions is to destroy the factorability of the translational partition function, at least partly. The translational energy, instead of being solely dependent on the molecular momenta, now has a potential energy term, dependent on the relative positions of the various molecules. This is a sum of terms, one for each pair of molecules. The force of interaction between molecule i and molecule j, to the first approximation, depends only on the distance r_{ij} between their centers of mass. It is zero when r_{ij} is large; as the molecules come closer together than their average distance the force is first weakly attractive until, at r_{ij} equal to twice the "radius" r_0 of each molecule, they "collide" and their closer approach is prevented by a strong repulsive force. Thus the potential energy $W_{ij}(r_{ij})$ of interaction between molecule i and molecule j has the form shown in Figure 21-1, with a small

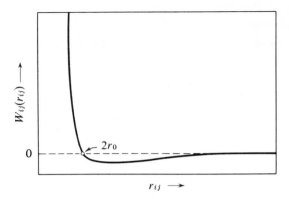

FIGURE 21-1

Potential energy of interaction between two molecules as a function of their distance apart.

positive slope (attractive force) for $r_{ij} > 2r_0$ and a large negative slope (repulsive force) for $r_{ij} < 2r_0$. By the time r_{ij} is as large as the average distance between molecules in the gas, W_{ij} is zero; in other words we still are assuming that the majority of the time the molecules do not affect each other.

The translational part of the Hamiltonian of the system is

$$H_{tr} = \tfrac{1}{2}M \sum_{i=1}^{N} p_i^2 + \sum_{\text{all pairs}} W_{ij}(r_{ij})$$

where the sum of the W_{ij}'s is over all the $\tfrac{1}{2}N(N-1) \simeq \tfrac{1}{2}N^2$ pairs of molecules in the gas. The translational partition function is then

$$Z_{tr} = \int \cdots \int e^{-H_{tr}/kT}\, dx_1\, dx_2\, \cdots\, dz_N\, dp_{x1}\, dp_{y1} \cdots dp_{zN}/h^{3N}$$

The integration over the momentum coordinates can be carried through as with Eq. (21-7) and, since the molecules are indistinguishable, we divide the result by $N!$. However, the integration over the position coordinates is not just V^N this time, because of the presence of the W_{ij}'s, so that $Z_{tr} = Z_p \cdot Z_q$

$$Z_p \simeq \frac{1}{N!}\left(\frac{2\pi MkT}{h^2}\right)^{(3/2)N} \simeq \left(\frac{e}{N}\right)^N \left(\frac{2\pi MkT}{h^2}\right)^{(3/2)N} \tag{21-16}$$

$$Z_q = \int \cdots \int \exp\left[-\sum_{\text{all pairs}} W_{ij}(r_{ij})/kT\right] dx_1\, dy_1 \cdots dy_N\, dz_N$$

Let us look at the behavior of the integrand for Z_q, as a function
of the coordinates x_1, y_1, z_1 of one molecule. The range of integration is over the volume V of the container. Over the great majority
of this volume the molecule will be far enough away from all other
molecules so that ΣW_{ij} is 0 and the exponential is 1; only when
molecule 1 comes close to another molecule (the jth one, say) does
r_{1j} become small enough for W_{1j} to differ appreciably from zero.
Of course if r_{1j} becomes smaller than $2r_0$, W_{1j} becomes very large
positive and the integrand for Z_q will vanish. The chance that two
molecules get closer together than $2r_0$ is quite small.

Thus it is useful to add and subtract 1 from the integrand,

$$Z_q = \int \cdots \int \{1 + [\exp(-\sum W_{ij}/kT) - 1]\}\, dx_1 \cdots dz_N$$

$$= V^N + \int \cdots \int [\exp(-\sum W_{ij}/kT) - 1]\, dx_1 \cdots dz_N \qquad (21\text{-}17)$$

where the first unity in the braces can be integrated as in Eq. (21-7)
and the second term is a correction to the perfect gas partition
function, to take molecular interaction approximately into account.
As we have just been showing, over most of the range of the position
coordinates the integrand of this correction term is zero. Only when
one of the r_{ij}'s is relatively small is any of the W_{ij}'s different from
zero. To the first approximation, we can assume that only one W_{ij}
differs from zero at a time, as the integration takes place.

Thus the integral becomes a sum of similar integrals, one for
each of the $\frac{1}{2}N(N-1) \simeq \frac{1}{2}N^2$ interaction terms W_{ij}. A typical one
is the integral for which W_{1j} is not zero; for this one the integrand
differs from zero only when (x_1, y_1, z_1) is near (x_j, y_j, z_j), so in the
integration over $dx_1\, dy_1\, dz_1 = dV_1$ we could use the relative
coordinates $r_{1j}, \theta_{1j}, \phi_{1j}$. Once this integral is carried out, the
integrand for the rest of the integrations is constant, so each of the
integrals over the other dV_i's is equal to V. Thus

$$Z_q = V^N + \tfrac{1}{2}N^2 \int_0^{2\pi} d\phi_{1j} \int_0^{\pi} \sin\theta_{1j}\, d\theta_{1j} \int_0^{\infty} (e^{-W_{1j}/kT} - 1)r_{1j}^2\, dr_{1j}$$

$$\times \int \cdots \int dV_2 \cdots dV_N$$

$$= V^N + \tfrac{1}{2}N^2 V^{N-1}\left[4\pi \int_0^{\infty} (e^{-W_{1j}/kT} - 1)r_{1j}^2\, dr_{1j}\right]$$

When $r_{1j} < 2r_0$, W_{1j} becomes very large positive and the integrand of the last term becomes -1, so this part of the quantity in brackets is just minus the volume of a sphere of radius $2r_0$, which we shall call -2β. For $r_{1j} > 2r_0$, W_{1j} is small and negative, so $(e^{-W_{1j}/kT} - 1) \simeq -(W_{1j}/kT)$, and this part of the quantity in brackets is roughly

$$-4\pi \int_{2r_0}^{\infty} (W_{1j}/kT)r_{1j}^2 \, dr_{1j}$$

which we shall call $2\alpha/kT$.

The van der Waals Equation of State

Therefore, to the first approximation, molecular interaction changes Z_{tr} from the simple expression of Eq. (21-13) to

$$Z_{tr} \simeq \left(\frac{eV}{N}\right)^N \left(\frac{2\pi MkT}{h^2}\right)^{(3/2)N}\left[1 - \left(\frac{N^2\beta}{V}\right) + \left(\frac{N^2\alpha}{VkT}\right)\right] \quad (21\text{-}18)$$

where $N\beta = N(8\pi r_0^3/3)$ is proportional to the total part of the volume V which is made unavailable to a molecule because of the presence of the other molecules, and where α is a measure of the attractive potential surrounding each molecule. The β and α terms in the bracket are both small compared to 1.

The Helmholtz function and the entropy for this partition function are

$$F_{tr} \simeq -\tfrac{3}{2}NkT \ln\left(\frac{2\pi MkT}{h^2}\right) - NkT \ln\left(\frac{eV}{N}\right)$$
$$- kT \ln\left(1 - \frac{N^2\beta}{V} + \frac{N^2\alpha}{VkT}\right)$$
$$\simeq -\tfrac{3}{2}NkT \ln\left(\frac{2\pi MkT}{h^2}\right) - NkT \ln\left(\frac{eV}{N}\right)$$
$$+ NkT\left(\frac{N\beta}{V}\right) - \left(\frac{N^2\alpha}{V}\right) \quad (21\text{-}19)$$
$$\simeq -\tfrac{3}{2}NkT \ln\left(\frac{2\pi MkT}{h^2}\right)$$
$$- NkT \ln\left[\frac{eV}{N}\left(1 - \frac{N\beta}{V}\right)\right] - \left(\frac{N^2\alpha}{V}\right)$$
$$S_{tr} \simeq \tfrac{3}{2}Nk + \tfrac{3}{2}Nk \ln\left(\frac{2\pi MkT}{h^2}\right) + Nk \ln\left[\frac{e}{N}(V - N\beta)\right]$$

Comparison with Eqs. (21-14) shows that U and C_v are unchanged, to this approximation, by the introduction of molecular interaction. However the equation of state becomes

$$P = -\left(\frac{\partial F}{\partial V}\right)_T \simeq \frac{NkT}{V - N\beta} - \frac{N^2\alpha}{V^2}$$

or

$$\left(P + \frac{N^2\alpha}{V^2}\right)(V - N\beta) \simeq NkT = nRT \qquad (21\text{-}20)$$

which is the van der Waals equation of state of Eq. (3-4), with $a = N_0^2\alpha$ and $b = N_0\beta$. The correction $N^2\alpha/V^2$ to P (which tends to decrease P for a given V and T) is caused by the small mutual attractions between molecules; the correction $N\beta$ to V (which tends to increase P for a given V and T) is the volume excluded by the presence of the other molecules. Thus measurement of a and b from the empirical equation of state can give us clues to molecular sizes and attractive forces; or else computation of the forces between like molecules can enable us to predict the van der Waals equation of state that a gas of these molecules should obey.

Integration of Eq. (21-16) to higher orders of approximation, taking into account regions where three or more particles are close together, may be worked out. The results fall naturally into the form of the virial equation of Eq. (3-5). Comparison between computed and measured values of the virial coefficients enable one to evaluate molecular interactions in some detail.

CHAPTER

22

A GAS OF DIATOMIC MOLECULES

In the molecular gas described in the preceding chapter, as long as kT is small compared to the energy spacing of rotational quantum levels of individual molecules, only Z_{tr} differs appreciably from unity and the gas behaves like a perfect gas of point atoms (if we neglect molecular interactions). To see for what temperature range this holds, we need to know the expression for the allowed energies of free rotation of a molecule. This expression is quite complicated for polyatomic molecules, so we shall go into detail only for diatomic molecules.

The Rotational Factor

If the two constituent nuclei have masses M_1 and M_2 and if they are held a distance R_0 apart at equilibrium, the moment of inertia of the molecule, for rotation about an axis perpendicular to R_0 through the center of mass, is $I = [M_1 M_2 R_0^2/(M_1 + M_2)]$. The moment of inertia about the R_0 axis is zero. The kinetic energy of rotation is then $1/2I$ times the square of the total angular momentum of the molecule.

This angular momentum is quantized, of course, the allowed values of its square being $\hbar^2 l(l + 1)$, where l is the rotational quantum number, and the allowed values of the component along some fixed direction in space are one of the $(2l + 1)$ values $-l\hbar$,

$-(l-1)\hbar,..., +(l-1)\hbar, +l\hbar$, for each value of l. Put another way, there are $2l + 1$ different rotational states which have the energy $(\hbar^2/2I)l(l+1)$, so the partition function for the rotational states of the gas system of N molecules is

$$Z_{\text{rot}} = \left\{ \sum_{l=0}^{\infty} (2l+1) \exp[-\theta_{\text{rot}}l(l+1)/T] \right\}^N \tag{22-1}$$

where $\theta_{\text{rot}} = \hbar^2/2Ik$. Therefore when T is very small compared to θ_{rot}, $Z_{\text{rot}} \simeq 1$ and, according to the discussion following Eq. (19-11), the rotational entropy and specific heat are negligible.

Values of θ_{rot} for a few diatomic molecules will indicate at what temperatures Z_{rot} begins to be important. For H_2, $\theta_{\text{rot}} = 85°K$; for HD, $\theta_{\text{rot}} = 64°K$; for D_2, $\theta_{\text{rot}} = 43°K$; for HCl, $\theta_{\text{rot}} = 15°K$; and for O_2, $\theta_{\text{rot}} = 2°K$. Therefore, except for protium (hydrogen), protium deuteride, and deuterium gases, T is appreciably larger than θ_{rot} in the temperature range where the system is a gas.

In these higher ranges of temperature we can change the sum for Z into an integral.

$$Z_{\text{rot}} = (z_{\text{rot}})^N$$

$$z_{\text{rot}} \simeq \int_0^{\infty} (2l+1) \exp[-\theta_{\text{rot}}(l^2+l)/T] \, dl = T/\theta_{\text{rot}}$$

so

$$Z_{\text{rot}} \simeq (T/\theta_{\text{rot}})^N = (8\pi^2 IkT/h^2)^N$$
$$F_{\text{rot}} \simeq -NkT \ln(T/\theta_{\text{rot}}); \qquad S_{\text{rot}} \simeq Nk \ln(eT/\theta_{\text{rot}})$$
$$U_{\text{rot}} \simeq NkT; \qquad C_v^{\text{rot}} \simeq Nk, \qquad T \gg \theta_{\text{rot}} \tag{22-2}$$

Thus for a gas of diatomic molecules at moderate temperatures, where both translational and rotational partition functions have their classical values, the total internal energy is $\frac{5}{2}NkT$ and the total heat capacity is $\frac{5}{2}Nk$, as mentioned in the discussion following Eq. (13-11). The rotational terms add nothing to the equation of state, however, for the effect of the neighboring molecules on a molecule's rotational states is negligible for a gas of moderate or low densities; consequently Z_{rot} and F_{rot} are independent of V. Therefore the equation of state is determined entirely by Z_{tr}, unless the gas density is so great that not even the van der Waals equation of state is valid.

For hydrogen and deuterium, a more careful evaluation of

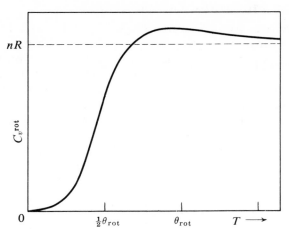

FIGURE 22-1

The rotational part of the heat capacity of a diatomic gas as a function of temperature.

Eq. (22-1) results in

$$
Z_{\rm rot} \to
\begin{cases}
(1 + 3e^{-\theta_{\rm rot}/T})^N, & T < \theta_{\rm rot} \\[2mm]
e^{N\theta_{\rm rot}/4T}\left(\dfrac{T}{\theta_{\rm rot}} + \dfrac{1}{12} + \dfrac{7\theta_{\rm rot}}{480T}\right)^N, & T > \theta_{\rm rot}
\end{cases}
\tag{22-3}
$$

A plot of the exact value of $C_v^{\rm rot}/Nk$, plotted against $T/\theta_{\rm rot}$ is shown in Figure 22-1. We note that $C_v^{\rm rot}$ rises somewhat above $Nk = nR$, as T increases, before it settles down to its classical value. The measured values of $C_v^{\rm rot}$ for HD fit this curve very well, from $T = 35°{\rm K}$ to several hundred degrees K, when molecular vibration begins to make itself felt. But the $C_v^{\rm rot}$ curves (i.e., $C_v - C_v^{\rm tr}$) for H_2 and D_2 do not match, no matter how one juggles the assumed values of $\theta_{\rm rot}$; for example, the curve for H_2 has no range of T for which $C_v^{\rm rot} > Nk$, and the peak for D_2 is not as large as Figure 22-1 would predict. The explanation of this anomaly lies again with the effects of indistinguishability of particles. The hydrogen and deuterium homonuclear molecules, H_2 and D_2, are the only ones with a low-enough boiling point so that these effects can be measured. The effects would not be expected for HD, for here the two nuclei in the molecule differ and are thus distinguishable. The calculations for H_2 and D_2 will be discussed later, in Chapter 27, after we take up in detail the effects of indistinguishability.

The Gas at Moderate Temperatures

Therefore, for all gases except H_2, HD, and D_2, over the temperature range from the boiling point of the gas to the temperature θ_{vib}, where vibrational effects begin to be noticeable, the only effective factors in the partition function are those for translation and rotation, and these factors can be computed classically, using Eqs. (21-14) and (22-2). In this range we can also calculate the partition function for polyatomic molecules. The classical Hamiltonian is $(p_1^2/2I_1) + (p_2^2/2I_2) + (p_3^2/2I_3)$, where I_1, I_2, I_3 are the moments of inertia of the molecule about its three principal axes and p_1, p_2, p_3 are the corresponding angular momenta. Therefore,

$$Z_{rot} \simeq \left\{ (8\pi^2/h^3) \int\int\int_{-\infty}^{\infty} \exp\left(\frac{-[(p_1^2/I_1) + (p_2^2/I_2) + (p_3^2/I_3)]}{2kT} \right) \times dp_1\, dp_2\, dp_3 \right\}^N$$

$$= [(8\pi^2/\sigma h^3) \sqrt{I_1 I_2 I_3}\, (2\pi kT)^{3/2}]^N \tag{22-4}$$

where $8\pi^2$ is the factor produced by the integration over the angles conjugate to p_1, p_2, p_3 and where σ is a symmetry factor, which enters when two or more indistinguishable nuclei are present in a molecule. This will be discussed further in Chapter 27.

We can now write the thermodynamic functions for a gas for which molecular interactions are negligible, for the temperature range where kT is large compared with rotational-energy-level differences but small compared with the vibrational-energy spacing. For monatomic gases, there is no Z_{rot} and, from Eq. (21-14),

$$F \simeq F_{tr} \simeq -NkT[\ln(V/N) + \tfrac{3}{2}\ln T + F_0]$$
$$U \simeq \tfrac{3}{2}NkT; \qquad C_v \simeq \tfrac{3}{2}Nk; \qquad P \simeq NkT/V$$

For diatomic gases, use Eq. (22-2) for Z_{rot}, and

$$F \simeq F_{tr} + F_{rot} \simeq -NkT[\ln(V/N) + \tfrac{5}{2}\ln T + F_0]$$
$$U \simeq \tfrac{5}{2}NkT; \qquad C_v \simeq \tfrac{5}{2}Nk; \qquad P \simeq NkT/V \tag{22-5}$$

For polyatomic gases, use Eq. (22-4) for Z_{rot}, and

$$F \simeq F_{tr} + F_{rot} \simeq -NkT[\ln(V/N) + 3\ln T + F_0]$$
$$U \simeq 3NkT; \qquad C_v \simeq 3Nk; \qquad P \simeq NkT/V$$

where the constant F_0 is a logarithmic function of k, h, the mass M of the molecule, and of its moments of inertia, the value of which can be computed from Eqs. (21-14) and (22-2) or (22-4). All these formulas are for perfect gases, in that the equation of state is $PV = NkT$ and the internal energy U is a function of T only. The specific heats depend on the nature of the molecule, whether it is monatomic, diatomic, or polyatomic.

We note that the result corresponds to the classical equipartition of energy for translational and rotational motion, U being $\frac{1}{2}kT$ times the number of "unfrozen" degrees of translational and rotational freedom. The effects of molecular interaction can be allowed for approximately by adding the factor in brackets in Eq. (21-18) to Z. These results check quite well with the experimental measurements, mentioned following Eq. (13-11).

The Vibrational Factor

When the temperature is high enough so that kT begins to equal the spacing between vibrational levels of the molecules, then Z_{vib} begins to depend on T and the vibrational degrees of freedom begin to "thaw out." In diatomic molecules there is just one such

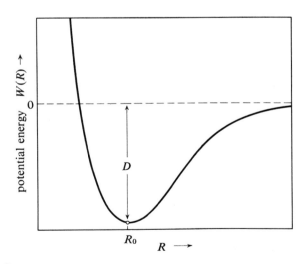

FIGURE **22-2**

Diatomic molecular energy $W(R)$ as a function of the separation R between nuclei.

degree of freedom, the distance R between the nuclei. The corresponding potential energy $W(R)$ has its minimum value at R_0, the equilibrium separation between the nuclei, and has a shape roughly like that shown in Figure 22-2. As $R \to \infty$, the molecule dissociates into separate atoms; the energy required to dissociate a molecule from equilibrium is D, the dissociation energy.

If the molecule is rotating there will be added a dynamic potential, corresponding to the centrifugal force, which is proportional to the square of the molecule's angular momentum and inversely proportional to R^3. Fortunately, for most diatomic molecules, this term, which would couple Z_{rot} and Z_{vib}, is small enough so we can neglect it here. For small-amplitude vibrations about R_0 the system acts like a harmonic oscillator, with a natural frequency $\omega/2\pi$ which is a function of the nuclear masses and of the curvature of the $W(R)$ curve near R_0. Thus the lower energy levels are $\hbar\omega(n + \frac{1}{2})$, where n is the vibrational quantum number.

Therefore, to the degree of approximation which neglects coupling between rotation and vibration and which considers all the vibrational levels to be those of a harmonic oscillator,

$$Z_{\text{vib}} \simeq \left(e^{-\varepsilon_0/kT} \sum_{n=0}^{\infty} e^{-n\hbar\omega/kT} \right)^N$$
$$= \left[e^{+\varepsilon_0/kT}(1 - e^{-\hbar\omega/kT}) \right]^{-N} \tag{22-6}$$

where $\varepsilon_0 = W(R_0) + \frac{1}{2}\hbar\omega$, and where we have used Eq. (20-2) to reduce the sum. The corresponding contributions to the Helmholtz function, entropy, etc., of the gas are

$$F_{\text{vib}} \simeq N\varepsilon_0 + NkT \ln(1 - e^{-\hbar\omega/kT})$$

$$U_{\text{vib}} \simeq N\varepsilon_0 + \frac{N\hbar\omega}{e^{+\hbar\omega/kT} - 1} \to \begin{cases} N\varepsilon_0 + N\hbar\omega e^{-\hbar\omega/kT}; & kT \ll \hbar\omega \\ N\varepsilon_0 + NkT; & kT \gg \hbar\omega \end{cases}$$

$$C_v^{\text{vib}} \simeq \frac{N\hbar^2\omega^2}{kT^2} \frac{e^{\hbar\omega/kT}}{(e^{\hbar\omega/kT} - 1)^2}$$
$$\to \begin{cases} (N\hbar^2\omega^2/kT^2)e^{-\hbar\omega/kT}; & kT \ll \hbar\omega \\ Nk; & kT \gg \hbar\omega \end{cases} \tag{22-7}$$

which are added to the functions of Eqs. (22-5) whenever the temperature is high enough (for T equal to or larger than $\theta_{\text{vib}} = \hbar\omega/k$).

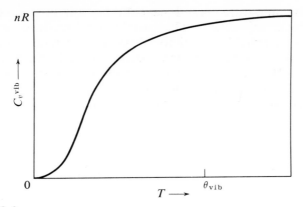

FIGURE 22-3

Vibrational part of the heat capacity of a diatomic molecular gas as a function of temperature.

As examples of the limits above which these terms become appreciable, the quantity θ_{vib} is equal to 2200°K for O_2, to 4100°K for HCl, and to 6100°K for H_2 (Figure 22-3). Therefore below roughly 1000°K the contribution of molecular vibration to S, U, and C_v of diatomic gases is small. Above several thousand degrees, the vibrational degree of freedom becomes "unfrozen," an additional energy kT is added per molecule, and an additional Nk to C_v [a degree of freedom with quadratic potential classically has energy kT; see Eq. (13-14)].

In the case of a polyatomic molecule with n nuclei, there are $3n - 6$ vibrational degrees of freedom, each with its fundamental frequency $\omega_j/2\pi$. The vibrational partition function is

$$Z_{vib} = e^{-N\varepsilon_0} Z_1 Z_2 \cdots Z_{3n-6}$$

where

$$Z_j \simeq (1 - e^{-\hbar\omega_j/kT})^{-N} \tag{22-8}$$

[compare this with Eq. (20-7)]. Again, for polyatomic gases, the vibrational contribution below about 1000°K is small, at higher T the ·contribution to U is $N(3n - 6)kT$. It often happens that the molecules are dissociated into their constituent atoms before the temperature is high enough for very many vibrations to "unfreeze."

The temperature would have to be still higher before Z_{el} began to have any effect. The usual electronic-level separation divided by k is roughly equal to $10,000\,°K$ at which temperatures most gases are dissociated and partly ionized. Such cases are important in the study of stellar interiors, but are too complex to discuss in this book. And before we can discuss the thermal properties of electrons we must return to first principles again.

CHAPTER

23

THE GRAND
CANONICAL ENSEMBLE

The canonical ensemble, representing a system of N particles kept at constant temperature T, has proved to be a useful model for such systems as the simple crystal and the perfect (or nearly perfect) gas. Many other systems, more complicated than these, can also be represented by the canonical ensemble, which makes it possible to express their thermodynamic properties in terms of their atomic structure. But in Chapter 21 we discovered a major defect, not in the accuracy of the canonical ensemble when correctly applied, but in its ease of manipulation in some important cases.

Whenever the N particles making up the system are identical and indistinguishable, the corresponding change in the multiplicity factors W_v has the result that the correct partition function does not separate into a product of N independent factors, even if the interaction between particles is negligible. In cases where kT is large compared to the separation between quantum levels of the system, we found we could take this effect into account approximately by dividing by $N!$. In this chapter we shall discuss a more general kind of ensemble, which will allow us to retain factorability of partition function and at the same time take indistinguishability into account exactly, no matter what value T has.

We have already found out how to devise an ensemble that will correspond to a specified thermodynamic situation. The canonical ensemble uses the variables T, V, and N as independent variables;

its partition function Z leads to the system's Helmholtz function $F = -kT \ln Z$, which is minimum at equilibrium when T, V, and N are held fixed. In Eq. (20-24) we used another ensemble, for which T, $\mathfrak{H}$, and N are the independent variables; the resulting partition function Z_m is related to the magnetic Gibbs function $G_m = -kT \ln Z_m$, which is minimal at equilibrium when T, $\mathfrak{H}$, and N are held fixed.

In the discussion of systems of indistinguishable particles it is inconvenient to use N, the total number of particles, as an independent variable. In Eq. (21-11), for example, the requirement that $\Sigma n_j = N$ makes Z_{ind} hard to calculate. We should change from N to μ, the chemical potential per particle, as the independent variable, and thus change from $F = U - TS$ to the grand potential $\Omega = U - TS - \mu N = -PV$ of Eq. (8-15), which is minimal at equilibrium when T, V, and μ are held fixed.

An Ensemble with Variable N

The new ensemble, which we shall call the *grand canonical ensemble*, is one in which we relax the requirement that we placed on the microcanonical and canonical ensembles—that each system in the ensemble has exactly N particles. We can imagine an infinitely large, homogeneous supersystem kept at constant T and P. The system the new ensemble will represent is that part of the supersystem contained within a volume V. We can imagine obtaining one of the sample systems of the ensemble by withdrawing that part of the supersystem that happens to be in a volume V at the instant of removal, and of doing this successively to obtain all the samples that make up the ensemble. Not only will each of the samples differ somewhat in regard to their total energy, but the number of particles N in each sample will differ from sample to sample. Only the average energy U and the average number of particles $\bar{N}$, averaged over the ensemble, will be specified.

The equations and subsidiary conditions serving to determine the distribution function are thus still more relaxed than for the canonical ensemble. A microstate of the grand canonical ensemble is specified by the number of particles N that the sample system has, and by the quantum numbers $v_N \equiv v_1, v_2,..., v_{3N}$, which the sample may have and which will specify its energy E_{Nv}. Thus for an equilibrium macrostate the distribution function f_{Nv} must satisfy the following requirements:

$$S = -k \sum_{N,v} f_{Nv} \ln f_{Nv} \text{ is maximum}$$

subject to

$$\sum_{N,v} f_{Nv} = 1; \qquad \sum_{N,v} E_{Nv} f_{Nv} = U; \qquad \sum_{N,v} N f_{Nv} = \bar{N}$$

$$(23\text{-}1)$$

where U, $\bar{N}$, and S are related by the usual thermodynamic relationships, such as $U = TS + \Omega + \bar{N}\mu$, for example, or any other of the Eqs. (8-21). Function Ω is the grand potential of Eq. (8-15). Note that, instead of n, the mean number of moles in the system, we now use $\bar{N}$, the mean number of particles, and therefore μ is now the chemical potential (the Gibbs function) *per particle*, rather than per mole, as it was in the first third of this book. We shall consistently use it thus henceforth, so it should not be confusing to use the same symbol, μ.

The Grand Partition Function

As before, we simplify the requirements by using Lagrange multipliers, and require that

$$-k \sum_{N,v} f_{Nv} \ln f_{Nv} + \alpha_1 \sum_{N,v} f_{Nv} + \alpha_e \sum_{N,v} E_{Nv} f_{Nv}$$
$$+ \alpha_n \sum_{N,v} N f_{Nv} \text{ be maximum}$$

with α_1, α_e, α_n chosen so that

$$\sum_{N,v} f_{Nv} = 1; \qquad \sum_{N,v} E_{Nv} f_{Nv} = U; \qquad \sum_{N,v} N f_{Nv} = \bar{N} \qquad (23\text{-}2)$$

The partials with respect to the f_{Nv}'s, which must be made zero, result in the equations

$$k \ln f_{Nv} + k = \alpha_1 + \alpha_n N + \alpha_e E_{Nv} \qquad (23\text{-}3)$$

To relate the Lagrange multipliers to the appropriate thermodynamic functions, as before, we multiply this equation by f_{Nv} and sum over v and N, relating the resulting sums to S, U, and $\bar{N}$ in accord with Eqs. (17-1) and (23-2),

$$-S = (\alpha_1 - k) + \alpha_e U + \alpha_n \bar{N}$$

which can be made to correspond to $-S = (\Omega - U + \mu\bar{N})/T$ if we set

$$\alpha_1 - k = (\Omega/T); \qquad \alpha_e = -(1/T); \qquad \alpha_n = (\mu/T)$$

Also the normalization condition $\Sigma f_{N\nu} = 1$ can be met by solving Eq. (23-3) for $f_{N\nu}$

$$f_{N\nu} = \exp[(1/k)(\alpha_1 - k + \alpha_n N + \alpha_e E_{N\nu})]$$
$$= \exp\left[\frac{\Omega + \mu N - E_{N\nu}}{kT}\right]$$

and then requiring that

$$1 = e^{\Omega/kT} \sum_{N\nu} e^{(\mu N - E_{N\nu})/kT} \quad \text{or} \quad e^{-\Omega/kT} = 3 = \sum_{N\nu} e^{(\mu N - E_{N\nu})/kT}$$

Thus the solution of Eq. (23-1) is

$$f_{N\nu} = (1/3) \exp\left(\frac{\mu N - E_{N\nu}}{kT}\right); \qquad 3 = \sum_{N,\nu} \exp\left(\frac{\mu N - E_{N\nu}}{kT}\right)$$

$$\Omega = -kT \ln 3 = -PV; \qquad (\partial\Omega/\partial\mu)_{TV} = -\bar{N}$$

$$(\partial\Omega/\partial T)_{V\mu} = -S; \qquad (\partial\Omega/\partial V)_{T\mu} = -P \qquad (23\text{-}4)$$

$$C_v = T(\partial S/\partial T)_{V\mu}; \qquad F = \Omega + \mu\bar{N};$$

$$U = F + ST = \Omega + ST + \mu\bar{N}$$

These are the equations for the grand canonical ensemble. The sum 3 is called the grand partition function; it is the sum of the canonical partition functions $Z(N)$ for ensembles with different N's, with weighting factors $e^{\mu N/kT}$,

$$3 = \sum_{N=0}^{\infty} e^{\mu N/kT} Z(N); \qquad Z(N) = \sum_{\nu} e^{-E_{N\nu}/kT} \qquad (23\text{-}5)$$

All the thermodynamic properties of the system can be obtained from Ω by differentiation, as with the canonical ensemble. We shall see that this partition function has even greater possibilities for factoring than does its canonical counterpart.

The Perfect Gas Once More

Just to show how this ensemble works we take up again the familiar theme of the perfect gas of point particles. From Eq. (21-13) we see that, if we take particle indistinguishability approximately into account, the canonical partition function for the gas of N particles is

$$Z(N) \simeq (1/N!)(V/l_t^3)^N; \qquad l_t = (h/\sqrt{2\pi MkT})$$

and therefore the grand partition function is, from Eq. (23-5),

$$3 = \sum_{N=0}^{\infty} (1/N!)[(V/l_t^3)e^{\mu/kT}]^N = \exp[(V/l_t^3)e^{\mu/kT}] \qquad (23\text{-}6)$$

where we have used the series expression

$$e^x = \sum_{n=0}^{\infty} (x^n/n!).$$

Then, from Eqs. (23-4)

$$\Omega = -kTV(2\pi MkT/h^2)^{3/2}e^{\mu/kT} = -PV$$

$$\bar{N} = V(2\pi MkT/h^2)^{3/2}e^{\mu/kT} = -(\Omega/kT) = PV/kT \qquad (23\text{-}7)$$

$$S = kV(2\pi MkT/h^2)^{3/2}e^{\mu/kT}\left(\frac{5}{2} - \frac{\mu}{kT}\right) = \bar{N}k\left(\frac{5}{2} - \frac{\mu}{kT}\right)$$

$$U = \Omega + ST + \mu\bar{N} = -\bar{N}kT + \bar{N}kT\left(\frac{5}{2} - \frac{\mu}{kT}\right)$$

$$+ \mu\bar{N} = \tfrac{3}{2}\bar{N}kT$$

$$F = \bar{N}(\mu - kT); \qquad \mu = -kT\ln[(V/\bar{N})(2\pi MkT/h^2)^{3/2}]$$

which, of course, present, in slightly different form, the same expressions for U, C_v, and the equation of state as did the other ensembles; only now $\bar{N}$ occurs instead of N. We also obtain directly an expression for the chemical potential per particle, μ, for the perfect gas.

The probability density that a volume V of such a gas, in equilibrium at temperature T and chemical potential μ, happens to contain N particles, and that these particles should have momenta $\mathbf{p}_1\mathbf{p}_2,...,\mathbf{p}_N$ and be located at the points specified by the vectors $\mathbf{r}_1, \mathbf{r}_2,..., \mathbf{r}_N$, is then

$$f_N(q, p) = \left(\frac{h^{-3N}}{N!}\right)\exp\left\{\frac{1}{kT}\left[\mu N - \sum_{j=1}^{N}\left(\frac{p_j^2}{2M}\right)\right]\right.$$

$$\left. - V\left(\frac{2\pi MkT}{h^2}\right)^{3/2}e^{\mu/kT}\right\} \qquad (23\text{-}8)$$

This is a generalization of the Maxwell distribution. The expression not only gives us the distribution in momentum of the N particles which happen to be in volume V at that instant (it is of course independent of their positions in V), but it also predicts the probability that there will *be* N molecules in volume V then. If we should wish to use P and T to specify the equilibrium state, instead of μ and T, this probability density would become

$$f_N(q, p) = \frac{1}{N!} \frac{(P/kT)^N}{(2\pi MkT)^{(3/2)N}} \exp -\left(\sum_{j=1}^{N} \frac{p_j^2}{2MkT} - \frac{PV}{kT} \right) \quad (23\text{-}9)$$

Density Fluctuations in a Gas

By summing f_{Nv} over v for a given N (or by integrating $f_N(q, p)$ over the q's and p's for a given N) we shall obtain the probability that a volume V of the gas, at equilibrium at pressure P and temperature T, will happen to have N molecules in it. From Eqs. (23-4) and (23-5) this is

$$f_N = \sum_v f_{Nv} = (1/3)e^{\mu N/kT}Z(N) = e^{(\Omega + \mu N)/kT}Z(N)$$

Using the expressions for $Z(N)$ and those for (Ω/kT) and (μ/kT), we obtain

$$f_N = e^{-\bar{N}} \left[\left(\frac{\bar{N}}{V} \right) \left(\frac{h^2}{2\pi MkT} \right)^{3/2} \right]^N \left(\frac{1}{N!} \right) \left[V \left(\frac{2\pi MkT}{h^2} \right)^{3/2} \right]^N$$

$$= (\bar{N}^N/N!)e^{-\bar{N}} = (1/N!)(PV/kT)^N e^{-PV/kT} \quad (23\text{-}10)$$

We could have obtained this directly by integrating $f_N(q, p)$ of Eq. (23-9) over phase space to obtain

$$f_N = \int \cdots \int f_N(q, p)\, dV_q\, dV_p$$

$$= \frac{1}{N!} \left(\frac{PV}{kT} \right)^N \exp(-PV/kT) \int \cdots \int \exp -\sum p_j^2/2MkT$$

$$\times \frac{dV_p}{(2\pi MkT)^{(3/2)N}} = \frac{(\bar{N})^N}{N!} e^{-\bar{N}}$$

where $\bar{N} = (PV/kT)$ is the expected value of N.

This is a Poisson distribution [see Eq. (11-5)] for the number of particles in a volume V of the gas. The mean number of particles is $\bar{N} = PV/kT$ and the probability f_N is greatest for N near $\bar{N}$ in value. But f_N is not zero when N differs from $\bar{N}$; it is perfectly possible to find a volume V in the gas which has a greater or smaller number of molecules in it than PV/kT. The variance of the number present is

$$(\Delta N)^2 = \sum_{N=0}^{\infty} (N - \bar{N})^2 f_N = \sum_{N=0}^{\infty} N^2 f_N - (\bar{N})^2$$
$$= \bar{N} = PV/kT \tag{23-11}$$

and the fractional deviation from the mean is

$$\Delta N/\bar{N} = 1/\sqrt{\bar{N}} = 1/\sqrt{kT/PV} \tag{23-12}$$

(It should be remembered that the system described by the grand canonical ensemble is not a gas of N molecules in a volume V, but that part of a supersystem that happens to be in a volume V, where V is much smaller than the volume occupied by the supersystem; thus the number of particles N that might be present can vary from zero to practically infinity.)

The smaller the volume of the gas looked at (the smaller the value of $\bar{N}$) the greater is this fractional fluctuation of number of particles present (or of density, for $\Delta N/\bar{N} = \Delta\rho/\rho$). Thus we have arrived at the result of Eq. (15-6), for the density fluctuations in various portions of a gas, by a quite different route.

CHAPTER

24

QUANTUM STATISTICS

But we still have not demonstrated the full utility of the grand canonical ensemble for handling calculations involving indistinguishable particles. The example in the previous chapter used the approximate correction factor $(1/N!)$, which we saw in Chapter 21 was not valid at low temperatures or high densities. We must now complete the discussion of the counting of states, which was begun there.

Occupation Numbers

In comparing the partition functions for distinguishable and indistinguishable particles, given in Eqs. (21-10) and (21-11) for the canonical ensemble, we saw that it was easier to compare the two if we talked about the number of particles occupying a given particle state rather than talking about which particle is in which state. In fact if the particles are indistinguishable it makes no sense to talk about which particle is in which state. We were there forced to describe the system state v by specifying the number n_j of particles that occupy the jth particle state, each of them having energy ε_j. The numbers n_j are called *occupation numbers*.

Of course if the interaction between particles is strong (as is the case with a crystal) we cannot talk about separate particle states; occupation numbers lose their specific meaning and we have to talk about normal modes instead of particles. But let us start with the particle interactions being small enough so we can talk about particle states and their occupation numbers. The results we obtain will turn out to be capable of extension to the strong-interaction case, as we shall show in Chapter 27.

We thus assume that, in the system of N particles, it makes sense to talk about the various quantum states of an individual particle, which we call particle states. These states are ranked in order of increasing energy, so that if ε_j is the energy of a particle in state j, then $\varepsilon_{j+1} > \varepsilon_j$. Instead of specifying the system state v by listing what state particle 1 is in, and so on for each particle, we specify it by saying how many particles are in state j (i.e., by specifying n_j). When the system is in state $v \equiv (n_1, n_2,..., n_j,...)$, the total number of particles and the total energy are

$$N = \sum_j n_j; \qquad E_N = \sum_j n_j \varepsilon_j \qquad (24\text{-}1)$$

Thus, for a system with no interaction between particles, the occupation numbers can be quantum numbers describing the state of the system.

For the canonical ensemble, we have to construct the partition function Z for a system with exactly N particles; the sum over v includes only those values of the n_j's for which their sum comes out to equal N. This restriction makes the calculation of a partition function such as that of Eq. (21-11) more difficult than it needs to be. With the grand canonical ensemble the limitation to a specific value of N is removed and the summation can be carried out over all the occupation numbers with no hampering restriction regarding their total sum.

Thus the grand partition function can be written in a form analogous to the Z of Eqs. (19-7) and (21-10),

$$\mathfrak{Z} = \sum_v W_v \exp\left[(1/kT) \sum_j n_j(\mu - \varepsilon_j)\right] \qquad (24\text{-}2)$$

by virtue of Eqs. (24-1). The multiplicities W_v, for each system state v (i.e., for each different set of occupation numbers n_j) are chosen according to the degree of distinguishability of the particles involved.

Indeed, this way of writing $\mathfrak{Z}$ is appropriate also when the "particles" are identical subsystems, such as the molecules of a gas. In such cases the "particle states" j are the molecular quantum states, specified by their translational, rotational, vibrational, and electronic quantum numbers, the allowed energies ε_j are the sums

$\varepsilon_{kmn}^{tr} + \varepsilon_{\lambda v}^{rot} + \varepsilon_n^{vib} + \varepsilon^{el}$ of Chapter 21, and the n_j's are the number of molecules that have the same totality of quantum numbers $j \equiv k, m, n, \lambda, v, n$, etc. However, we shall postpone discussion of this generalization until Chapter 27.

Maxwell-Boltzmann Particles

At present we wish to utilize the grand canonical ensemble to investigate systems of "elementary" particles, such as electrons or protons or photons or the like, sufficiently separated so that their mutual interactions are negligible. Each particle in such a system will have the same mass m and will be subject to the same conservative forces, so that the Schrödinger equation for each will be the same. Therefore, the total set of allowed quantum numbers, represented by the index j, will be the same for each particle (although at any instant different particles may have different values of j). The allowed energy corresponding to the set of quantum numbers represented by j is ε_j and the number of particles in this state is n_j. The grand partition function can then be written as in Eq. (24-2).

When the elementary particles are distinguishable, the discussion leading up to Eq. (21-9) indicates that W_v, the number of different ways N particles can be assigned to the various particle states, n_j of them being in the jth particle state, is $[N!/n_1!n_2!\cdots]$. Since $0! = 1$, we can consider all the n's as being represented in the denominator, even those for states unoccupied; an infinite product of 1's is still 1. The grand partition function will thus have the same kind of terms as in Eq. (21-10), but the restriction on the summation is now removed; all values of the n_j's are allowed. Therefore, for *distinguishable particles*,

$$\mathfrak{Z}_{dist} = \sum_{n_1, n_2, \ldots} \frac{(n_1 + n_2 + \cdots)!}{n_1!n_2!\cdots} \exp\left[\frac{1}{kT}\sum_j n_j(\mu - \varepsilon_j)\right] \quad (24-3)$$

In contrast to Eq. (21-10), this sum is not separable into a simple product of one-particle factors; because of the lack of limitation on N it reduces to the sum $\sum_N [\sum_j e^{(\mu - \varepsilon_j)/kT}]^N$, each term of which is a product of one-particle partition functions.

In the classical limit, when there are many particle states in the range of energy equal to kT, the chance of two particles being in the same particle state is vanishingly small, and the preponderating terms in series (24-3) are those for which no n_j is larger than 1. In this case we may correct for the indistinguishability of particle by the "shot-gun" procedure, used in Chapter 21, of dividing every term by the total number of ways in which N particles can be arranged in N different states. The resulting partition function

$$3_{MB} = \sum_{n_1, n_2, \ldots} \frac{1}{n_1! n_2! \cdots} \exp\left[(1/kT) \sum_j n_j(\mu - \varepsilon_j)\right]$$

$$= \sum_{n_1} \frac{1}{n_1!} e^{n_1(\mu - \varepsilon_1)/kT} \sum_{n_2} \frac{1}{n_2!} e^{n_2(\mu - \varepsilon_2)/kT} \cdots$$

$$= \exp\left[\sum_j e^{(\mu - \varepsilon_j)/kT}\right] = 3_1 3_2 \cdots \qquad (24\text{-}4)$$

$$3_j = \exp\left[e^{(\mu - \varepsilon_j)/kT}\right]$$

can be separated, being a product of factors 3_j, one for each particle *state j, not* one for each particle. This is the partition function we used to obtain Eqs. (23-6) and (23-7). As was demonstrated there and earlier, this way of counting states results in the thermodynamics of a perfect gas. It results also in the Maxwell–Boltzmann distribution for the mean number of particles occupying a given particle state j.

This last statement can quickly be shown by obtaining the grand potential Ω from 3 and then, by partial differentiation by μ, obtaining the mean number of particles in the system,

$$\Omega_{MB} = -kT \ln(3_{MB}) = -kT \sum_j e^{(\mu - \varepsilon_j)/kT} = -PV$$

$$\bar{N} = e^{\mu/kT} \sum_j e^{-\varepsilon_j/kT} = \sum_j \bar{n}_j; \qquad \bar{n}_j = e^{(\mu - \varepsilon_j)/kT} \qquad (24\text{-}5)$$

where $\bar{N}$ is equal to PV/kT, thus fixing the value of

$$\mu = -kT \ln\left[(kT/PV) \sum_j e^{-\varepsilon_j/kT}\right].$$

In fact μ acts like a magnitude parameter in the grand canonical ensemble; its value is adjusted to make $-(\partial\Omega/\partial\mu)_{TV}$ equal to the specified value of $\bar{N}$. The quantity $\bar{n}_j$, the mean value of the occupation number for the jth particle state for this ensemble, takes the place of the particle probabilities (for we can no longer ask what state a given particle is in). We see that the mean number of particles in state j, with energy ε_j, is

$$\bar{n}_j = \left(\frac{PV}{kT}\right)\left[\sum_i e^{-\varepsilon_i/kT}\right]^{-1} e^{-\varepsilon_j/kT} \qquad (24\text{-}6)$$

which is proportional to the Maxwell–Boltzmann factor $e^{-\varepsilon_j/kT}$.

Therefore particles that correspond to this partition function may be called *Maxwell–Boltzmann particles* (MB particles for short). No actual system of particles corresponds exactly to this distribution for all temperatures and densities. But all systems of particles approach this behavior in the limit of high-enough temperatures, whenever the classical phase-space approximation is valid.

Before the advent of the quantum theory the volume of phase space occupied by a single microstate was not known; in fact it seemed reasonable to assume that every element of phase space, no matter how small, represented a separate microstate. If this were the case, the chance that two particles would occupy the same state was of the second order in the volume element and could be neglected. Thus for classical statistical mechanics the procedure of dividing by $N!$ was valid. Now we know that the magnitude of phase-space volume occupied by a microstate is finite, not infinitesimal; it is apparent that there can be situations in which the system points are packed closely enough in phase space so that two or more particles are within the volume that represents a single microstate; in such cases MB statistics is not an accurate representation.

Bosons and Fermions

Actual particles are of two types. Both types are indistinguishable and thus, according to Eq. (20-11), have multiplicity factors $W_v = 1$, rather than $(N!/n_1! n_2! \cdots)$. A state of the system is specified by specifying the values of the occupation numbers n_j. Each such state is a single one; it has no meaning to try to distinguish which particle is in which state; all we can specify are the numbers n_j in each state.

In addition to their indistinguishability, different particles obey different rules regarding the maximum value of n_j. One set of particles can pack as many into a given particle state as the distribution will allow; n_j can take on all values from 0 to ∞. Such particles are called *bosons*; they are said to obey the *Bose–Einstein statistics* (BE for short). Photons and helium atoms are examples of bosons. For these particles the W_v of Eq. (24-2) are all unity and the grand partition function is

$$\mathfrak{Z}_{BE} = \sum_{n_1, n_2, \ldots} \left[\exp (1/kT) \sum_j n_j(\mu - \varepsilon_j) \right]$$
$$= \sum_{n_1} e^{n_1(\mu - \varepsilon_1)/kT} \sum_{n_2} e^{n_2(\mu - \varepsilon_2)/kT} \cdots = \mathfrak{z}_1 \mathfrak{z}_2$$

$$\mathfrak{z}_1 = [1 - e^{(\mu - \varepsilon_j)/kT}]^{-1} \tag{24-7}$$

where we have used Eq. (20-2) to consolidate the factor sums $\mathfrak{z}_j$.

Here again the grand partition function separates into factors, one for each particle state, rather than one for each particle. We note that the series for the jth factor does not converge unless μ is less than the corresponding energy ε_j.

Here again we can calculate the grand potential and the mean number of particles in the system of bosons,

$$\Omega_{BE} = kT \sum_j \ln[1 -- e^{(\mu - \varepsilon_j)/kT}] = -PV$$

$$\bar{N} = \sum \bar{n}_j; \qquad \bar{n}_j = [e^{(\varepsilon_j - \mu)/kT} - 1]^{-1} \tag{24-8}$$

where $\bar{n}_j$ is the mean number of particles in the jth particle state. In this case there is no simple equation fixing the value of μ in terms of $\bar{N}$ (or of PV) and T, nor is the relationship between $\bar{N}$ and $\Omega = -PV$ as simple as it was with Eq. (24-5). Nevertheless, knowing the allowed energy values ε_j and the temperature T, we can adjust μ so the sum of $1/[e^{(\varepsilon_j - \mu)/kT} - 1]$ over all j is equal to $\bar{N}$. This value of μ is then used to compute the other thermodynamic quantities.

Note the difference between the occupation number n_j for the boson and that of Eq. (24-5) for the MB particle. For higher states, where $\varepsilon_j - \mu \gg kT$, the two values do not differ much, but for the lower states, at lower temperatures, where $\varepsilon_j - \mu$ is equal to or smaller than kT, the n_j for the boson is appreciably greater than that for the MB particle (shall we call it a maxwellon?). Bosons tend to "condense" into their lower states, at low temperatures, more than do maxwellons.

The other kind of particle encountered in nature has the idiosyncrasy of refusing to occupy a state that is already occupied by another particle. In other words the occupation numbers n_j for such particles can be 0 or 1, but not greater than 1. Particles exhibiting such unsocial conduct are said to obey the *Pauli exclusion principle*. They are called *fermions* and are said to obey *Fermi–Dirac statistics* (FD for short). Electrons, protons, and other elementary particles with spin $\frac{1}{2}$ are fermions. For these particles $W_v = 1$, but the sum over each n_j omits all terms with $n_j > 1$. Therefore,

$$\mathfrak{z}_{FD} = \sum_{n_1, n_2, \ldots} \exp\left[(1/kT) \sum_j n_j(\mu - \varepsilon_j)\right] = \mathfrak{z}_1 \cdot \mathfrak{z}_2 \cdots$$

$$\mathfrak{z}_j = [1 + e^{(\mu - \varepsilon_j)/kT}] \tag{24-9}$$

Again the individual factors $\mathfrak{z}_j$ are for each quantum state, rather than for each particle. The mean values of the occupation numbers can be obtained from $\mathfrak{Z}$ as before,

$$\Omega_{\text{FD}} = -kT \sum_j \ln[1 + e^{(\mu - \varepsilon_j)/kT}] = -PV$$

$$\bar{N} = \sum_{j=1}^{\infty} \bar{n}_j; \qquad \bar{n}_j = [e^{(\varepsilon_j - \mu)/kT} + 1]^{-1} \tag{24-10}$$

where again the relation between $\bar{N}$ and PV (i.e., the equation of state) is not so simple as it is for maxwellons, and again there is no simple relationship that determines μ in terms of $\bar{N}$; the equation for $\bar{N}$ must be inverted to find μ as a function of $\bar{N}$.

Comparing the mean number $\bar{n}_j$ of particles in state j for fermions with the $\bar{n}_j$ for MB particles [Eq. (24-5)], we see that for the higher states, where $\varepsilon_j - \mu \gg kT$, the two values are roughly equal, but for the lower states the $\bar{n}_j$ for fermions is appreciably smaller (for a given value of μ) than the $\bar{n}_j$ for maxwellons. Fermions tend to stay away from the lower states more than do maxwellons, and thus much more than do bosons. In fact, fermions cannot enter a state already occupied by another fermion; according to the Pauli principle $\bar{n}_j$ cannot be larger than 1.

Comparison among the Three Statistics

The differences between the BE, MB, and FD statistics can be most simply displayed by comparing the factors multiplying the exponentials in the sums over the n_j's in Eqs. (24-4), (24-7), and (24-9). For MB statistics this factor (call it b_j) is $(1/n_j!)$, for BE statistics it is 1 for all n_j and for FD statistics it is 1 for $n_j = 0$ or 1, zero otherwise. The three sets of values are

$$b_j(n_j) = \begin{cases} 1(n_j = 0 \text{ or } 1) & = 1(n_j > 1) \text{ BE statistics} \\ 1(n_j = 0 \text{ or } 1) & = \dfrac{1}{n_j!}(n_j > 1) \text{ MB statistics} \\ 1(n_j = 0 \text{ or } 1) & = 0(n_j > 1) \text{ FD statistics} \end{cases}$$
$$\tag{24-11}$$

The b's are identical for $n_j = 0$ or 1; they differ for the higher values of the occupation numbers. Bosons have $b_j = 1$ for all values of n_j; they don't care how many others are in the same state. Fermions have $b_j = 0$ for $n_j > 1$; they are completely unsocial. The approximate statistics we call MB has values intermediate between 0 and 1 for $n_j > 1$; these particles are moderately unsocial; the b_j tend toward zero as n_j increases.

In terms of the energy ε_j of the jth particle state and the value of the normalizing parameter μ, the mean number of particles in state j is

$$\bar{n}_j = \begin{cases} 1/[e^{(\varepsilon_j - \mu)/kT} - 1] & \text{BE statistics} \\ 1/e^{(\varepsilon_j - \mu)/kT} & \text{MB statistics} \\ 1/[e^{(\varepsilon_j - \mu)/kT} + 1] & \text{FD statistics} \end{cases} \qquad (24\text{-}12)$$

For FD statistics, $\bar{n}_j$ can never be larger than 1; for MB statistics $\bar{n}_j$ can be larger than 1 for those states with μ larger than ε_j; for BE statistics μ cannot be larger than ε_1 [see discussion of Eq. (24-7)] but $\bar{n}_j$ can be much larger than 1 if $(\varepsilon_j - \mu)/kT$ is small.

In each case the value of μ is determined by requiring that the sum of the $\bar{n}_j$'s, over all values of j, be equal to the mean number $\bar{N}$ of particles in the system. If kT is large compared to the energy spacings $\varepsilon_{j+1} - \varepsilon_j$, then $\bar{n}_{j+1}$ will not differ much from $\bar{n}_j$ and the sum for $\bar{N}$ will consist of a large number of $\bar{n}_j$'s, of slowly diminishing magnitude. Therefore much of the sum for $\bar{N}$ will be "carried" by the $\bar{n}_j$'s for the higher states $(j > 1)$. If, at the same time, $\bar{N}$ is small, then all the $\bar{n}_j$'s must be small; even $\bar{n}_1$ must be less than 1. For this to be so, $(\varepsilon_1 - \mu)$ must be larger than kT, so that the terms $\exp[(\varepsilon_j - \mu)/kT]$ must all be considerably larger than 1. In this case the values of the $\bar{n}_j$'s, for the three statistics, are nearly equal, and we might as well use the intermediate MB values, since these provide us with a simpler set of equations for μ, S, P, C_v, etc. [Eqs. (24-5)].

In other words, in the limit of high temperature and low density, both bosons and fermions behave like classical Maxwell–Boltzmann particles. For this reason, the fact that classical statistical mechanics is only an approximation did not become glaringly apparent until systems of relatively high density were studied at low temperatures (except in the case of photons, which are a special case, since $\mu = 0$ for them).

When kT is the same size as $\varepsilon_2 - \varepsilon_1$ or smaller, the three statistics display markedly different characteristics. For *bosons* μ becomes very nearly equal to $\varepsilon_1 (\mu = \varepsilon_1 - \delta$, where $\delta \ll kT)$ so that

$$\bar{n}_1 = [e^{(\varepsilon_1 - \mu)/kT} - 1]^{-1} \simeq kT/\delta$$

and

$$\bar{n}_j \simeq [e^{(\varepsilon_j - \varepsilon_1)/kT} - 1]^{-1} \qquad \text{for} \quad j > 1$$

which is considerably smaller than $\bar{n}_1$ if $\varepsilon_2 - \varepsilon_1 > kT$. Therefore at low temperatures and high densities, most of the bosons are in the lowest state $(j = 1)$ and

$$\bar{n}_1 \simeq \bar{N} - \sum_{j=2}^{\infty} \left[e^{(\varepsilon_j - \varepsilon_1)/kT} - 1 \right]^{-1} \to \bar{N}, \qquad kT \to 0 \qquad (24\text{-}13)$$

which serves to determine δ, and therefore $\mu = \varepsilon_1 - \delta$. At very low temperatures bosons "condense" into the ground state. The "condensation" is not necessarily one in space, as with the condensation of a vapor into a liquid. The ground state may be distributed all over position space but may be "condensed" in momentum space. This will be illustrated later.

For *fermions* such a condensation is impossible; no more than one fermion can occupy a given state. As $T \to 0$, μ must approach ε_N, so that $\bar{n}_j = \{\exp[(\varepsilon_j - \mu)/kT] + 1\}^{-1}$ is practically equal to 1 for $j < N$ (since $\exp[(\varepsilon_j - \mu)/kT] \simeq \exp[(\varepsilon_j - \varepsilon_N)/kT]$ is then very much smaller than 1) and is much smaller than 1 for $j > N$ (since $\exp[(\varepsilon_j - \mu)/kT]$ is then very large compared to 1 for $\varepsilon_j > \varepsilon_N$). Thus at low temperatures the lowest N particle states are completely filled with fermions (one per state) and the states above this "Fermi level" ε_N are devoid of particles.

The behavior of *MB particles* differs from that of either bosons or fermions at low temperatures and high densities. The lower states are populated by more than one particle, in contrast to the fermions, but they don't condense exclusively and suddenly in just the ground state, as do bosons.

These differences can be shown graphically by changing from summation over the different quantum states j to integration over the allowed energy values ε_j. The relationship between the particle-state energy ε_j and the particle-state quantum number (or numbers) j depend on the forces acting on the particles in the particular system under study; the symmetry of the system determines the multiplicity $g(\varepsilon_j)$ [see Eq. (19-9)], the number of different particle states that have the same value ε_j of energy. In the sum $\bar{N} = \Sigma \bar{n}_j$, if $\bar{N}$ is large enough so that many states are occupied, the sum can be approximated by the integral

$$\bar{N} = \sum_j \bar{n}_j \simeq \int \bar{n}_j \, dj = \int \bar{n}_j g(\varepsilon_j)(dj/d\varepsilon_j) \, d\varepsilon_j = \int D(\varepsilon)\bar{n}(\varepsilon) \, d\varepsilon$$

where $D(\varepsilon_j) \, d\varepsilon = g(\varepsilon_j)(dj/d\varepsilon_j) \, d\varepsilon$ is the mean number of *particle states* having energy between ε_j and $\varepsilon_j + d\varepsilon$ and where $\bar{n}_j D(\varepsilon_j) \, d\varepsilon$

is the mean number of *particles* occupying the states with energy between ε_j and $\varepsilon_j + d\varepsilon$. For free particles in a gas, $D(\varepsilon_j)$ is proportional to $\sqrt{\varepsilon_j}$ [see Eqs. (25-17) and (26-1)]; for particles held in a three-dimensional harmonic-oscillator potential well (a simplified model of a nucleus) $D(\varepsilon_j)$ is proportional to ε_j^2.

No matter which statistics is to be used, the value of the chemical potential μ, entering into the formula for $\bar{n}_j$, is fixed by setting the area under the curve of $\bar{n}_j D(\varepsilon_j)$ versus ε_j equal to $\bar{N}$. For a given form of $D(\varepsilon_j)$ and a given value of μ, the area under the curve of $\bar{n}D$ for BE statistics is larger than the area under the $\bar{n}D$ curve for MB statistics (for the same μ), and this again is larger than the area under the $\bar{n}D$ curve for FD statistics. For all three cases, the smaller the value of μ the smaller is $\bar{n}D$. Combining these two facts, we see that, for a given value of $\bar{N}$ the μ for the FD statistics must be larger than the μ for the MB statistics and this must in turn be larger than the μ for the BE statistics. In fact the μ for the BE statistics must never be larger than the lowest allowed energy ε_1; otherwise $\bar{n}_j$ for this case would become infinite for $\varepsilon_j = \mu$. By comparison, μ for a degenerate FD system can be much larger than ε_1.

As a simple illustration we plot in Figure 24-1 curves of $(kT/D)(d\bar{N}/d\varepsilon) = \bar{n}(\varepsilon)$ versus (ε/kT), for the three statistics, for a system for which $D(\varepsilon_j)$ is a constant D. The upper trio of curves are for a small value of $(\bar{N}/D)$, such as 0.5. In this case none of the energy levels are very full ($\bar{n}$ is less than 0.5 for all ε for MB and FD statistics, and is less than 0.5 for BE statistics for $\varepsilon > 0.2\,kT$) and thus the curves for the three statistics do not differ much. With smaller values of $(\bar{N}/D)$ these differences would be even less.

On the other hand, for $(\bar{N}/D)$ as large as 5 the distributions differ considerably, for the two quantum cases are becoming *degenerate*. For the BE statistics the curve for $\bar{n}$ rises to nearly 150 at $\varepsilon = 0$; nearly all the $\bar{N}$ bosons have squeezed themselves into the lowest few states. The FD case is just the opposite; the fermions cannot all crowd into the lowest states, so they fill the lowest $\bar{N}$ levels nearly continually and leave the levels above these sparsely occupied ($\bar{n}$ is greater than 0.5 for ε less than $5kT$ and drops rapidly below 0.5 for ε greater than $5kT$).

Another comparison, for the case of a gas in a container of volume V, where $D(\varepsilon)$ is proportional to $\sqrt{\varepsilon}$, is given in Figure 26-1. Here the quantity plotted is $2\bar{n}(\varepsilon)\sqrt{[y/\pi]}$, where y, the ordinate, is (ε/kT), as in Figure 24-1. We see again that when η (which is proportional to $\bar{N}$) is small, so that $\bar{n}(\varepsilon) < 1$, the three curves do not

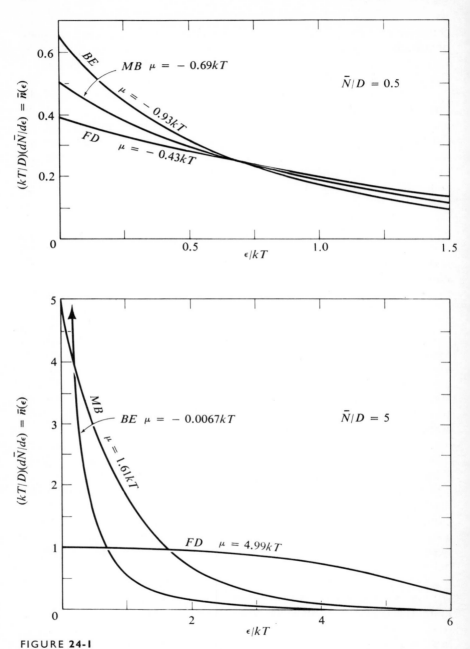

FIGURE 24-1

Comparison of distribution in energy, among the three statistics, for a system with level density D independent of ε. Area under each curve equals $(\bar{N}/D)$.

differ much. For η large, however, degeneracy sets in; with BE statistics most of the particles squeeze into the lowest states; with FD statistics all they can do is to fill all the lower states ($\bar{n} \simeq 1$) up to $\varepsilon = \mu$.

Distribution Functions and Fluctuations

With indistinguishable particles there is no sense in asking the probability that a specific particle is in state j; all we can ask for is the probability $f_j(n_j)$ that n_j particles are in state j. These probabilities can be obtained from the distribution function of the ensemble, given in Eq. (23-4). For

$$f_{Nv} = (W_v/\mathfrak{Z}) \exp\left[\sum_{j=1}^{\infty} n_j(\mu - \varepsilon_j)/kT\right] = f_1(n_1) \cdot f_2(n_2) \cdots$$

$$f_j(n_j) = (b_j/\mathfrak{Z}_j)e^{n_j(\mu - \varepsilon_j)/kT} \tag{24-14}$$

where the factor $\mathfrak{Z}_j$ of the partition function, for the jth particle state, is given by Eq. (24-4), (24-7), or (24-9), depending upon whether the particles in the system are MB, BE, or FD particles.

To be specific, the probability that n particles are in state j (we can leave the subscript off n without producing confusion here) for the three statistics, is

For bosons:

$$f_j(n) = e^{n(\mu - \varepsilon_j)/kT} - e^{(n+1)(\mu - \varepsilon_j)/kT}$$

For MB particles:

$$f_j(n) = \frac{1}{n!} \exp\left[n\frac{\mu - \varepsilon_j}{kT} - e^{(\mu - \varepsilon_j)/kT}\right]$$

For fermions:

$$f_j(n) = \begin{cases} e^{n(\mu - \varepsilon_j)/kT}/[1 + e^{(\mu - \varepsilon_j)/kT}] & n = 0, 1 \\ 0 & n > 1 \end{cases} \tag{24-15}$$

Reference to Eqs. (24-12) shows that the mean value of n is given by the usual formula,

$$\bar{n}_j = \sum_{n=0}^{\infty} nf_j(n) \tag{24-16}$$

With a bit of algebraic juggling, we can then express the probability $f_j(n)$ in terms of n and of its mean value $\bar{n}_j$ (we can call it $\bar{n}$ without confusion here):

$$f_j(n) = \begin{cases} (\bar{n})^n/(\bar{n} + 1)^{n+1} & \text{for bosons} \\ [(\bar{n})^n/n!]e^{-\bar{n}} & \text{for MB particles} \\ 1 - \bar{n} \text{ if } n = 0, \ = \bar{n} \text{ if } n = 1, \\ \qquad = 0 \text{ if } n > 1 & \text{for fermions} \end{cases}$$

$$(24\text{-}17)$$

The distribution function $f_j(n)$ for bosons is a *geometric* distribution. The ratio $f_j(n)/f_j(n - 1)$ is a constant, $\bar{n}/(\bar{n} + 1)$; the chance of adding one more particle to state j is the same, no matter how many bosons are already in the state. The MB distribution is the familiar Poisson distribution of Eqs. (11-15) and (23-10), with ratio $f_j(n)/f_j(n - 1) = \bar{n}/n$, which decreases as n increases. The presence of maxwellons in a given state discourages the addition of others, to some extent. On the other hand, $f_j(n)$ for FD statistics is zero for $n > 1$; if a fermion occupies a given state, no other particle can join it (the Pauli principle).

Using these expressions for $f_j(n)$ we can calculate the variance $(\Delta n_j)^2$ of the occupation number n_j for state j, for each kind of statistics:

$$(\Delta n_j)^2 = \sum_{n=0}^{\infty} (n - \bar{n}_j)^2 f_j(n) = \sum_n n^2 f_j(n) - (\bar{n}_j)^2$$

$$= \begin{cases} \bar{n}_j(\bar{n}_j + 1) & \text{for bosons} \\ \bar{n}_j & \text{for MB particles} \\ \bar{n}_j(1 - \bar{n}_j) & \text{for fermions} \end{cases} \qquad (24\text{-}18)$$

and, from this, obtain the fractional fluctuation $\Delta n_j/\bar{n}_j$ of the occupation numbers,

$$\Delta n_j/\bar{n}_j = \begin{cases} \sqrt{1 + (1/\bar{n}_j)} & \text{for bosons} \\ \sqrt{1/\bar{n}_j} & \text{for MB particles} \\ \sqrt{(1/\bar{n}_j) - 1} & \text{for fermions} \end{cases} \qquad (24\text{-}19)$$

The fractional fluctuation is greatest for the least-occupied states ($\bar{n}_j \ll 1$). As the mean occupation number increases the fluctuation decreases, going to zero for fermions as $\bar{n}_j \to 1$ (the degenerate state) and to zero for maxwellons as $\bar{n}_j \to \infty$. But the standard deviation Δn_j for bosons is never less than the mean occupancy $\bar{n}_j$. We shall see later that the local fluctuations in intensity of thermal radiation (photons are bosons) are always large, of the order of magnitude of the intensity itself, as predicted by Eq. (24-19).

CHAPTER

25

BOSE-EINSTEIN STATISTICS

The previous chapter has indicated that, as the temperature is lowered or the density is increased, systems of bosons or of fermions enter a state of degeneracy, wherein their thermodynamic properties differ considerably from those of the corresponding classical system, subject to Maxwell–Boltzmann statistics. These differences are apparent even when the systems are perfect gases, where the interaction between particles is limited to the few collisions needed to bring the gas to equilibrium. Indeed, in some respects, the differences between the three statistics are more apparent when the systems are perfect gases than when they are more complex in structure. Therefore it is useful to return once again to the system we started to study in Part II, this time to analyze in detail the differences caused by differences in statistics. In this chapter we take up the properties of a gas of bosons. Two different cases will be considered; a gas of photons (electromagnetic radiation) and a gas of material particles, such as helium atoms.

General Properties of a Boson Gas

Using Eqs. (24-8) et seq., we compute the distribution function, mean occupation numbers, and thermodynamic functions for the gas of bosons:

$$f_j(n) = \left[1 - e^{(\mu - \varepsilon_j)/kT} \right] e^{n(\mu - \varepsilon_j)/kT} = \left[\bar{n}_j^n / (\bar{n}_j + 1)^{n+1} \right]$$

$$\bar{n}_j = \sum_{n=0}^{\infty} n f_j(n) = \left[e^{(\varepsilon_j - \mu)/kT} - 1 \right]^{-1}$$

$$\Omega_{\text{BE}} = -PV = -kT \ln \mathfrak{Z} = kT \sum_{j=1}^{\infty} \ln\left[1 - e^{(\mu - \varepsilon_j)/kT} \right]$$

$$\bar{N} = -(\partial\Omega/\partial\mu)_{TV} = \sum_{j=1}^{\infty} \bar{n}_j = \sum_{j=1}^{\infty} \left[e^{(\varepsilon_j - \mu)/kT} - 1 \right]^{-1}$$

$$S = -(\partial\Omega/\partial T)_{\mu V} = (U - \bar{N}\mu - \Omega)/T$$

$$U = \sum_{j=1}^{\infty} \varepsilon_j \bar{n}_j; \qquad C_v = T(\partial S/\partial T)_{\mu V}$$

(25-1)

where μ must be less than the lowest particle energy ε_1 in order that the series expansions converge. All these quantities are functions of the chemical potential μ. For systems in which the mean number of particles $\bar{N}$ is specified, the value of μ, as a function of $\bar{N}$, V, and T, is determined implicitly by the equation for $\bar{N}$ given above. The value obtained by inverting this equation is then inserted in the other equations, to give S, U, P, and C_v as functions of $\bar{N}$, V, and T.

In the case of the photon gas, in equilibrium at temperature T in a volume V (black-body radiation), the number of photons $\bar{N}$ in volume V is not arbitrarily specified; it adjusts itself so that the radiation is in equilibrium with the constant-temperature walls of the container. Since, at constant T and V, the Helmholtz function $F = \Omega + \mu\bar{N}$ comes to a minimal value at equilibrium [see the discussion following Eq. (8-10)], if $\bar{N}$ is to be varied to reach equilibrium at constant T and V, we must have

$$(\partial F/\partial\bar{N})_{TV} = \frac{\partial}{\partial\bar{N}}(\Omega + \mu\bar{N}) = \mu \text{ equal to zero} \qquad (25\text{-}2)$$

Therefore, for a photon gas at equilibrium, at constant T and V, the chemical potential of the photons must be zero [see the discussion following Eq. (7-8)].

Classical Statistics of Black-Body Radiation

At this point the disadvantages of a "logical" presentation of the subject become evident; a historical presentation would bring out more vividly the way experimental findings forced a revision of classical statistics. It was the work of Planck, in trying to explain the frequency distribution of electromagnetic radiation which first exhibited the inadequacy of the Maxwell–Boltzmann statistics and pointed the way to the development of quantum statistics. A purely logical demonstration that quantum statistics does conform with observation leaves out the atmosphere of struggle which permeated the early development of quantum theory— struggle to find a theory that would fit the many new and unexpected measurements.

Experimentally, the energy density of black-body radiation having frequency between $\omega/2\pi$ and $(\omega + d\omega)/2\pi$ was found to fit an empirical formula

$$de = \frac{\hbar}{\pi^2 c^3} \frac{\omega^3\, d\omega}{e^{\hbar\omega/kT} - 1} \rightarrow \begin{cases} (\omega^2 kT/\pi^2 c^3)\, d\omega & kT \gg \hbar\omega \\[2mm] (\hbar\omega^3/\pi^2 c^3) e^{-\hbar\omega/kT}\, d\omega & kT \ll \hbar\omega \end{cases}$$

$$(25\text{-}3)$$

where, at the time, $\hbar$ was an empirical constant, adjusted to fit the formula to the experimental curves. Classical statistical mechanics could explain the low-frequency part of the curve ($kT \gg \hbar\omega$) but could not explain the high-frequency part (Figure 25-1).

Classically, each degree of freedom of the electromagnetic radiation should possess a mean energy kT [see the discussion of Eq. (15-5)], so determining the formula for de should simply involve finding the number of degrees of freedom of the radiation between ω and $\omega + d\omega$. Since the radiation is a collection of standing waves, it can proceed exactly as was done in Chapter 20, in finding the number of standing waves in a crystal with frequencies between $\omega/2\pi$.

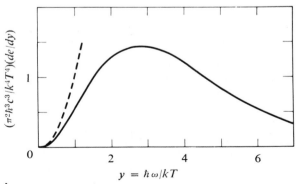

FIGURE 25-1

The Planck distribution of energy density of black-body radiation per frequency range. Dashed line is Rayleigh–Jeans distribution.

and $(\omega + d\omega)/2\pi$ [see Eqs. (20-10) and (20-11)]. In a rectangular enclosure of sides l_x, l_y, l_z the allowed values of ω are

$$\omega_j = \pi c[(k/l_x)^2 + (m/l_y)^2 + (n/l_z)^2]^{1/2} \qquad (25\text{-}4)$$

where k, m, n are integers and where c is the velocity of light. Each different combination of k, m, n corresponds to a different electromagnetic wave, a different degree of freedom or, in quantum language, a different quantum state j for a photon.

By methods completely analogous to those used in Chapter 20, we find that the number of different degrees of freedom having allowed values of ω_j between ω and $\omega + d\omega$ are

$$dj = (V/\pi^2 c^3)\omega^2 \, d\omega, \qquad V = l_x l_y l_z \qquad (25\text{-}5)$$

which is twice the value given in Eq. (20-11) because light can have two mutually perpendicular polarizations, so there are two different standing waves for each set of values of k, m, and n. As mentioned before, this formula is valid for nonrectangular enclosures of volume V.

Now, if each degree of freedom carries a mean energy kT, then the total energy within V, between ω and $\omega + d\omega$, is $(kT) \, dj$ and the energy density of radiation with frequency between $\omega/2\pi$ and $(\omega + d\omega)/2\pi$ is

$$de = (kT/V) \, dj = (\omega^2 kT/\pi^2 c^3) \, d\omega$$

which is called the Rayleigh–Jeans formula. We see that it fits the empirical formula (25-3) at the low-frequency end (see the dashed curve of Figure 25-1) but not for high frequencies.

As a matter of fact it is evident that the Rayleigh–Jeans formula cannot hold over the whole range of ω from 0 to ∞, for the integral of de would then diverge. If this were the correct formula for the energy density then, to reach equilibrium with its surroundings, a container filled with radiation would have to withdraw an infinite amount of energy from its surroundings; all the thermal energy in the universe would drain off into high-frequency electromagnetic radiation. This outcome was dramatized by calling it the *ultraviolet catastrophe*. There is no sign of such a fate, so the Rayleigh–Jeans formula cannot be correct for high frequencies. In fact the empirical curve has the energy density de dropping down exponentially, according to the factor $e^{-\hbar\omega/kT}$, when $\hbar\omega \gg kT$, so that the integral of the empirical expression does not diverge.

Parenthetically, a similar catastrophe cannot arise with waves in a crystal because a crystal is not a continuous medium; there can only be as many different standing waves in a crystal as there are atoms in the crystal; integration over ω only goes to ω_m [see Eq (26-13)] not to ∞. In contrast, the electromagnetic field is continuous, not atomic, so there is no lower limit to wavelength, no upper limit to the frequency of its standing waves.

A satisfactory exposition (to physicists, at any rate) would be to proceed from empirical formula (25-3) to the theoretical mode that fits it, showing that the experimental findings lead inexorably to the conclusion that photons obey Bose–Einstein statistics. We have not the space to do this; we shall show instead that assuming photons are bosons (with $\mu = 0$) leads directly to the empirical formula (25-3) and by identifying the empirical constant $\hbar = h/2\pi$ with Planck's constant, joins the theory of black-body radiation to all the rest of quantum theory.

The Planck Distribution Formula

As we have already pointed out in Eq. (25-2), photons are a rather special kind of boson; their chemical potential is zero when they are in thermal equilibrium in volume V at temperature T. Formulas (25-1) thus simplify. For example, the mean number of photons in state j is $\bar{n}_j = [\exp(\varepsilon_j/kT) - 1]^{-1}$. But state j has been defined as the state that has frequency $\omega_j/2\pi$, where ω_j is given in Eq. (25-4) in terms of its quantum numbers. Since a photon of frequency $\omega_j/2\pi$ has energy $\varepsilon_j = \hbar\omega_j$, the mean occupation number becomes

$$\bar{n}_j = 1/(e^{\hbar\omega_j/kT} - 1) \tag{25-6}$$

Since there are $(V/\pi^2 c^3)\omega^2\,d\omega = dj$ different photon states (different standing waves) with frequencies between $\omega/2\pi$ and $(\omega + d\omega)/2\pi$, the mean number of photons in this frequency range in the container is

$$d\bar{n} = \frac{V}{\pi^2 c^3}\frac{\omega^2\,d\omega}{e^{\hbar\omega/kT} - 1} \tag{25-7}$$

The mean energy density de of black-body radiation in this frequency range is $d\bar{n}$ times the energy $\hbar\omega$ per photon, divided by V, which turns out to be identical with the empirical formula for de given in Eq. (25-3). Thus the assumption that photons are bosons with $\mu = 0$ leads directly to agreement with observation.

The frequency distribution of radiation given in Eq. (25-3) is called the *Planck distribution*. The energy density per unit frequency band increases proportionally to ω^2 at low frequencies; it has a maximum at $\omega = 2.82(kT/h)$ [where $x = 2.82$ is the solution of the equation $(3 - x)e^x = 3$] and it drops exponentially to zero as ω increases beyond this maximum. Measurements have checked all these details; in fact this was the first way by which the value of h was determined. The mean number of photons, and the mean energy density, for all frequencies, can be obtained from the following formulas:

$$\int_0^\infty \frac{x^2 \, dx}{e^x - 1} = 2.404; \qquad \int_0^\infty \frac{x^3 \, dx}{e^x - 1} = \frac{\pi^4}{15} = 6.494 \qquad (25\text{-}8)$$

For example, the mean energy density is

$$e(T) = \int de = \frac{(kT)^4}{\pi^2 c^3 \hbar^3} \int_0^\infty \frac{x^3 \, dx}{e^x - 1} = aT^4, \; a = \frac{\pi^2 k^4}{15\hbar^3 c^3} \qquad (25\text{-}9)$$

which is the same as Eq. (7-8), of course, only now we have obtained an expression for Stefan's constant a in terms of k, h, and c (which checks with experiment).

The grand potential Ω (which also equals F, since $\mu = 0$) is

$$\Omega = kT \int dj \ln(1 - e^{-\hbar\omega/kT}) = \frac{VkT}{\pi^2 c^3} \int_0^\infty \omega^2 \, d\omega \ln(1 - e^{-\hbar\omega/kT})$$

$$= -\frac{k^4 T^4 V}{3\pi^2 c^3 \hbar^3} \int_0^\infty \frac{x^3 \, dx}{e^x - 1} = -\tfrac{1}{3}aVT^4 = -\tfrac{1}{3}Ve(T) = F \qquad (25\text{-}10)$$

where we have integrated by parts. The other thermodynamic quantities are obtained by differentiation,

$$S = -\left(\frac{\partial F}{\partial T}\right)_v = \tfrac{4}{3}aVT^3; \qquad P = -\left(\frac{\partial F}{\partial V}\right)_T = \tfrac{1}{3}aT^4 = \tfrac{1}{3}e(T) \qquad (25\text{-}11)$$

which also check with Eqs. (7-8). The mean number of photons of any frequency in volume V is

$$\bar{N} = \int dn = \frac{Vk^3 T^3}{\pi^2 c^3 \hbar^3} \int_0^\infty \frac{x^2 \, dx}{e^x - 1} = \frac{2.404}{\pi^2}\left(\frac{kT}{c\hbar}\right)^3 \times V$$

$$= \left(\frac{0.625k}{c\hbar}\right)^3 VT^3 \qquad (25\text{-}12)$$

Statistical Properties of a Photon Gas

We saw earlier that the assumption of classical equipartition of energy for each degree of freedom of black-body radiation leads to the nonsensical conclusion that each container of radiation has an infinite heat capacity. The assumption that photons are bosons, with $\mu = 0$, leads to the Planck formula, rather than the Rayleigh–Jeans formula, and leads to the conclusion that the mean energy carried per degree of freedom of the thermal radiation is

$$\hbar\omega\bar{n}_j = \hbar\omega_j/(e^{\hbar\omega/kT} - 1) \rightarrow \begin{cases} kT & \hbar\omega_j \ll kT \\ \hbar\omega_j e^{-\hbar\omega_j/kT} & \hbar\omega_j \gg kT \end{cases} \quad (25\text{-}13)$$

which equals kT, the classical value, only for low-frequency radiation. At high frequencies each photon carries so much energy that, even in thermal equilibrium, very few can be excited (just as with other quantized oscillators that have an energy level spacing large compared to kT), and the mean energy possessed by these degrees of freedom falls off proportionally to $e^{-\hbar\omega/kT}$. Thus, as we have seen, the mean energy $e(T)$ is not infinite as classical statistics had predicted.

As was pointed out at the end of Chapter 24, the fluctuations in a boson gas are larger than those in a classical gas. For a photon gas the standard deviation Δn_j of the number of photons in a particular state j, above and below the mean value $\bar{n}_j$, is $\sqrt{[\bar{n}_j(\bar{n}_j + 1)]}$ and consequently the fractional fluctuation is

$$\Delta n_j/\bar{n}_j = \sqrt{(\bar{n}_j + 1)/\bar{n}_j} = e^{\hbar\omega/2kT} \quad (25\text{-}14)$$

This is also equal to the fractional fluctuation of energy density $\Delta e_j/e_j$ or of intensity $\Delta I_j/I_j$ of the standing wave having frequency $\omega_j/2\pi$. This quantity is always greater than unity, indicating that the standard deviation of the intensity of a standing wave is equal to or greater than its mean intensity.

Such large fluctuations may be unusual for material gases; they are to be expected for standing waves. If the jth wave is fairly steadily excited (i.e., if $\bar{n}_j > 1$, i.e., if $e^{\hbar\omega_j/kT} < 2$) then it will be oscillating more or less sinusoidally and its intensity will vary more or less regularly between zero and twice its mean value, which corresponds to a standard deviation roughly equal to its mean value. If, on the other hand, the standing wave is excited only occasionally, the sinusoidal oscillation will occur only occasionally and the amplitude will be zero in between times. In this case the standard deviation will be larger than the mean. Thus Eq. (25-14) is not as anomalous as it might appear at first.

Statistical Mechanics of a Boson Gas

When the bosons comprising the gas are material particles, rather than photons, μ is not zero but is determined by the mean particle density. The particle energy ε is not $\hbar\omega$ but is the kinetic energy $p^2/2m$ of the particle if m is its mass. We have already shown [see Eqs. (19-13) and (21-7)] that, for elementary particles in a box of "normal" size, the translational levels are spaced closely enough so that we can integrate over phase space instead of summing over particle states. The number of particle states in an element $dx\,dy\,dz \times dp_x\,dp_y\,dp_z = dV_q\,dV_p$ of phase space is $g(dV_q\,dV_p/h^3)$ where g is the multiplicity factor caused by the particle spin. If the spin is s and no magnetic field is present, $g = (2s + 1)$ different spin orientations have the same energy ε. Therefore the sum for $\bar{N}$ of Eq. (25-1) becomes

$$\bar{N} = (g/h^3) \int \cdots \int \left[e^{(\varepsilon - \mu)/kT} - 1 \right]^{-1} dV_q\,dV_p$$

$$= (gV/h^3) \int_0^{2\pi} d\beta \int_0^\pi \sin\alpha\,d\alpha \int_0^\infty \left[e^{(\varepsilon - \mu)/kT} - 1 \right]^{-1} p^2\,dp \quad (25\text{-}15)$$

where angles α and β are the spherical angles in momentum space of Eq. (12-1).

We can integrate over α and β and, since $\varepsilon = (p^2/2m)$ or $p = \sqrt{(2m\varepsilon)}$, we can change to ε for the other integration variable, so

$$\bar{N} = 2\pi g V \left(\frac{2m}{h^2}\right)^{3/2} \int_0^\infty \frac{\sqrt{\varepsilon}\,d\varepsilon}{e^{(\varepsilon - \mu)/kT} - 1}$$

$$= g V \left(\frac{2\pi m k T}{h^2}\right)^{3/2} f_{1/2}(-\mu/kT) \quad (25\text{-}16)$$

where

$$f_m(x) = \frac{1}{m!} \int_0^\infty \frac{z^m\,dz}{e^{z+x} - 1} = \sum_{n=1}^\infty (e^{-nx}/n^{m+1}) \to e^{-x}, \qquad x \to \infty$$

$$(df_m/dx) = -f_{m-1}(x), \qquad m > 0 \quad (25\text{-}17)$$

[See Eq. (18-14) for a definition of $m!$, for m a half-integer.] The series for f_m converges if x is positive. However we recollect that with Bose–Einstein statistics μ must be less than the lowest energy level, which is zero for gas particles. Therefore μ is negative and $x = -(\mu/kT)$ is positive, and the series does converge.

It should be pointed out that the change from summation to integration has one defect; it leaves out the ground state $\varepsilon = 0$. This term, in the sum of particle states, is the largest term of all; in the integral approximation it is completely left out because the density function $\sqrt{\varepsilon}$ goes to zero there. Ordinarily this does not matter, for there are so many terms in the sum for $\bar{N}$ for ε small compared to kT (which *are* included in the integral) that the omission of this one term makes a negligible difference in the result. At low temperatures, however, bosons "condense" into this lowest state [see Eq. (24-13)] and its population becomes much greater than that for any other state. We shall find that above a limiting temperature T_B the ground state is no more densely populated than many of its neighbors and that it can then be neglected without damage. Below T_B, however, the lowest state begins to collect more than a normal share of particles, and we have to add an extra term to the integral for $\bar{N}$, corresponding to the number of particles that have "condensed" into the zero-energy state.

We should have mentioned this complication when we were discussing a photon gas, of course, for the integrals of (25-9) to (25-11) also have left out the zero-energy state. But a photon of zero energy has zero frequency, so this lowest energy state represents a static electromagnetic field. We do not usually consider a static field as an assemblage of photons and, furthermore, the exact number of photons present is not usually of interest; the measurable quantities are the energy density and intensity. For more-material bosons, however, the mean number of particles can be measured directly, so we must account for the excess of particles in the zero-energy state when the gas is degenerate.

Thermal Properties of a Boson Gas

The value of $-\mu$ is determined implicitly by the equation

$$\eta \equiv \bar{N}l_t^3/gV = f_{1/2}(x); \qquad x = -\mu/kT; \qquad l_t = h/\sqrt{2\pi mkT}$$

$$(25\text{-}18)$$

which can be inverted to obtain $-\mu$ as a function of T and η. When the parameter η is small (low density and/or high temperature), $f_{1/2}$ has its limiting form $e^{-x} = e^{\mu/kT}$ and

$$\mu \to -kT \ln(g V/\bar{N}l_t^3) = kT \ln \eta, \qquad \eta \to 0$$

which is the value for a classical, perfect gas of Eqs. (23-7). A better approximation can be obtained by inverting the series for $f_{1/2}(x)$,

$$\eta = f_{1/2}(x) = e^{-x} + 2^{-3/2}e^{-2x} + 3^{-3/2}e^{-3x} + \cdots$$

or
$$e^{-x} = \eta - 2^{-3/2}\eta^2 + (2^{-2} - 3^{-3/2})\eta^3 - \cdots$$

$$x = -\ln \eta + 2^{-3/2}\eta - 3(2^{-4} - 3^{-5/2})\eta^2 \cdots \qquad (25\text{-}19)$$

From these series we can compute $x = -(\mu/kT)$ as a function of η, which is proportional to the particle density and inversely proportional to the $\frac{3}{2}$ power of T. The first two columns of Table 25-1 give x for a few values of η, up to $\eta = 2.612$, for which value $x = 0$; the integral diverges for x negative, as was pointed out earlier.

Over the range $0 \le \eta < 2.612$ the integral for the internal energy U differs from that for $\bar{N}$ only by containing an extra factor ε inside the integral; thus U is proportional to $f_{3/2}(x)$. The integral for the grand potential Ω can be shown, by integration by parts, to be proportional to U. The expression for the entropy can be obtained from the equation $S = (U - \Omega - \mu\bar{N})/T$, or else by differentiation of Ω with respect to T,

TABLE 25-1

Functions for a Boson Gas

η	x	T/T_B	$PV/\bar{N}kT$	$S/\bar{N}k$	$2C_v/3\bar{N}k$	$N_c/\bar{N}$
0	∞	∞	1.000	∞	1.00	0
0.1	2.342	8.803	0.977	4.784	1.01	0
1	0.358	1.897	0.818	2.403	1.09	0
2	0.033	1.195	0.637	1.625	1.19	0
2.5	0.001	1.030	0.536	1.341	1.26	0
2.612	0	1.000	0.513	1.282	1.28	0
3	0	0.912	0.447	1.116	1.12	0.129
10	0	0.409	0.134	0.335	0.33	0.739
30	0	0.196	0.045	0.112	0.11	0.913
∞	0	0	0	0	0	1.000

$$U = 2\pi g V\left(\frac{2m}{h^2}\right)^{3/2} \int_0^\infty \frac{\varepsilon^{3/2}\, d\varepsilon}{e^{(\varepsilon-\mu)/kT} - 1} = \frac{3}{2}\left(\frac{\bar{N}kT}{\eta}\right) f_{3/2}(x)$$

$$= \tfrac{3}{2}\bar{N}kT[1 - 2^{-5/2}\eta - 2(3^{-5/2} - 2^{-4})\eta^2 - \cdots]$$

$$-\Omega = P_tV = -2\pi kTg V\left(\frac{2m}{h^2}\right)^{3/2} \int_0^\infty \sqrt{\varepsilon}\, d\varepsilon$$

$$\times \ln[1 - e^{(\mu-\varepsilon)/kT}] = \tfrac{2}{3}U$$

$$= \bar{N}kT[1 - 2^{-5/2}\eta - 2(3^{-5/2} - 2^{-4})\eta^2 - \cdots] \qquad (25\text{-}20)$$

$$S = -\left(\frac{\partial\Omega}{\partial T}\right)_{V\mu} = \frac{5}{3}\left(\frac{U}{T}\right) - \mu\,\frac{\bar{N}}{T} = \bar{N}k\left[x + \frac{5}{2}\frac{f_{3/2}(x)}{f_{1/2}(x)}\right]$$

$$= \bar{N}k[-\ln\eta + \tfrac{5}{2} - 2^{-7/2}\eta - 2(3^{-5/2} - 2^{-4})\eta^2 - \cdots]$$

$$C_v = \left(\frac{\partial U}{\partial T}\right)_{\bar{N}V} = \tfrac{3}{2}\bar{N}k[1 + 2^{-7/2}\eta + 4(3^{-5/2} - 2^{-4})\eta^2 + \cdots]$$

Values of these quantities are given in Table 25-1 for a few values of η between 0 and 2.612. We see that when η is small (particle density $\bar{N}/V$ is small or T is large or both) the expressions for U, S, C_v, etc., do not differ much from the corresponding quantities for a perfect gas of MB particles. As η increases, however, these quantities diverge from the classical values; U, PV, and S all dropping below the perfect-gas values. The bosons tend to congregate, more than maxwellons, in the lower energy levels, which contribute less to pressure and entropy. On the other hand the heat capacity is some-what larger than the classical value of $\tfrac{3}{2}\bar{N}k$. When η reaches the value 2.612, μ equals zero; for any larger values of η the boson gas becomes degenerate, exhibiting properties which differ markedly from those of a MB gas.

Before discussing the degenerate state, we should point out that the relation between P, T, and V for the adiabatic expansion of a boson gas is exactly the same as for a maxwellon gas. Since

$$(S/\bar{N}k) = \tfrac{5}{2}[f_{3/2}(-\mu/kT)/f_{1/2}(-\mu/kT)] + (-\mu/kT)$$

and

$$(\bar{N}/VT^{3/2}) = g(2\pi mk/h^2)^{3/2} f_{1/2}(-\mu/kT)$$

are functions of $(-\mu/kT)$ alone and since, in adiabatic expansion, both S and $\bar{N}$ are constant, we see that $(-\mu/kT)$ and $VT^{3/2}$ remain constant during an adiabatic expansion of a boson gas. But the equation $VT^{3/2} = $ constant is the equation for the adiabatic expansion of a MB perfect gas [see Eq. (4-12)]. Also since

$$PV^{5/3} = V^{2/3}(\bar{N}kT/\eta)f_{3/2}(-\mu/kT)$$
$$= gk\left(\frac{2\pi mk}{h^2}\right)^{3/2}(V^{2/3}T)(VT^{3/2})f_{3/2}(-\mu/kT)$$

this product also is constant during an adiabatic expansion.

The Degenerate Boson Gas

As the density of particles is increased and/or the temperature is decreased, η increases, $x = \mu/kT$ decreases and the thermal properties of the gas depart farther and farther from those of a classical perfect gas, until at $\eta = 2.612$, μ becomes zero. If η becomes larger than this, Eq. (25-16) no longer can be satisfied. For the maximum value of $f_{1/2}(x)$ is 2.612, for $\mu = 0$, and μ cannot become positive. The only way the additional particles can be accommodated is to "condense" them into the hitherto-neglected zero-energy state mentioned several pages back.

If $\bar{N}$ is held constant and T is reduced, the condensation starts when $\eta = 2.612$, and thus when T reaches the value

$$T_B = (h^2/2\pi mk)(\bar{N}/2.612gV)^{2/3} = 3.31(h^2/mk)(\bar{N}/gV)^{2/3}$$

$$= 0.1124\rho^{2/3} \,^\circ K \qquad \text{for} \quad \text{He}^4, \text{ with } \rho \text{ in kg/m}^3 \qquad (25\text{-}21)$$

Any further reduction of T will force some of the particles to condense into the zero-energy state. In fact the number N_x of particles that can stay in the upper states are those which satisfy Eq. (25-16) with $\mu = 0$.

$$N_x = 2.612gV(2\pi mkT/h^2)^{3/2} = \bar{N}(T/T_B)^{3/2} \qquad (25\text{-}22)$$

and the rest,

$$N_c = \bar{N}[1 - (T/T_B)^{3/2}]$$

are condensed in the ground state, exerting no pressure and carrying no energy. Therefore, the thermodynamic functions for the gas in this partly condensed state are

$$PV = -\Omega = \tfrac{2}{3}U = 0.513\,\bar{N}kT(T/T_B)^{3/2}$$

$$= 0.086\frac{m^{3/2}gV}{h^3}(kT)^{5/2}$$

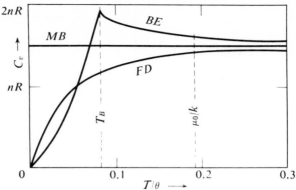

FIGURE 25-2

Heat capacity of a gas (according to the three statistics) versus temperature in units of $\theta = (h^2/mk)(\bar{N}/gV)^{2/3}$.

or

$$P = 0.086(m^{3/2}g/h^3)(kT)^{5/2}$$

$$S = 5U/3T = 1.28\,\bar{N}k(T/T_B)^{3/2} = \tfrac{2}{3}C_v \qquad (25\text{-}23)$$

The pressure is independent of volume, because this is all the pressure the uncondensed particles can withstand. Further reduction of volume simply condenses more particles into the ground state, where they contribute nothing to the pressure. The heat capacity of the gas as a function of T has a discontinuity in slope at T_B, as shown in Figure 25-2. At high temperatures the gas is similar to an MB gas of point particles, with $C_v = \tfrac{3}{2}\bar{N}k$. As T is diminished C_v rises until, at $T = T_B$, it has its largest value, $C_v = 1.92\bar{N}k$. For still smaller values of T, C_v decreases rapidly, to become zero at $T = 0$.

The "condensed" particles are not condensed in position space, as with the usual change of phase from gas to liquid or solid; they are condensed in momentum space, at $p = 0$, a set of stationary particles, distributed at random throughout the volume V, having zero energy and zero entropy. As such they resemble the particles of superfluid in helium II, discussed in Chapter 7.

We shall indicate, in Chapter 27, that deuterium, H^2, with 1 proton and 1 neutron, and He^4, with 2 protons and 2 neutrons in its nucleus, both should obey the BE statistics, whereas protium, H^1, with a single proton, and He^3, with 2 protons and 1 neutron, should obey FD statistics. Hydrogen gas condenses into a solid before the temperature reaches the transition temperature T_B, where the boson condensation is exhibited. But helium atoms have less mutual attraction, so helium is the only substance that is still liquid at temperatures low enough to exhibit BE condensation (we shall see in Chapter 27 that hydrogen exhibits other quantum statistical effects at temperatures that are low but not so low as to turn the hydrogen into a liquid).

As described in Chapter 7, He^4 liquid does go through a phase change of the second kind at about 2.2°K, which may correspond to a BE condensation. Below this temperature the liquid behaves as though part of it were a superfluid, with zero entropy and energy, the fraction of superfluid present increasing as T approaches zero. It is thus tempting to assume that the superfluid corresponds to the condensed part of the BE gas, which also has zero entropy and energy. This assumption is somewhat strengthened when we compute that for He^4 liquid, of density approximately 145 kg/m^3, the value of T_B, given by Eq. (25-21) is 3.1°K, and still further strengthened when we find that He^3 liquid (which should not obey BE statistics) does not have a phase change at low temperatures and does not exhibit the presence of a superfluid fraction.

Of course the calculations of this chapter were made for a gas of point bosons, whereas helium at these low temperatures is a liquid, so we cannot expect the theoretical model of the preceding pages to correspond too closely to the actual behavior of HeII. It is not surprising that the $T_B = 3.1°$ of the gas model is not in exact agreement with the experimental value of 2.2° for the phase change, or that the heat capacity curve for HeII is not exactly the shape of the curve of Figure 25-2. The measured heat capacity has a more pronounced peak at the λ point and drops more rapidly to zero than does Figure 25-2. Although many improvements have been made in the theoretical model, to make it represent more nearly a boson liquid rather than a gas, and although some of these improvements have improved the fit between theory and experiment, no completely satisfactory theory of HeII has yet been devised. It is generally agreed, however, that the superfluid does correspond, in some sense, to the condensed portion of a degenerate boson fluid.

CHAPTER

26

FERMI-DIRAC STATISTICS

Fermi–Dirac statistics is appropriate for electrons and other elementary particles that are subject to the Pauli exclusion principle. The occupation numbers n_j can only be zero or unity and the mean number of particles in state j is the $\bar{n}_j$ of Eq. (24-10). In this chapter we shall work out the thermal properties of a gas of fermions, to compare with those of a gas of bosons and with those of a perfect gas of MB particles, particularly in the region of degeneracy. There are no FD analogues to photons, with $\mu = 0$.

General Properties of a Fermion Gas

For a gas of fermions, at temperature T in a volume V, the particle energy is $\varepsilon = p^2/2m$ as before, and the number of allowed translational states in the element of phase space $dV_q\, dV_p$ is $g(dV_q\, dV_p/h^3)$ as before (g is the spin multiplicity $2s + 1$). Integrating over dV_q and over all directions of the momentum vector, we find the number of states with kinetic energy between ε and $\varepsilon + d\varepsilon$ is $2\pi g V(2m/h^2)^{3/2}\sqrt{\varepsilon}\, d\varepsilon$, as before. Multiplying this by n_j [Eq. (24-10)] gives us the mean number of fermions with kinetic energy between ε and $\varepsilon + d\varepsilon$,

$$dN = 2\pi g V(2m/h^2)^{3/2}\frac{\sqrt{\varepsilon}\, d\varepsilon}{e^{(\varepsilon - \mu)/kT} + 1} \tag{26-1}$$

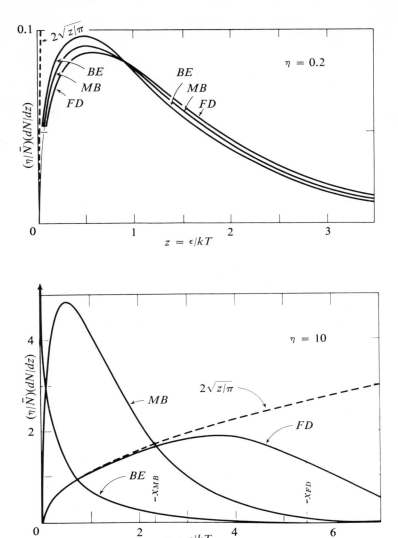

FIGURE **26-1**

    Density of particles per energy range for a gas, according to the three statistics, for nondegenerate and degenerate conditions. Dashed curve $2(z/\pi)^{1/2}$ *corresponds to a density of one particle per particle state. Area under each curve equals* η.

which is to be compared with the integrand of Eq. (25-16) for the boson gas and with $dN = 2\pi g V (2m/h^2)^{3/2} e^{(\mu - \varepsilon)/kT} \sqrt{\varepsilon}\, d\varepsilon$ for a perfect gas of MB particles.

Figure 26-1 compares plots of $dN/d\varepsilon$ for these three statistics for two different degrees of degeneracy. As $\eta = (N/gV)(h^2/2\pi mkT)^{3/2}$ varies, the MB distribution changes scale, but not shape. For small values of η, the values of $\mu = -xkT$ for the three cases are all negative and do not differ much in value, nor do the three curves differ much in shape. In this region the MB approximation is satisfactory.

For large values of η the curves differ considerably, and the values of the chemical potential μ differ greatly for the three cases. For bosons, as we saw in Chapter 25, μ is zero and a part of the gas has "condensed" into the ground state, making no contribution to the energy or pressure of the gas, and being represented on the plot by the vertical arrow at $z = 0$. For fermions, μ is positive, and the states with ε less than μ are practically completely filled, whereas those with ε greater than μ are nearly empty. Because of the Pauli principle, no more than one particle can occupy a state; at low temperatures and/or high densities the lowest states are filled, up to the energy $\varepsilon = \mu$, and the states for $\varepsilon > \mu$ are nearly empty, as shown by the curve (which is the parabolic curve $2\sqrt{(z/\pi)}$ for z less than $-x$ and which drops to zero shortly thereafter).

The dotted parabola $2\sqrt{(z/\pi)}$ corresponds to the level density $dN/d\varepsilon = 2\pi g V (2m/h^2)^{3/2}\sqrt{\varepsilon}$, corresponding to one particle per state. We see that the curve for MB particles rises above this for η large, corresponding to the fact that some of the lower levels have more than one particle apiece. The BE curve climbs still higher at low energies. The FD curve, however, has to keep below the parabola everywhere.

The conduction electrons in a metal are the most accessible example of a fermion gas. In spite of the fact that these electrons are moving through a dense lattice of ions, they behave in many respects as though the lattice were not present. Their energy distribution is more complicated than the simple curves of Figure 26-1 and, because of the electric forces between them and the lattice ions, the pressure they exert on the container is much less than that exerted by a true gas; nevertheless their heat capacity, entropy, and mean energy are remarkably close to the Fermi-gas values. Measurements on conduction electrons constitute most of the verifications of the theoretical model to be described in this chapter.

Intermediate Temperatures

In correspondence with Eq. (25-16) we can set the equation relating the total number $\bar{N}$ of fermions to the chemical potential μ per fermion, into the form

$$\bar{N} = 2\pi g V \left(\frac{2m}{h^2}\right)^{3/2} \int_0^\infty \frac{\sqrt{\varepsilon}\, d\varepsilon}{e^{(\varepsilon - \mu)/kT} + 1} \qquad \text{or} \qquad \eta = F_{1/2}(y)$$

$$\eta = (\bar{N}/gV)(h^2/2\pi mkT)^{3/2}; \qquad y = (\mu/kT) \tag{26-2}$$

$$F_m(y) = \frac{1}{m!} \int_0^\infty \frac{z^m\, dz}{e^{(z-y)} + 1} = e^y - 2^{-m-1}e^{2y} + 3^{-m-1}e^{3y} - \cdots$$

where the last series is valid for negative values of y. Note the sign difference between x and y, and in the denominator of the integrand, as compared with Eqs. (25-16).

From this we can obtain series expansions for the thermodynamic functions of the fermion gas, which converge when $\eta < 1$,

$$e^y = \eta + 2^{-3/2}\eta^2 + (2^{-2} - 3^{-3/2})\eta^3 + \cdots$$

$$y = (\mu/kT) = \ln \eta + 2^{-3/2}\eta + 3(2^{-4} - 3^{-5/2})\eta^2 + \cdots$$

$$PV = -\Omega = \tfrac{2}{3}U = (\bar{N}kT/\eta)F_{3/2}(y) \tag{26-3}$$

$$= \bar{N}kT[1 + 2^{-5/2}\eta + 2(2^{-4} - 3^{-5/2})\eta^2 + \cdots]$$

$$S = [(U - \Omega - \mu\bar{N})/T] = \bar{N}k[-\ln \eta + \tfrac{5}{2} + 2^{-7/2}\eta$$

$$+ 2(2^{-4} - 3^{-5/2})\eta^2 + \cdots]$$

$$C_v = \tfrac{3}{2}\bar{N}k[1 - 2^{-7/2}\eta - 4(2^{-4} - 3^{-5/2})\eta^2 - \cdots]$$

These formulas show that for intermediate temperatures and densities the pressure and internal energy, for a given temperature and density, is somewhat larger than the corresponding values for a MB gas, and still larger than those for a BE gas. The constraint of the Pauli principle already shows up in forcing more particles into states with greater momentum, which produce greater pressure. Values for these quantities are tabulated in Table 26-1. The quantity η is equal to $A\rho T^{-3/2}$, where ρ is the density of the gas in kg/m^3, T is in degrees Kelvin, and where $A = 1.59$ for protons or neutrons and $A = 2.36 \times 10^8$ for electrons.

TABLE 26-1

Functions for a Fermion Gas

η	y	kT/μ_0	PV/NkT	S/Nk	$2C_v/3Nk$
0	$-\infty$	∞	1.000	∞	1.000
0.01	-4.60	17.81	1.001	7.1	0.997
0.1	-2.26	3.841	1.017	4.8	0.989
1	$+0.35$	0.827	1.174	2.6	0.919
10	$+5.46$	0.178	2.521	0.85	0.529
100	$+26.0$	0.038	10.48	0.18	0.145
316	$+56.0$	0.008	22.48	0.09	0.084
∞	$+\infty$	0	∞	0	0

The Completely Degenerate Fermion Gas

As T approaches zero the FD distribution takes on its fully degenerate form, with all states up to the $\bar{N}$th completely filled and all states beyond the $\bar{N}$th completely empty. In other words, the limiting value of μ (call it μ_F) is large and positive, and

$$dN = \begin{cases} 2\pi g V(2m/h^2)^{3/2}\sqrt{\varepsilon}\,d\varepsilon & 0 \leq \varepsilon \leq \mu_F \\ 0 & \varepsilon > \mu_F \end{cases} \qquad (26\text{-}4)$$

where μ_F has the value that allows the integral of dN to equal $\bar{N}$,

$$\bar{N} = 2\pi g V(2m/h^2)^{3/2} \int_0^{\mu_F} \sqrt{\varepsilon}\,d\varepsilon$$

$$\mu_F = (h^2/2m)(3\bar{N}/8\pi V)^{2/3} \qquad (26\text{-}5)$$

for particles of spin $\frac{1}{2}$ (electrons, protons, or neutrons, for example) for which $g = 2$. Numerically

$$\mu_F = \begin{cases} 0.625 \times 10^{-17}\rho^{2/3}\,\text{joule}, & \text{or } 39\rho^{2/3}\,\text{ev for electrons} \\ 0.227 \times 10^{-22}\rho^{2/3}\,\text{joule}, & \text{or } 1.42 \times 10^{-4}\rho^{2/3}\,\text{ev} \\ & \qquad\qquad\qquad\qquad \text{for protons} \end{cases}$$

Alternatively the top of the occupied levels can be expressed in terms of an equivalent temperature, the Fermi temperature

$$T_F = (\mu_F/k) = \begin{cases} 4.52 \times 10^5 \rho^{2/3} \ ^\circ\text{K for electrons} \\ 1.64\rho^{2/3} \ ^\circ\text{K for protons or neutrons} \end{cases}$$

In all cases the density ρ of the gas is in kg per m^3 (the density of water is 1000 kg/m^3).

Even at absolute zero most of the fermions are in motion, some of them moving quite rapidly. For an electron gas of density $\rho = \bar{N}m/V$ kg/m^3, the kinetic energy of the fastest, μ_F, is roughly equal to $40\rho^{2/3}$ electron volts; μ_F/k is approximately equal to $4.5 \times 10^5 \rho^{2/3}\,^\circ\text{K}$. In other words the top of the occupied levels (the Fermi level) corresponds to the mean energy $[=\frac{3}{2}kT]$ of a MB particle in a gas at the temperature $3 \times 10^5 \rho^{2/3}\,^\circ\text{K}$. For the conduction electrons in metals, where $\rho_{el} \simeq 0.1$ kg/m^3, this corresponds to about 100,000°K; for free electrons in a white-dwarf star, where $\rho_{el} > 1000$, it corresponds to more than 5×10^7 degrees. Until the actual temperature of a Fermi gas is larger than this value, it remains degenerate. The parameter $\eta = 0.752(T_F/T)^{3/2}$ is a good index of the onset of degeneracy (when $\eta > 1$ there is degeneracy).

The internal energy of the completely degenerate gas (which, like the boson gas, is equal to $-\frac{3}{2}\Omega$ at all temperatures), is

$$U_0 = \int_0^{\mu_0} \varepsilon \, dN = \tfrac{3}{5}\bar{N}\mu_F = -\tfrac{3}{2}\Omega; \qquad S_0 = 0 \qquad (26\text{-}6)$$

Therefore the pressure of the degenerate fermion gas is

$$P_0 = \left(\frac{-\Omega}{V}\right) = \tfrac{2}{5}\mu_F\left(\frac{\bar{N}}{V}\right) = \begin{cases} 2.71 \times 10^7 \rho \text{ atm for electrons} \\ 5.36 \times 10^{-2}\rho \text{ atm for protons} \end{cases}$$

Thus even at absolute zero a fermion gas exerts a pressure. If the electrons in copper were electrically neutral they would exert a pressure of about 5×10^5 atm. Because of the strong electrical attraction to the ions of the crystal lattice this pressure is largely counterbalanced by the forces holding the crystal together.

Low Temperatures

When T is not zero but is small compared to T_F, so that $\eta \gg 1$ and the gas is degenerate, the series of Eqs. (26-3) do not converge and a different series must be used. This approximation is obtained by using the fact that for the degenerate fermion gas the mean occupation number $\bar{n}(\varepsilon) = [e^{(\varepsilon - \mu)/kT} + 1]^{-1}$ changes very slowly with ε except in the region $\varepsilon \simeq \mu$, where it changes very rapidly. Put another way, $-(\partial \bar{n}/\partial \varepsilon)$, for $\eta \gg 1$, resembles a delta function, being negligibly small except when $|\varepsilon - \mu|$ is small and being large enough at $\varepsilon = \mu$ so that the area under the peak is unity. Therefore any integral of a function $w(\varepsilon)$ times $\bar{n}(\varepsilon)$ can be integrated by parts to obtain

$$\int_0^{\infty} w(\varepsilon)\bar{n}(\varepsilon)\, d\varepsilon = \left[W(\varepsilon)\bar{n}(\varepsilon) \right]_0^{\infty} - \int_0^{\infty} W(\varepsilon)\left(\frac{d\bar{n}}{d\varepsilon} \right) d\varepsilon$$

$$\simeq W(\mu_F) \tag{26-7}$$

when $\eta \gg 1$. Function

$$W(\varepsilon) = \int_0^{\varepsilon} w(x)\, dx$$

and we have assumed that $W(\varepsilon)\bar{n}(\varepsilon) \to 0$ as $\varepsilon \to \infty$.

For example, Eq. (26-5) can be written

$$\bar{N} = 2\pi g V(2m/h^2)^{3/2} \int_0^{\infty} \sqrt{\varepsilon}\,\bar{n}(\varepsilon)\, d\varepsilon \simeq -\pi g V(2m\mu_F/h^2)^{3/2}$$

But we need another term in our series to obtain expressions for the heat capacity and entropy. To do this we can manipulate the integral for the function $F_m(y)$. Integrating by parts, we have

$$F_m(y) = \frac{1}{m!} \int_0^{\infty} \frac{z^m\, dz}{e^{z-y} + 1} = \frac{1}{(m+1)!} \int_0^{\infty} \frac{z^{m+1}e^{z-y}\, dz}{(e^{z-y} + 1)^2}$$

The factor

$$e^{z-y}(e^{z-y} + 1)^{-2} = \left[4\cosh^2\left(\frac{z-y}{2} \right) \right]^{-1}$$

has a sharp maximum at $z = y$ [as mentioned in connection with Eq. (26-7)]. We therefore expand z^{m+1} around $z = y$, setting $z = u + y$,

$$z^{m+1} = (u + y)^{m+1} = y^{m+1} + (m+1)u y^m$$
$$+ \tfrac{1}{2}m(m+1)u^2 y^{m-1} + \cdots$$

The first term in the series is independent of u, so the first term in the series for F_m is

$$-\frac{y^{m+1}}{(m+1)!}\int_0^\infty \frac{d}{dz}(e^{z-y}+1)^{-1}\,dz = \frac{y^{m+1}}{(m+1)!}\frac{1}{e^{-y}+1}$$

$$\simeq \frac{y^{m+1}}{(m+1)!} \qquad y \gg 1$$

The integrals involving odd powers of u are zero since the factor $[1/4\cosh^2(u/2)]$ is symmetric about $u = 0$. Thus the next nonzero term in the series for F_m is the one with u^2. This will be a small correction to the first term as long as $\eta \gg 1$ and $y \gg 1$. Therefore the integral over u from $-y$ to ∞ can be changed to an integral over u from $-\infty$ to ∞ without much error, and the second term in the series for F_m is approximately

$$\frac{y^{m-1}}{2(m-1)!}\int_{-\infty}^\infty \frac{u^2 e^u\,du}{(e^u+1)^2} = \frac{y^{m-1}}{(m-1)!}\int_0^\infty \frac{u^2 e^{-u}\,du}{(1-e^{-u})^2}$$

$$= \frac{y^{m-1}}{(m-1)!}\int_0^\infty u^2[e^{-u} - 2e^{-2u}$$

$$+ 3e^{-3u} - \cdots]\,du$$

$$= \frac{2y^{m-1}}{(m-1)!}[1 - 2^{-2} + 3^{-2} - \cdots]$$

$$= \frac{\pi^2}{6}\frac{y^{m-1}}{(m-1)!}$$

since the sum in bracket equals $(\pi^2/12)$. Therefore to the second order in the small quantity y^{-1} we can write

$$F_m(y) \simeq \frac{y^{m+1}}{(m+1)!}[1 + \tfrac{1}{6}\pi^2 m(m+1)y^{-2} + \cdots] \qquad (26\text{-}8)$$

With this formula we can now write out the various thermodynamic functions for the Fermi gas. Since

$$\eta = F_{1/2}(y) \simeq \frac{4}{3\sqrt{\pi}}y^{3/2} + \frac{\pi^{3/2}}{6}y^{-1/2}$$

then

$$y^{3/2} \simeq \frac{3\sqrt{\pi}}{4}\eta - \frac{\pi^2}{8}\left(\frac{4}{3\sqrt{\pi}\eta}\right)^{1/3}$$

and

$$y = (T_F/T) - \frac{\pi^2}{12}(T/T_F)$$

where the formula for T_F is given below Eq. (26-5). Therefore

$$PV = -\Omega = -U = (\bar{N}kT/\eta)F_{3/2}(y)$$

$$\simeq \bar{N}kT_F\left[\frac{2}{5} + \frac{\pi^2}{6}(T/T_F)^2\right] \qquad T \ll T_F \qquad (26\text{-}9)$$

$$S = (1/T)(U - \Omega - \mu\bar{N}) \simeq \tfrac{1}{2}\pi^2 \bar{N}k(T/T_F) \simeq C_v$$

These results are very different from those for the classical MB gas. For $T \ll T_F$ the particles are packed into the lower states up to their full capacity of one per state. The number of filled states is large enough so that the mean energy and pressure are large, but the degree of disorder is small since only the levels near $\varepsilon = \mu_F$ are partly filled and disorder is possible only for these levels. Addition of heat cannot move most particles from the states they occupy since there are no states, near them in energy, that are unoccupied and thus open to receive them. Thus most of the particles are "frozen in" and cannot be affected by external forces; only the "top fringe" of the particles, those with energy nearly equal to μ_F, can be disturbed easily. The heat capacity is considerably smaller than the $\tfrac{3}{2}\bar{N}k$ of a perfect MB gas, when $T \ll T_F$.

Electrons in Metals

At first consideration it seems quite unlikely that the electrons in a crystal lattice could be considered to be equivalent to a perfect gas, Fermi or MB. Each atom in the crystal has around it a number of electrons equal to the atom's nuclear charge. When the atoms are separated, as in a gas, all the electrons are held in bound states, with separations between their allowed energies so large that they are unaffected by thermal motion. Thus at normal temperatures the electrons in each gas atom are in their lowest states; the probability of finding one in an upper state is vanishingly small.

This is shown in Figure 26-2a, where the curved lines are the electronic potential energy as a function of distance along a line through the nucleus and the horizontal lines represent levels occupied by a pair of electrons, one for each spin orientation. The lowest level is for the K shell of 2 electrons, the next four are for the L shell, and so on. For sodium, with 11 electrons, the K and L shells are completely filled at ordinary temperatures and the extra electron half-occupies the lowest level of the M shell, which is represented by the dashed line.

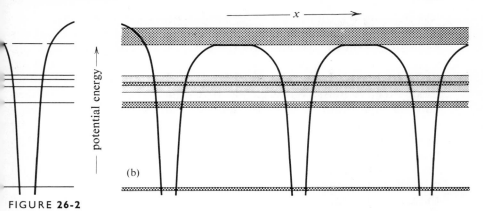

FIGURE 26-2
Plots of electronic potential energy and of filled energy levels for (a) a single atom and (b) a crystal lattice of atoms.

When N atoms are brought together to form a crystal lattice the potential energies of the nuclei overlap and each allowed atomic energy spreads out into a band of N levels. The band arising from the tightly bound K shell is very narrow, those for the L shell spread out somewhat more, and some of them may overlap in energy. In sodium, however, these levels are all filled, so these electrons cannot take part in thermal behavior unless they are pulled all the way up to an upper level. At the height of the M shell, however, the overlap has spread the levels into a wide band, which begins to be similar to the energy distribution of free electrons. The potential overlap is such that, at this level, there is no potential barrier between the lattice atoms.

The corresponding potential distribution, along a line between nuclei, is shown by the curved lines in Figure 26-2b. The energy bands, arising from each atomic level, are also shown. The potential overlap is such that there is a region of nearly constant potential between the nuclei; in the three-dimensional lattice the volume in which the potential is nearly constant is rather larger than the volume occupied by the potential "wells" around each nucleus. The M-shell electrons in sodium move freely in this interspace, avoiding the closed shells around each ion. The electrons in the ionic K and L shells are in filled bands, which have not spread out much, are still tightly bound to their individual nuclei, and contribute nothing to the heat capacity or the electrical or thermal conductivity of the crystal. The M-shell electrons, one per atom in sodium, are more or less free and only half-fill the M band, so they can contribute to the crystal's thermal and electrical properties. Their electronic charge is neutralized, on the average, by the net positive charge on each ion.

In a very approximate way the energy difference between individual levels in a band can be said to be inversely proportional to an effective mass. With free electrons the momentum is quantized and the separation between the allowed values of the kinetic energy $(p^2/2m)$ is inversely proportional to the particle's mass m. In the lower, narrow bands in a crystal the separation between levels is much smaller than those for free electrons, so that these electrons behave as though they had a much larger mass. In the upper band of sodium the spread is nearly as large as for free electrons, so that the usual electronic mass can be used in computing their behavior. For semiconductors, the upper band is not so widely spread and the electrons act as though they had greater mass. In the non-conductors the upper occupied band is completely filled so that even these electrons cannot contribute to the crystal's thermal behavior unless they are given enough energy to raise one or more to a higher, unoccupied band, an energy much larger than the usual kT. In this book we cannot do more than sketch the behavior of electrons in good conductors, where the upper electrons are more or less free to travel in the spaces between the ions and where the upper conduction band is not completely filled.

Conduction Electrons in Good Conductors

Before the advent of the quantum theory an attempt was made to explain the electrical, magnetic, and thermal properties of good conductors by assuming that their conduction electrons constituted a MB gas. In this case the electrons should have a heat capacity, over and above that of the lattice [as given by Eq. (20-16)], which should equal $\frac{3}{2}N_c k$. If the number N_c of conduction electrons were equal to the number N of atoms in the crystal, this should constitute one-third of the crystal's total heat capacity at room temperatures and should constitute the majority of its heat capacity at low temperatures, where C_v for the lattice goes to zero as T^3.

This prediction is not borne out by measurement. The heat capacity of a metal at room temperatures is very little larger than $3Nk$, the value for the lattice alone, and this quantity drops off, as T is reduced, as predicted by the Debye formula, Eq. (20-16), until T is somewhat below 10°K, where a small term, linear in T, becomes apparent. For nonconductors the T^3 dependence seems to be valid down to $T = 0$. These measurements hint that the heat capacity of the conduction electrons is linearly proportional to T, with a proportionality constant small enough so that this capacity is still much smaller than $\frac{3}{2}N_c k$ at room temperatures. We cannot assume that the electrons become bound to the ion at very low temperatures because they still conduct electric current at low temperatures—better, in fact, than they do at room temperatures.

Other properties of metals also disagree with the assumption that the conduction electrons constitute a MB gas. For example, each conduction electron has a magnetic moment equal to a Bohr magneton $(eh/4\pi m) = 0.93 \times 10^{-23}$ amp-m^2 = $\mathfrak{m}_B$. If it satisfies the MB statistics the magnetization of these electrons should be that given by Eqs. (13-26), with $J = \frac{1}{2}$. For fields of moderate intensity the Curie law

$$\mathfrak{M} = [N_c(\mu_0 \mathfrak{m}_e)^2/3kT] \tag{26-10}$$

should hold, with $\mathfrak{m}_e = \sqrt{(\frac{3}{4})}\mathfrak{m}_B$. Experimentally, however, the magnetization of a good conductor is several hundred times smaller than this (except for ferromagnetic metals); also it does not vary proportionally to the reciprocal of T (in fact it is nearly independent of T).

Likewise the electrical conductivity of the MB gas of electrons should be predicted by a formula similar to Eq. (14-5). The collisions which disrupt the free motion of these electrons are not electron-electron collisions, but are collisions with the lattice ions. Thus, according to Eq. (12-12) the electronic mean free path λ should equal the reciprocal of the cross section σ for collision between an electron and a lattice ion, divided by the number (N/V) of atoms per unit volume. The electric conductivity, from Eq. (14-5), should then be

$$(I/\mathfrak{E}) \simeq (N_e/V)(e^2\lambda/m)\langle 1/v \rangle = (N_e/N)(e^2/m\sigma)\langle 1/v \rangle \tag{26-11}$$

where, if there is one conduction electron per atom, the first factor in the last expression should be unity. Quantity $\langle 1/v \rangle$ is the mean value of the reciprocal of the electron's velocity, averaged over the Maxwell distribution; we showed it to be equal to $\sqrt{(2m/\pi kT)}$.

This prediction also is not confirmed by experiment. The dependence on temperature is not proportional to $T^{-1/2}$ but more nearly proportional to T^{-1} at room temperatures and to T^{-5} at temperatures below 50°K (not to mention the superconductive state, which cannot be explained by any simple transport theory). If we assume that somehow the formula (26-11) is correct we have to conclude that the mean free path, and thus the cross section, change markedly with temperature. For example, for silver at room temperatures the conductivity is about 7×10^7 mho/m and with $(N_c/V) \simeq 6 \times 10^{28}$ m^{-3}, the value of λ to satisfy Eq. (26-11) is about 10^{-8} m, more than 50 times the distance between the atoms in the silver lattice. At $T = 50°K$ the conductivity of silver is such that λ would have to be about 10^{-4} m to satisfy Eq. (26-11).

Finally the thermal conductivity of the metal is also a property which can be used to explore the properties of the conduction electrons. It can be shown that the conduction electrons are much more efficient at transporting heat than the lattice vibrations; in fact the thermal conductivity of metals is almost proportional to their electrical conductivity. For good conductors the lattice vibration carries less than 1 per cent of the total heat flow, so the measured heat conductivity should be approximately equal to that computed for the conduction electrons. But, if these electrons constitute a MB gas, the heat conductivity should be given by Eq. (14-22).

$$\kappa \simeq \tfrac{5}{2}(N_c/V)(\lambda/m)k^2 T \langle 1/v \rangle \tag{26-12}$$

which should vary proportional to $T^{1/2}$ if λ is independent of T. This formula also disagrees with measured values. However, the *ratio* between thermal and electrical conductivity, which from Eqs. (26-11) and (26-12) should come out independent of λ and of $\langle 1/v \rangle$, has the simple form

$$[\kappa/(I/\mathfrak{E})] = \tfrac{5}{2}(k/e)^2 T = LT; \qquad L = 1.85 \times 10^{-8} \tag{26-13}$$

where L is called the Lorentz number.

The constancy of the Lorentz number was demonstrated experimentally by Wiedemann and Franz, but the experimental values of L are not exactly equal to those predicted from Eq. (26-13). Some experimental values are given in Table 26-2 to show how remarkably independent L is of the material as well as of the temperature (at temperatures well below the Debye temperature, how-

TABLE 26-2

Values of the Lorentz Number L, times 10^{-8}

Metal	$T = 0°C$	$T = 100°C$	Metal	$T = 0°C$	$T = 100°C$
Ag	2.31	2.37	Pb	2.47	2.56
Cu	2.23	2.33	Sn	2.52	2.49

ever, L reduces sharply in value). Somehow, although the MB formulas for the electrical and thermal conductivity of the electrons do not fit the data, their ratio fits quite well, except for a constant error of about 15 per cent in magnitude. Perhaps the general form of Eqs. (26-11) and (26-12) is correct although our assumptions regarding the way λ and $\langle 1/v \rangle$ change with T is not correct.

Electrons as a Fermion Gas

Our earlier discussion has indicated what is wrong with Eqs. (26-11) and (26-12). The conduction electrons in a metal are packed densely enough so that they exhibit their fermion properties, as distinguished from those of MB particles. This was first pointed out by Sommerfeld, who derived Eq. (26-9) for the electronic heat capacity. Since T_F is of the order of $10^5\,°\mathrm{K}$ for metals we see that

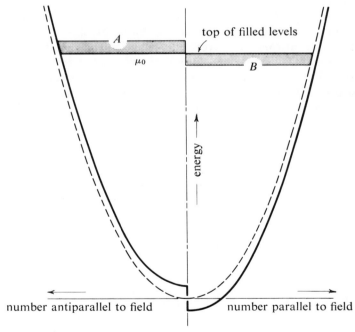

FIGURE **26-3**

Effect of magnetic field on the degenerate Fermi distribution of conduction electrons.

this equation is in agreement with the experimental results, which require that C_v for the electrons be proportional to T and be considerably smaller than $\frac{3}{2}Nk$ at $T \simeq 400°\text{K}$, in contrast to the MB model, which predicts C_v to be $\frac{3}{2}Nk$, for all values of T.

Next, the magnetization of the electron gas cannot be that predicted using MB statistics, for the great majority of the electrons cannot reorient themselves along the magnetic field because there is already another electron in the reoriented state; only near the top of the Fermi band are there free states that allow reorientation. We can show this graphically in Figure 26-3 for $T \ll T_F$, by plotting the number of electrons between ε and $\varepsilon + d\varepsilon$ which have orientation along $\mathfrak{H}$ by distances to the right and the number having orientations opposed to $\mathfrak{H}$ by distances to the left. The dashed curve shows the distribution when $\mathfrak{H} = 0$; equal numbers of each orientation from $\varepsilon = 0$ to $\varepsilon = \mu_F$, so that $\mathfrak{M} = 0$. When $\mathfrak{H} > 0$ the energy of those parallel to $\mathfrak{H}$ is lowered by $\mu_0 \mathfrak{m}_B \mathfrak{H}$ [see Eq. (13-29)] and the energy of those antiparallel to $\mathfrak{H}$ is raised by $\mu_0 \mathfrak{m}_B \mathfrak{H}$, as is indicated by the solid parabolas. (Note that μ_F is the energy of the top of the Fermi band, whereas $\mu_0 = 4\pi \times 10^{-7}$ is the magnetic permeability of vacuum.) The top of the occupied band must still be the same for both orientations, however, for the electrons in the region A find themselves with energy larger than those parallel to $\mathfrak{H}$ and promptly fill in the empty levels in region B by changing their orientation. Thus there is now an excess of electrons with moments parallel to $\mathfrak{H}$, and thus a net magnetization,

$$\mathfrak{M} = \mu_0 \mathfrak{m}_B (\bar{N}_p - \bar{N}_a)$$

where $\bar{N}_p$ is the number of electrons parallel to the field and $\bar{N}_a$ the number antiparallel. When $T \ll T_F$ these numbers are given by the integrals [see Eq. (26-5)]

$$\bar{N}_p \simeq 2\pi g V (2m/h^2)^{3/2} \int_0^{\mu_F} \tfrac{1}{2}\sqrt{\varepsilon + \mathfrak{m}_B \mathfrak{B}}\; d\varepsilon$$

$$\bar{N}_a \simeq 2\pi g V (2m/h^2)^{3/2} \int_0^{\mu_F} -\sqrt{\varepsilon - \mathfrak{m}_B \mathfrak{B}}\; d\varepsilon$$

If $\mathfrak{m}_B \mathfrak{B} \ll \mu_F$ we can replace the difference $\sqrt{(\varepsilon + \mathfrak{m}_B \mathfrak{B})} - \sqrt{(\varepsilon - \mathfrak{m}_B \mathfrak{B})}$ by the differential $2\mathfrak{m}_B \mathfrak{B}[d\sqrt{(\varepsilon)}/d\varepsilon]$ and

$$\mathfrak{M} \simeq (\mu_0 \mathfrak{m}_B)^2 \mathfrak{H} 2\pi g V (2m/h^2)^{3/2} \int_0^{\mu_F} d\sqrt{\varepsilon}$$

$$= (\mu_0 \mathfrak{m}_B)^2 \mathfrak{H} \cdot \tfrac{3}{2} N_c \mu_F^{-3/2} \cdot \mu_F^{1/2}$$

$$= [3 N_c (\mu_0 \mathfrak{m}_B)^2 \mathfrak{H} / 2\mu_F] = [2 N_c (\mu_0 \mathfrak{m}_e)^2 \mathfrak{H} / k T_F]$$

which is to be compared with Eq. (26-10) for a MB gas. The difference is in the numerical factor and in the fact that T_F is substituted for T. The numerical factor is modified by the fact that the free-electron translational motion is also affected by the magnetic field, the resulting circular motion producing an opposed magnetization $-[2 N_c (\mu_0 \mathfrak{m}_e)^2 \, \mathfrak{H} / 3\mu_F]$. The net electronic magnetization is thus

$$\mathfrak{M} \simeq \frac{4 N_c (\mu_0 \mathfrak{m}_e)^2}{3 k T_F} \tag{26-14}$$

which is the same as the Curie formula, except that the constant $\tfrac{1}{4} T_F \simeq 1.1 \times 10^5 \rho^{2/3}\,°\mathrm{K}$ is inserted instead of the actual temperature T, thus reducing $\mathfrak{M}$ by a factor of about 100 for room temperatures and for $\rho \simeq 0.1$.

In exactly the same way, calculation of the electrical and thermal conductivity of a FD electron gas results in formulas similar to those of Eqs. (26-11) and (26-12), except that the speed of the electrons at the top of the Fermi band,

$$v_F = \sqrt{2\mu_F/m} \simeq 3.7 \times 10^6 \rho^{1/3}\ \mathrm{m/sec}$$

instead of the mean speed $\langle v \rangle = 2 \times 10^4 \sqrt{T}\ \mathrm{m/sec}$ of the Maxwell distribution, is present in the formulas,

$$(I/\mathfrak{E}) \simeq (N_c/V)(e^2 \lambda/m)\sqrt{m/2\mu_F}$$

$$\kappa \simeq \frac{\pi^2}{3}(N_c/V)(k^2 \lambda/m) T \sqrt{m/2\mu_F} \tag{26-15}$$

The Wiedemann–Franz ratio of Eq. (26-13) is thus still proportional to T, but the new value of the Lorentz factor $L = \tfrac{1}{3}\pi^2 (k/e)^2 = 2.43 \times 10^{-8}$ is more nearly equal to the experimental values given in the table.

When it comes to predicting numerical values of the conductivities themselves, however, the simple fermion gas model is not impressively successful, though much better than the MB gas model.

The complicating factor is the fact that the impediment to the motion of the electrons is not collisions with all the lattice ions, but only with those which deviate from their equilibrium position in the lattice. In other words the electrons would suffer no collisions at all in a completely stationary, perfect crystal lattice. They are only scattered, and thus retarded, by an atom of impurity in the lattice or by a crystal ion which is displaced from its equilibrium position by thermal vibration or by lattice imperfection. The Debye theory of Chapter 20 expresses the lattice vibrations in terms of phonons, quanta of acoustic vibrations; we can say that the electrons are scattered by phonons, and vice versa.

Therefore the effective mean free path of a conduction electron in a metal is quite sensitive to impurities and imperfections of the crystal and is also strongly dependent on temperature, particularly at low temperatures, where the number of phonons diminishes markedly as T decreases. When, in addition, the differences between the energy bands of Figure 26-2b and the simple energy levels of a perfect gas are taken into account, the formulas for conductivity become quite complicated; they also begin to correspond fairly well with measured values. Some late developments of the theory even promise to "explain" superconductivity.

CHAPTER

INTERPARTICLE
FORCES

In the comparison between the two quantum statistics, given in Eq. (24-11), we might have wondered why only two varieties of statistics were observed in nature. One variety (FD) does not allow more than one particle to occupy a given quantum state, the other variety (BE) allows any number to be present. One might wonder why there is no other variety, one which allows no more than two particles in a given state, for example—or no more than three or four.

A more serious question lies in the extension to systems of particles that interact with each other. All our formulas, in the past three chapters, have been for perfect gases, in which we assumed that we could neglect the forces between particles. In such simplified cases the total energy of the system is just the sum $\Sigma n_j \varepsilon_j$ of products of the allowed single-particle energy ε_j times the number n_j of particles which have this energy, as was indicated in Eq. (24-1).

In "real life," however, the elementary particles do interact with each other and the energy of the system is not just a sum of single-particle energies. The quantum numbers specifying the state of the system are not simply related to individual particles; in particular the occupation numbers n_j are no longer appropriate quantum numbers for the system. The question then arises as to how we are to distinguish between the two quantum statistics. If every quantum state involves the cooperative motion of all the particles, how can we decide which particle is in which state, if we wish to apply the Pauli principle?

Particles and Wave Functions

The answer can only be given in terms of the quantum theory, and can most easily be discussed in terms of Schrödinger wave functions. As any text on quantum mechanics will state, a wave function is a solution of a Schrödinger equation; its square is a probability density. For a single particle of mass m in a potential field $\phi(\mathbf{r})$ the equation is $H\psi = \varepsilon\psi$, where H is the differential operator,

$$H \equiv -\frac{\hbar^2}{2m}\left(\frac{\partial^2}{\partial x^2} + \frac{\partial^2}{\partial y^2} + \frac{\partial^2}{\partial z^2}\right) + \phi(\mathbf{r}) \tag{27-1}$$

which is applied to the wave function $\psi(\mathbf{r})$. The values of the energy factor ε, for which the equation can be solved to obtain a continuous, single-valued, and finite ψ, are the allowed energies ε_j for the particle. The square of the corresponding solutions $\psi_j(\mathbf{r})$ (the square of its magnitude, if ψ is complex) is equal to the probability density that the particle is at the point $\mathbf{r}$. Therefore ψ must be *normalized*,

$$\int\int\int |\psi_j(\mathbf{r})|^2 \, dx \, dy \, dz = 1 \tag{27-2}$$

The mathematical theory of such equations easily proves that wave functions for different states i and j are *orthogonal*,

$$\int\int\int \psi_i(\mathbf{r})\psi_j(\mathbf{r}) \, dx \, dy \, dz = 0 \qquad \text{unless} \qquad i = j \tag{27-3}$$

The wave function $\psi_j(\mathbf{r})$ embodies what we can know about the particle in state j. According to quantum theory, we cannot know the particle's exact position; so we cannot expect to obtain a solution of its classical motion by finding x, y, and z as functions of time. All we can expect to obtain is the probability that the particle is at $\mathbf{r}$ at time t, which is $|\psi|^2$. The relation between classical and quantum mechanics is the relation between the operator H of Eq. (27-1) and the Hamiltonian function $H(q, p)$ of Eqs. (13-9) and (16-4). For a single particle (the kth one, say)

$$H_k(q, p) = \frac{1}{2m}(p_{kx}^2 + p_{ky}^2 + p_{kz}^2) + \phi(\mathbf{r}_k)$$

We see that the quantum-mechanical operator is formed from the classical Hamiltonian by substituting $(\hbar/i)(\partial/\partial q)$ for each p. For this reason we call the H of Eq. (27-1) a *Hamiltonian operator*.

The generalization to a system of N similar particles is obvious. If there is no interaction between the particles, the Hamiltonian for the system is the sum of the single-particle Hamiltonians,

$$H(p, q) = \sum_{k=1}^{N} H_k(q, p)$$

and the Schrödinger equation for the system is

$$H\Psi = E\Psi; \qquad H = \sum_{k=1}^{N} H_k; \qquad H_k = -\frac{\hbar^2}{2m}\nabla_k^2 + \phi(\mathbf{r}_k)$$

$$(27\text{-}4)$$

where

$$\nabla_k^2 = \frac{\partial^2}{\partial x_k^2} + \frac{\partial^2}{\partial y_k^2} + \frac{\partial^2}{\partial z_k^2}$$

The values of E for which there is a continuous, single-valued, and finite solution $\Psi_v(\mathbf{r}_1, \mathbf{r}_2,..., \mathbf{r}_N)$ of Eq. (27-4) are the allowed values E_v of the energy of the system. They are, of course, the sums of the single-particle energies ε_j, one for each particle. We have used these facts in previous chapters [see Eqs. (19-8) and (24-1), for example].

A possible solution of Eq. (27-4) is a simple product of single-particle wave functions,

$$\Psi_v(\mathbf{r}_1, \mathbf{r}_2,..., \mathbf{r}_N) = \psi_{j_1}(\mathbf{r}_1) \cdot \psi_{j_2}(\mathbf{r}_2) \cdots \psi_{j_N}(\mathbf{r}_N)$$

$$E_v = \sum_{k=1}^{N} \varepsilon_{j_k} \qquad\qquad (27\text{-}5)$$

where j_k stands for the set of quantum numbers of the kth particle. This would be an appropriate solution for distinguishable particles, for it has specified the state of each particle; state j_1 for particle 1, state j_2 for particle 2, and so on. The square of Ψ_v is a product of single-particle probability densities $|\psi_{j_k}(\mathbf{r}_k)|^2$ that particle k, which is in the j_kth state, is at $\mathbf{r}_k$. We should note that for particles with spin, each ψ has a separate factor which is a function of the spin coordinate, a different function for each different spin state. Thus coordinate $\mathbf{r}_k$ represents not only the position of the particle but also its spin coordinate, and the quantum numbers represented by j_k include the spin quantum number for the particle.

Symmetric Wave Functions

This product wave function, however, will not do for indistinguishable particles. What is needed for them is a probability
density that will have the same value if particle 1 is placed at point
r (including spin) as it has if particle k is placed there. To be more
precise, we wish a probability density $|\Psi_v(\mathbf{r}_1, \mathbf{r}_2,..., \mathbf{r}_N)|^2$ which is
unchanged in value when we interchange the positions (and spins)
of particle 1 and particle k, or any other pair of particles. The simple
product wave function of Eq. (27-5) does not provide this; if j_1
differs from j_k, then interchanging $\mathbf{r}_1$ and $\mathbf{r}_k$ produces a different
function.

However, other solutions of Eq. (27-4), having the same value
E_v of the energy of the system as does solution (27-5), can be obtained
by interchanging quantum numbers and particles. For example,

$$\Psi_v(\mathbf{r}_N, r_{N-1},..., \mathbf{r}_1) = \psi_{j_N}(\mathbf{r}_1) \cdot \psi_{j_{N-1}}(\mathbf{r}_2) \cdots \psi_{j_1}(\mathbf{r}_N)$$

is another solution with energy E_v. There are $N!$ possible permutations of N different quantum numbers among N different particle
wave functions. If several different particles have the same quantum
numbers, if n_j particles are in state j, for example, then there are
$(N!/n_1!n_2! \cdots)$ [compare with Eq. (21-9)] different wave functions
Ψ_v which can be obtained from (27-5) by permuting quantum numbers and particles.

Therefore a possible solution of Eq. (27-4), for the allowed
energy E_v, would be a *sum* of all the different product functions that
can be formed by permuting states j among particles k. Use of
Eqs. (27-2) and (27-3) can show that for such a sum to be normalized,
it must be multiplied by $\sqrt{(n_1!n_2! \cdots /N!)}$. However, such details need
not disturb us here; what is important is that this sum is a solution
of Eq. (27-4) for the system state v with energy E_v, which is unchanged in value when any pair of particle coordinates is interchanged (the change rearranges the order of functions in the sum
but does not introduce new terms). Therefore, its square is unchanged
by such interchange and the wave function is an appropriate one
for indistinguishable particles. For such a wave function it is no
longer possible to talk about *the* state of *a* particle; all particles
participate in all states; all we can say is that n_j particles are in
state j at any time. Such a wave function is said to be *symmetric*
to interchange of particle coordinates.

A few examples of symmetric wave functions for two par-
ticles are $\psi_1(\mathbf{r}_1)\psi_1(\mathbf{r}_2)$ and $(1/\sqrt{2})[\psi_1(\mathbf{r}_1)\psi_2(\mathbf{r}_2) + \psi_1(\mathbf{r}_2)\psi_2(\mathbf{r}_1)]$; a

few for three particles are $\psi_1(\mathbf{r}_1)\psi_1(\mathbf{r}_2)\psi_1(\mathbf{r}_3)$ or $(1/\sqrt{3})[\psi_1(\mathbf{r}_1)$ $\times \psi_1(\mathbf{r}_2)\psi_2(\mathbf{r}_3) + \psi_1(\mathbf{r}_1)\psi_1(\mathbf{r}_3)\psi_2(\mathbf{r}_2) + \psi_1(\mathbf{r}_3)\psi_1(\mathbf{r}_2)\psi_2(\mathbf{r}_1)]$; and so on.

Antisymmetric Wave Functions

However, since our basic requirement is that of symmetry for the *square* of Ψ, we have an alternative choice, that of picking a wave function *antisymmetric* with respect to interchange of particle coordinates, which changes its sign but not its magnitude when the coordinates of any pair are interchanged. The square of such a Ψ also is unchanged by the interchange. Such an antisymmetric solution can be formed out of the product solutions of Eq. (27-5), but *only* if all particle ψ's are for *different states*. If every j_k differs from every other j, then an antisymmetric solution of Eq. (27-4), with energy E_v, is the determinant

$$\Psi_v = \frac{1}{\sqrt{N!}} \begin{vmatrix} \Psi_{j_1}(\mathbf{r}_1) & \Psi_{j_1}(\mathbf{r}_2) & \cdots & \cdots & \Psi_{j_1}(\mathbf{r}_N) \\ \Psi_{j_2}(\mathbf{r}_1) & \Psi_{j_2}(\mathbf{r}_2) & \cdots & \cdots & \Psi_{j_2}(\mathbf{r}_N) \\ \cdots & \cdots & \cdots & \cdots & \cdots \\ \cdots & \cdots & \cdots & \cdots & \cdots \\ \Psi_{j_N}(\mathbf{r}_1) & \Psi_{j_N}(\mathbf{r}_2) & \cdots & \cdots & \Psi_{j_N}(\mathbf{r}_N) \end{vmatrix} \quad (27\text{-}6)$$

The properties of determinants are such that an interchange of any two columns (interchange of particle coordinates) or of any two rows (interchanging quantum numbers) changes the sign of Ψ_v. The proof that $1\sqrt{N!}$ must be used to normalize this function is immaterial here. What is important is that another whole set of wave functions, satisfying the requirements of particle indistinguishability, is the set of functions that are antisymmetric to interchange of particle coordinates. For this set, no state can be used more than once (a determinant with two rows identical is zero).

By now it should be apparent that the two types of wave functions correspond to the two types of quantum statistics. Wave functions for a system of *bosons* are *symmetric* to interchange of particle coordinates; any number of particles can occupy a given particle state. Wave functions for *fermions* are *antisymmetric* to interchange of particle coordinates; because of the antisymmetry, no two particles can occupy the same particle state (which is the Pauli exclusion principle). Both sets of wave functions satisfy the indistinguishability requirement—that the square of Ψ be symmetric to interchange of particle coordinates. A simple application of quantum theory will prove that no system of forces which are the same for all particles will change a symmetric wave function into an antisymmetric one, or vice versa. Once a boson, always a boson, and likewise for fermions. It is an interesting indication of the way that all parts of quantum theory "hang together" that the fact that the quantity of physical importance is the *square* of the wave function should not only allow, but indeed *demand*, two different kinds of statistics, one for symmetric wave functions, the other for antisymmetric. Also, from this point of view, *only* two alternatives are possible.

We have introduced the subject of symmetry of wave functions by talking about a system with no interactions between particles. But this restriction is not binding. Suppose we have a system of N particles, having a classical Hamiltonian $H(q, p)$ which includes potential energies of interaction $\phi(r_{kl})$, which is symmetric to interchange of particles (as it must be if the particles are identical in behavior). Corresponding to this H is a Hamiltonian operator H, which includes the interaction terms $\phi(r_{kl})$ and in which each p_k is changed to $(\hbar/i) \times (\partial/\partial q_k)$; this operator is also symmetric to interchange of particle coordinates. Of the many possible solutions of the general Schrödinger equation

$$H\Psi(\mathbf{r}_1,..., \mathbf{r}_i,..., \mathbf{r}_j,..., \mathbf{r}_N) = E\Psi$$

there are only two subsets which satisfy the indistinguishability requirement that the square of Ψ be unchanged when a pair of particle coordinates (such as $\mathbf{r}_i$ and $\mathbf{r}_j$) are interchanged. One set is *symmetric* to interchange of coordinates,

$$\Psi(\mathbf{r}_1,..., \mathbf{r}_j,..., \mathbf{r}_i,..., \mathbf{r}_N) = \Psi(\mathbf{r}_1,..., \mathbf{r}_i,..., \mathbf{r}_j,..., \mathbf{r}_N)$$

If the particles are *bosons* this set of solutions is the one which must be used; all other solutions of the general Schrödinger equation are invalid, for bosons.

On the other hand, if the particles under study are *fermions* the subset of solutions, which must be used, are those *antisymmetric* in interchange of particle coordinates,

$$\Psi(\mathbf{r}_1,..., \mathbf{r}_j,..., \mathbf{r}_i,..., \mathbf{r}_N) = -\Psi(\mathbf{r}_1,..., \mathbf{r}_i,..., \mathbf{r}_j,..., \mathbf{r}_N)$$

for all pairs i, j. This set automatically satisfies the Pauli principle and thus has the characteristics of the FD statistics. When there are particle interactions the two sets usually have different allowed energies E_v. As long as H is symmetric to interchange of particle coordinates, a wave function of one symmetry cannot change into one of the opposite symmetry.

Thus we have transferred the specification of the particle statistics from one involving the occupation numbers n_j, which breaks down when there is particle interaction and the occupation numbers are no longer appropriate quantum numbers, to one involving the symmetry of the wave function of the system, which is applicable to all systems encountered in nature.

Little more would need to be said if we could obtain exact wave-function solutions for all situations of interest. We would work out the solutions having the symmetry appropriate for the statistics of the elementary particles involved, and that would be that. However, when interparticle interactions are involved, very few problems can be solved exactly and we usually must be satisfied with approximate solutions. In some cases we would be satisfied to consider tightly bound groups of particles (such as the electrons and nuclei making up a molecule, or the elementary nucleons making up an atomic nucleus) as being single units, ignoring their internal structure—suppressing some particle coordinates. We then must determine what sort of symmetry our approximate wave functions must have, in terms of the unsuppressed coordinates. The rest of this chapter discusses how this is done and illustrates the discussion with a few simple examples, some of which are amenable to experimental verification of the formulas.

Bound States and Free States

Ordinary matter is a combination of three kinds of elementary particles: protons, neutrons, and electrons (each of them are fermions). The protons and neutrons are held together by nuclear forces to form atomic nuclei. The nuclei in turn are surrounded by electrons, which are held to the nuclei by longer-range electric forces. Nuclei and molecules are thus quite complex systems. Let us start our discussion of the effects of particle interaction with a greatly simplified example. Suppose we have two identical, one-dimensional fermions, subject to no force except a mutual attraction. The displacement of one fermion is coordinate x_1, displacement of the other is x_2, as shown in Figure 27-1a. Thus the potential energy of the interaction can be written as $\phi(X_{12})$, where $X_{12} = (x_1 - x_2)/\sqrt{2}$. The shape of ϕ, chosen in the example, corresponds to a strong attraction between the particles when they are closer than a distance δ apart, negligible interaction when $X_{12} > \delta$, somewhat analogous to the short-range nuclear interaction.

The Schrödinger equation

$$\left[-\frac{\hbar^2}{2m}\left(\frac{\partial^2}{\partial x_1^2} + \frac{\partial^2}{\partial x_2^2}\right) + \phi(X_{12})\right]\Psi = E\Psi$$

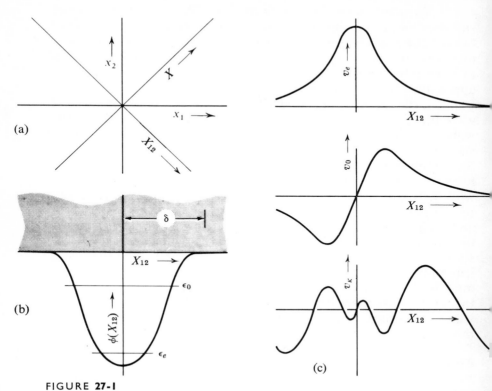

FIGURE 27-1
Coordinates, interaction potential ϕ, bound and free wave functions for two mutually attracting, one-dimensional particles.

can be rewritten as

$$\left[-\frac{\hbar^2}{2m} \frac{\partial^2}{\partial X_{12}^2} + \phi(X_{12}) - \frac{\hbar^2}{2m} \frac{\partial^2}{\partial X^2} \right] \Psi = E\Psi$$

which separates into

$$\left[-\frac{\hbar^2}{2m} \frac{\partial^2}{\partial X_{12}^2} + \phi(X_{12}) \right] v_\nu(X_{12}) = \varepsilon_\nu v_\nu(X_{12})$$

and   (27-7)

$$-\frac{\hbar^2}{2m} \frac{\partial^2}{\partial X^2} u_k(X) = \frac{\hbar^2 k^2}{2m} u_k(X)$$

where $\Psi = v_v(X_{12}) \cdot u_k(X)$ and $E = \varepsilon_v + (\hbar^2 k^2/2m)$. The factor for the $X = (x_1 + x_2)/\sqrt{2}$ direction, perpendicular to the attractive force, is that for free translation of the center of mass of the pair of particles, with a momentum component $\hbar k$, kinetic energy $(\hbar^2 k^2/2m)$, and wave-function factor $u_k(X) = e^{\pm ikX}$.

We choose a binding potential which has two bound levels, ε_e and ε_o, as shown in the figure, and a continuum of "free states," in which the two particles are moving apart or together with mutual momentum $(\hbar\kappa)$ and kinetic energy $(\hbar^2\kappa^2/2m)$. The corresponding wave-function factors are v_e and v_o for the bound states, and, for the free states, a solution which has the asymptotic form $e^{\pm i\kappa X_{12}}$ when $|X_{12}| \to \infty$. The allowed free states are indicated by the shading, from $\phi = 0$ up, in Figure 27-1b.

The wave-function factors $v_e(X_{12})$ and $v_o(X_{12})$ for the two bound states are plotted in Figure 27-1c. The factor for the most tightly bound state, the one with the lowest energy, is symmetric about $X_{12} = 0$; the upper bound state is antisymmetric, i.e., has a node at $X_{12} = 0$. (In general, the more nodes the wave function has, the more times it changes sign, the higher is the energy.) Reversing the sign of X_{12} is equivalent to interchanging the values of x_1 and x_2, i.e., to interchanging the two particles. If these particles are bosons then only the factors which are symmetric in X_{12} are allowed; energy ε_e is allowed but not energy ε_o. Alternatively if the particles are fermions, ε_e is not allowed but ε_o, corresponding to the antisymmetric factor $v_o(X_{12})$, is allowed. The free-state factors $v_\kappa(X_{12})$ for two fermions would be the antisymmetric combinations of the exponentials, which would have the asymptotic form $A \sin(\kappa X_{12} - \beta)$ for $|X_{12}| \to \infty$, where the value of the phase angle β is determined by the form of the potential function $\phi(X_{12})$. This function is also shown in Figure 27-1c, for one value of κ. The solution for bosons would be symmetric about $X_{12} = 0$ and would thus have the asymptotic form $A \cos(\kappa X_{12} - \beta)$.

For fermions, therefore, the allowed wave functions and corresponding energies are

$$\Psi_{bk}(x_1, x_2) = Bv_o(X_{12})e^{ikX}; \qquad E_{bk} = \varepsilon_o + (\hbar^2 k^2/2m)$$

$$\Psi_{\kappa k}(x_1, x_2) = Av_\kappa(X_{12})e^{ikX} \qquad\qquad (27\text{-}8)$$

$$\to \frac{A}{2i}\left[e^{ik_1 x_1 + ik_2 x_2} - e^{ik_2 x_1 + ik_1 x_2}\right] \qquad |X_{12}| \to \infty$$

$$E_{\kappa k} = (\hbar^2/2m)(\kappa^2 + k^2) = (\hbar^2/2m)(k_1^2 + k_2^2)$$

where $k_1 = (k + \kappa)/\sqrt{2}$ and $k_2 = (k - \kappa)/\sqrt{2}$. Because of the antisymmetry κ cannot be zero, i.e., k_1 cannot equal k_2, where k_1 is the wave number for one particle and k_2 that for the other. Each of these wave functions change sign if x_1 and x_2 are interchanged (i.e., if X_{12} is changed to $-X_{12}$) as is required for fermions. The constants A and B are adjusted so that the integral of $|\psi|^2$ over the allowed range of x_1 and x_2 is unity. We must also adjust the values of k and κ so the boundary conditions at the edge of the allowed region are satisfied, but if the allowed range of x_1 and x_2 is much larger than δ the spacing between the allowed levels of $(\hbar^2 k^2/2m)$ and $(\hbar^2 \kappa^2/2m)$ is so much smaller than the magnitude of ε_o that the translational energy can be considered to be a continuous variable.

Thus, in this very simple example of two particles with mutual interaction, the wave function appropriate for fermions turns out to be a product of two factors. One is a function of the distance X_{12} between the particles; it must be antisymmetric about $X_{12} = 0$ if the particles are fermions. The other is a simple exponential (or sine or cosine) of X, representing free motion of the center of mass of the particles and having no particular symmetry about $X = 0$. The bound-state wave functions Ψ_{bk} have an X_{12} factor v_o which is concentrated near the line $x_1 = x_2$, representing the fact that the two particles travel together as a unit; the X factor corresponds to the free motion of this unit. The antisymmetric requirements are satisfied by the bound-state factor $v_o(X_{12})$, since it is antisymmetric about $X_{12} = 0$. The free-state wave functions correspond to separate motions of each particle; except in the relatively small area near the line $x_1 = x_2$ the solution is an antisymmetric combination of traveling (or standing) waves for each particle separately, of the form of Eq. (27-6); it represents, over most of the allowed area, a perfect gas of two fermions.

The discussion for three-dimensional fermions is similar to that for the one-dimensional case, except that it cannot be pictured as easily. If the interaction potential $\phi(r_{12})$ is such as to permit a bound state, the bound-state wave functions would be most easily represented in terms of the coordinates X, Y, Z (or $\mathbf{R}$) of the center of mass of the pair and the coordinates r_{12}, ϑ, φ, specifying the magnitude and direction of the vector distance $\mathbf{r}_{12}$ between the two particles, as shown in Figure 27-2a. The bound-state wave function $\Psi = u(\mathbf{R})v(\mathbf{r}_{12})$ would thus have two factors. One, $u(X, Y, Z) = u(\mathbf{R})$, would represent the free motion of the two-particle combination. The other factor, $v(\mathbf{r}_{12})$, would be large only when the

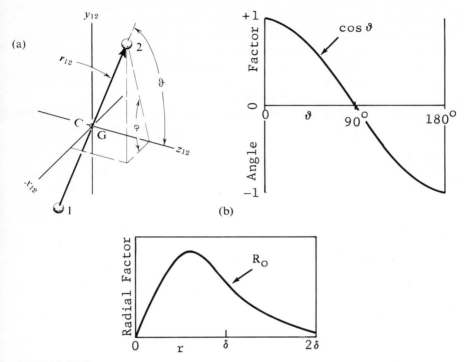

FIGURE 27-2

Vector interparticle distance and direction angles. Angle and radial dependence of antisymmetric wave function.

two particles were close together ($r_{12} < \delta$) and would have to be antisymmetric to interchange of $\mathbf{r}_1$ and $\mathbf{r}_2$, i.e., to reversal of direction of $\mathbf{r}_{12}$. A function which would accomplish this would be one which is positive for $\vartheta < \frac{1}{2}\pi$ and negative for $\vartheta > \frac{1}{2}\pi$, as shown in Figure 27-2b.

The functional form for such an antisymmetric, bound-state wave function could thus be $v_o(\mathbf{r}_{12}) = \cos \vartheta R_o(r_{12})$, with the antisymmetry about the $x_{12} - y_{12}$ plane. Other possibilities could be $\sin \vartheta \cos \varphi R_o(r_{12})$ or $\sin \vartheta \sin \varphi R_o(r_{12})$, with the antisymmetry about the $z_{12} - x_{12}$ or the $z_{12} - y_{12}$ planes, respectively. The radial factor $R_o(r_{12})$ rises to a maximum for r_{12} less than δ and drops to zero for r_{12} large, corresponding to the fact that the two particles stay close together in the bound state. The energy of the combined-particle system is ε_o, the energy of the bound state, plus $(\hbar^2 k^2/2m)$, the kinetic energy of the translational motion of the two-particle unit. If we were computing the thermal behavior of the system at temperatures small enough so that kT is small compared to the magnitude of ε_o, we could neglect the fact that the unit is a bound pair and consider it as a particle of mass $2m$, with kinetic energy $(\hbar^2 k^2/2m)$ and momentum $\hbar k \sqrt{2}$.

On the other hand, it is possible for the particles to be in free states, in independent motion. The corresponding wave function is an antisymmetric combination of products of free-particle wave functions, as in Eq. (27-6), except in the small region where r_{12} is less than δ where the wave function has a more complicated form, as suggested by the third curve of Figure 27-1c. Because of the antisymmetry, the translational states of the two particles cannot be exactly the same. Thus the possible wave functions for such a pair of interacting, three-dimensional fermions are

$$\Psi_{bi}(12) = u_i(\mathbf{R})v_o(\mathbf{r}_{12}) = u_i(\mathbf{R})R_o(r_{12}) \begin{cases} \cos \vartheta \\ \sin \vartheta \cos \varphi \\ \sin \vartheta \sin \varphi \end{cases}$$

$$E = \varepsilon_o + (\hbar^2 k^2/2m) \tag{27-9}$$

$$\Psi_{ij}(12) \simeq \frac{1}{\sqrt{2}}[u_i(\mathbf{r}_1)u_j(\mathbf{r}_2) - u_j(\mathbf{r}_1)u_i(\mathbf{r}_2)] \qquad r_{12} > \delta$$

$$E = (\hbar^2/2m)(k_i^2 + k_j^2)$$

which is to be compared with Eqs. (27-8) for the one-dimensional case.

If the particles are bosons the bound-state wave function would be symmetric with respect to reversal of direction of $\mathbf{r}_{12}$. A possible solution, $v_e(\mathbf{r}_{12})$, would be independent of the direction of r_{12}. The binding energy of this state, ε_e, would usually be greater than the binding energy ε_o for the antisymmetric wave function v_o of Eq. (27-9). In spite of this, there is no chance that a pair of fermions would simultaneously change into bosons, just to become bound more tightly together; wave functions cannot change their symmetry.

The Effect of Particle Spin

A particle with spin is a four-dimensional particle; the coordinates specifying its wave function must be the three space coordinates plus the spin coordinate; its state is specified by the three quantum numbers, determining its spatial distribution, plus a fourth number specifying the orientation of its spin. If the spin magnitude is $\frac{1}{2}$, as is the case with elementary fermions, there are only two possible spin states, one which can be called "plus" and the other "minus" [see discussion of Eqs. (13-24) and (20-21)]. The spin factor in the wave function for each particle is analogous to a vector; the factor $\beta(1)$, corresponding to the "minus" state of particle 1, is "perpendicular" to the factor $\alpha(1)$ for the same particle, corresponding to the "plus" state. Therefore $\alpha^2(1) = \beta^2(1) = 1$ and $\alpha(1)\beta(1) = \beta(1)\alpha(1) = 0$. Note that the spin factor for one particle does not combine with that of another particle, so that $\alpha(1)\beta(2)$ is not zero.

For fermions having spin $\frac{1}{2}$ the wave function must be anti-symmetric for interchange of all four dimensions, the spin as well as the space dimensions. For example, an appropriate wave function for two fermions with no mutual interaction is

$$(1/\sqrt{2})[u_i(1)\alpha(1)u_j(2)\beta(2) - u_i(2)\alpha(2)u_j(1)\beta(1)] \tag{27-10}$$

representing one particle with spatial state i and spin state $+$ and the other particle with spatial state j and spin $-$. This function changes sign when we interchange 1 and 2, not only the space part of the function but also the spin part.

We notice that with the particles represented by these wave functions, it is possible for the spatial state of the two particles to be the same $(i = j)$, as long as the spin states are different. The function

$$(1/\sqrt{2})[u_i(1)u_i(2)][\beta(1)\alpha(2) - \alpha(1)\beta(2)]$$

is still antisymmetric to interchange of particles 1 and 2 in their entirety. Thus we can have a spatial part of the combined wave function which is symmetric, as long as the spin part is anti-symmetric. Vice versa, if the space part is antisymmetric, the spin part must be symmetric, as with the functions

$$\frac{1}{\sqrt{2}}[u_i(1)u_j(2) - u_i(2)u_j(1)] \begin{cases} \alpha(1)\alpha(2) \\ \frac{1}{\sqrt{2}}[\alpha(1)\beta(2) + \alpha(2)\beta(1)] \\ \beta(1)\beta(2) \end{cases}$$

There are three possible symmetric spin factors, only one possible antisymmetric one.

Returning to the particles with interaction potential $\phi(r_{12})$, we see that fermions with spin can occupy both kinds of bound state, the even and the odd, depending on the symmetry of their spin factors. Suppose the interaction potential is such that both types are possible; the most tightly bound state would be the one with symmetric $\mathbf{r}_{12}$ factor, $v_e(\mathbf{r}_{12}) = v_e(-\mathbf{r}_{12})$, with energy ε_e. For the whole wave function to be antisymmetric, the spin factor must be the single antisymmetric one. On the other hand, the antisymmetric space state, with $v_o(\mathbf{r}_{12}) = -v_o(-\mathbf{r}_{12})$, with energy ε_o, would require one of the three symmetric spin factors. The lowest possible bound state of the fermion pair would thus be a single level antisymmetric in the spin part, and the next lowest would be three equal levels, symmetric in their spin parts,

$$\Psi_{si} = u_i(\mathbf{R})v_e(\mathbf{r}_{12})[\alpha(1)\beta(2) - \alpha(2)\beta(1)]/\sqrt{2}$$

for $\quad E = \varepsilon_e + (\hbar^2 k_i^2/2m)$

$$\Psi_{ti} = u_i(\mathbf{R})v_o(\mathbf{r}_{12}) \begin{cases} \alpha(1)\alpha(2) \\ [\alpha(1)\beta(2) + \alpha(2)\beta(1)]/\sqrt{2} \\ \beta(1)\beta(2) \end{cases} \qquad (27\text{-}11)$$

for $E = \varepsilon_0 + (\hbar^2 k_i^2/2m)$. The lowest state, symmetric in its space part, has only one spin arrangement possible, with the two spins opposed in direction and thus a zero total spin. The next lowest state, with an antisymmetric space part, can have three different spin factors; one with both spins plus, one with two opposed, and one with both spins negative, corresponding to the three components $+1$, 0, -1 of a system with total spin $J = 1$. For obvious reasons the state with the antisymmetric spin part is called a *singlet* state; the three states with symmetric spin factors comprise the *triplet* state. Unless the particles are charged or a magnetic field is present the three triplet states all have the same energy.

It is not hard to see that if the particles were bosons, with spin $\frac{1}{2}$, the triplet spin factor would have to go with the symmetric space factor and binding energy ε_e, the singlet state would correspond to the antisymmetric space factor and energy ε_o.

The unbound states for the two fermions with spin can be written in the form of Eq. (27-10), except in the small region where $r_{12} < \delta$. They can also be rearranged as singlet and triplet states, if desired; here there is no energy difference between symmetric and antisymmetric space factors.

Thus the additional degree of freedom represented by the particle spin makes possible states which are not possible for spinless particles, whether they be fermions or bosons. It also increases the multiplicity of some of the states. The energy difference between the singlet and the triplet bound states is not an energy of electromagnetic coupling between spin and particle motion; it is the consequence of the difference in symmetry requirements of the FD or the BE statistics.

Larger Numbers of Particles

To see what happens when we have more than two particles, let us again take the simplest situation. We assume that each pair of particles (particles n and m, for example) has a mutual attraction, expressible in terms of a potential energy $\phi(r_{nm})$, where r_{nm} is the distance between the two particles. We further assume that $\phi(r)$ is large and negative only when r is small, going to zero when $r > \delta$. Also, its shape and depth are such that the only bound state possible is one for which the r_{nm} factor in the wave function is independent of the orientation of $\mathbf{r}_{nm}$, is $v_e(\mathbf{r}_{nm})$ which is symmetric to interchange of particles n and m. The corresponding energy is ε_e, which is negative in value, representing a bound state, and has a magnitude large compared to the spacing between the allowed energies of translation of the bound pair.

The potential $\phi(r_{nm})$ also allows free states, where particles n and m move freely throughout the allowed volume with momenta $\hbar k_n$ and $\hbar k_m$ (and thus kinetic energies $\hbar^2 k_n^2/2m$ and $\hbar^2 k_m^2/2m$). This wave function is an antisymmetric combination of single-particle wave functions of the form of Eq. (27-10), except in the restricted region, $r_{nm} < \delta$, where the form of the function is modified by the mutual attraction.

Suppose there are three such fermions in volume V. With the interaction potential ϕ, just described, any two of the three could be bound together (1 and 2, for example) with a wave function

$$\psi_{bi}(12) = U_i(\mathbf{R}_{12})v_e(\mathbf{r}_{12})[\alpha(1)\beta(2) - \alpha(2)\beta(1)]/\sqrt{2} \qquad (27\text{-}12)$$

where, since v_e is symmetric with respect to reversal of $\mathbf{r}_{12}$, the spin factor must be antisymmetric (singlet state). Factor U_i is a function of $\mathbf{R}_{12} = \frac{1}{2}(\mathbf{r}_1 + \mathbf{r}_2)$, the position of the center of mass of the two particles and represents the translational motion of the pair as a unit. The energy of the pair in this state is ε_e, the binding energy, plus the kinetic energy $(p_i^2/4m)$ of the ith translational state of the bound pair.

The Pauli principle rules out the binding of three particles together, for the potential function assumed here. Since the symmetric function v_e corresponds to the only bound state, three fermions could be bound together by this potential only if an antisymmetric spin function could be formed for three particles. This is impossible for particles with spin $\frac{1}{2}$ (i.e., having only two spin states). The third particle must therefore be in an unbound state, moving freely around in the allowed volume V and being affected by the other pair only when it comes close. Its wave function is a free-particle wave function

$$\psi_{j\sigma}(3) = u_j(\mathbf{r}_3) \begin{cases} \alpha(3) & \text{if} \quad \sigma = + \\ \beta(3) & \text{if} \quad \sigma = - \end{cases} \tag{27-13}$$

where u_j is the translational space part for a single particle, with momentum p_j and energy $(p_j^2/2m)$.

Therefore a possible solution of the Schrödinger equation

$$-(\hbar^2/2m)[\nabla_1^2 + \nabla_2^2 + \nabla_3^2]\Psi$$
$$+ [\phi(r_{12}) + \phi(r_{13}) + \phi(r_{23})]\Psi = E\Psi \tag{27-14}$$

is

$$\Psi = \psi_{bi}(12)\psi_{j\sigma}(3) \qquad \text{for} \qquad E = \varepsilon_e + (p_i^2/4m) + (p_j^2/2m)$$

for particles 1 and 2 bound and particle 3 free. Another set of states corresponds to all three particles being free,

$$\Psi = \psi_{i\sigma}(1)\psi_{j\tau}(2)\psi_{l\eta}(3) \qquad \text{for} \qquad E = (1/2m)(p_i^2 + p_j^2 + p_l^2)$$

Neither of these solutions has the antisymmetry property required for fermions, however. The correct function for the three free particles must be the function which, when r_{12}, r_{13}, and r_{23} are large compared to δ, has the determinantal form

$$\Psi_f \simeq \frac{1}{\sqrt{6}} \begin{vmatrix} \psi_{i\sigma}(1) & \psi_{i\sigma}(2) & \psi_{i\sigma}(3) \\ \psi_{j\tau}(1) & \psi_{j\tau}(2) & \psi_{j\tau}(3) \\ \psi_{l\eta}(1) & \psi_{l\eta}(2) & \psi_{l\eta}(3) \end{vmatrix} \tag{27-15}$$

whereas the antisymmetric form for one pair bound, the third free, is

$$\Psi_b = \frac{1}{\sqrt{3}}[\psi_{bi}(12)\psi_{j\sigma}(3) - \psi_{bi}(13)\psi_{j\sigma}(2) + \psi_{bi}(23)\Psi_{j\sigma}(1)]$$

$$\tag{27-16}$$

which changes sign when any pair of particles is interchanged, as required. This is easily checked if we remember that $\psi_{bi}(mn)$ $= -\psi_{bi}(nm)$, from Eq. (27-12).

There is an interesting difference between these two wave-function forms. In the free-particle determinant Ψ_f, all the three particles are on an equal footing and the determinantal form enforces the Pauli principle, forbidding them to share a quantum state, by vanishing if any one of the quantum number sets, $i\sigma$, $j\tau$, or $l\eta$, is equal to any of the others. In the bound-particle function Ψ_b the particles in any term of the wave-function sum are on a different footing; the two-particle, bound-state function $\psi_{bi}(12)$ has quite a different form from the single-particle, free-state function $\psi_{j\sigma}(3)$, so there is no possibility of Ψ_b going to zero when $i = j$; the (12) pair is a different "particle" from the single (3). The antisymmetry has "done its work" by requiring that the bound-state wave function $\psi_{bi}(12)$ be itself antisymmetric (and thus, in this case, by requiring that the bound state be a singlet spin state). The determinantal form of Ψ_f is essential to ensure the fermion behavior of the particles; the extended form of Ψ_b is not as essential, the important requirement of the antisymmetry has been satisfied by the requirement that ψ_{bi} be antisymmetric. Once this is ensured we could just as well have used the simple product $\psi_{bi}(12)\psi_{j\sigma}(3)$ to compute the properties of the system.

If the temperature is low enough so that kT is small compared to the binding energy ε_e, the bound pair can be considered to be a permanent unit (like a molecule or a complex nucleus). This unit is a different kind of particle from the remaining single particle; the two are distinguishable particles and can be so handled statistically. For example, deuterium nuclei are distinguishable from protons. Wave functions relating a proton and a deuteron need not be antisymmetric for interchange between the two—nor need they be symmetric. However, the wave function for the internal structure of the deuteron must be antisymmetric.

Deuterium and Helium Nuclei

To see what sort of symmetry is required between pairs of deuterons, we study the system of four particles, with the same simple interaction potential as before. Here again there can be no three- or four-particle bound state. For the potential we have assumed to join three or four particles in a bound state would require that more than one of them would have to occupy the same spatial and spin state, which is forbidden for fermions. However, it is possible for two separate pairs to be formed.

The wave function for two combined pairs of particles would be a function of their twelve spatial degrees of freedom and of their four spin coordinates. In the small region of this space where all particles are close together the wave function must be small and complex in form in order to satisfy the symmetry requirements (it must be zero when $\mathbf{r}_1 = \mathbf{r}_2 = \mathbf{r}_3 = \mathbf{r}_4$). But, in the region when r_{13} and r_{14} are large, the function would have the form of a product of two two-particle wave functions such as that of Eq. (27-12). The $\psi_{bi}(12)$ for the pair (12) would be antisymmetric for this pair, and the factor $\psi_{bj}(34)$ would be antisymmetric for 3 and 4, and probably its translational factor U_j would be for a different momentum $\mathbf{p}_j$.

The appropriately antisymmetrized combination for two bound pairs would thus have the form

$$\Psi_b(1234) = \frac{1}{\sqrt{6}}\{[\psi_{bi}(12)\psi_{bj}(34) + \psi_{bi}(34)\psi_{bj}(12)]$$
$$+ [\psi_{bi}(23)\psi_{bj}(14) + \psi_{bi}(14)\psi_{bj}(23)] \qquad (27\text{-}17)$$
$$- [\psi_{bi}(13)\psi_{bj}(24) + \psi_{bi}(24)\psi_{bj}(13)]\}$$

except in the small region where all four particles are close together, when Ψ_b becomes quite small. This function is antisymmetric with respect to interchange of any single particle with another. For example $\Psi_b(2134) = \Psi_b(3214) = \Psi_b(4231) = -\Psi_b(1234)$, etc. If each bound pair is considered to be a particle of twice the mass of the elementary particle, then interchange of 12 with 34, for example, does not change the value or sign of Ψ. The wave function is *symmetric* with respect to interchange of *pairs* of particles. The bound-together pairs behave toward each other as though they were *bosons*.

If the energy of binding of a pair of particles is very large compared to kT at ordinary temperatures then it is quite unlikely that a pair will dissociate into two free, elementary particles, and we might consider the bound pairs as being indivisible but non-elementary particles of mass $2m$, in which case we could disregard the internal structure of the pair and omit the constant binding energy ε_e from the energy equations, dealing only with the kinetic energies of the bound pairs. In other words we would then use the factor $U_i(\mathbf{R})$, representing the translational motion of the center of mass of the pair, as the wave function for the combined particle, with $(p_i^2/2M)$ $(M = 2m)$ as its energy. We would then find that a wave function for two such pairs would have to be symmetric for interchange of the pairs and thus that each bound pair acted as though it were a boson. The inherent fermion character of the constituent elementary particles has "gone underground" in the internal structure of each pair and in the fact that the two pairs tend to avoid interpenetration in a way not typical of elementary bosons. If the range of action δ of the interaction potential ϕ is small (as it is for nucleons) this latter peculiarity may not be readily apparent.

The nucleus of deuterium, which contains two elementary nucleons, can be considered to be an individual boson at ordinary temperatures. The bound state of two nucleons differs from the simple case we have studied in that the forces are spin-dependent, so that the deuteron bound state is the triplet instead of the singlet. Our conclusions in regard to the statistics of the deuterons, as particles, however, follow from the same symmetry arguments which were used with our simplified example. If interchange of elementary particles changes the sign of the wave function, interchange of pairs of particles leaves the sign unchanged, which is the hallmark of the boson. The underlying fermion nature of the elementary nucleons only becomes apparent when we apply high enough temperatures or voltages, so that an appreciable number of deuterons are dissociated.

Liquid Helium II, Metastable Equilibrium

Extending the argument, we see that, if elementary nucleons (protons and neutrons) are fermions, then nuclei with an *even* number of nucleons will act as though they are *bosons* and nuclei with an *odd* number of nucleons will act as if they are *fermions*. An important experimental corroboration of this theory is the fact that liquid helium exhibits the low-temperature behavior typical of bosons if the helium is He^4 (with four nucleons per nucleus) but does not exhibit this behavior if the helium is He^3 (with three nucleons per nucleus). (See the latter parts of Chapters 7 and 25.)

Nuclear forces, in contrast to the simplified binding force used in our earlier example, allow more than two nucleons to be bound together to form nuclei of various masses and electric charges. For example, the He^4 nucleus contains two neutrons and two protons, the equivalent of two deuterons, bound together. The energy of binding of the helium is quite large; it would require about 24 Mev of energy to convert a helium nucleus into two deuterons. This being the case, we might wonder why all the deuterium in the universe has not already been converted into helium. Which brings us back to a point, touched on in Chapter 9, concerning thermodynamic equilibrium and rates of transition.

Glass is a supercooled liquid; its Gibbs function is large than that for the corresponding crystal at the same temperature By rights, it should spontaneously change to crystalline form, t reach thermodynamic equilibrium. As a matter of fact it is changing but its rate of transition is so slow, at room temperatures, that w would have to wait several centuries for it to reach its equilibrium phase, a heap of crystals. Because of its very slow rate of transition we can measure and compute the thermodynamic properties o glass as though it were in equilibrium. We say it is in *metastable equilibrium*.

Likewise, at ordinary temperatures, the rate of combination of deuterium nuclei into helium nuclei is negligible. The minute siz of the deuterons, and the strong repulsion between them becaus of their electric charge, makes it exceedingly unlikely that tw deuterons will come close enough to stick together, at room tempera tures. The deuterium in the world thus is in metastable equilibrium In all our thermodynamic measurements and calculations we ca consider it to be a separate substance, basically different from th helium in the universe, not just a bunch of fragments of helium nuclei which have not yet recombined.

Nuclei are extreme cases of metastable equilibrium. Other less extreme examples will be encountered as we examine furthe the implications of wave-function symmetry.

Homonuclear Diatomic Molecules

Homonuclear diatomic molecules, because of the duplicatio of elementary particles in their makeup, thus behave as bosons witl respect to each other. Their internal structure, however, must reflec the fact that the elementary particles, of which they are built, ar fermions. As with other manifestations of quantum statistics, th effects are most noticeable at low temperatures. They can best b demonstrated with hydrogen, which is gaseous at lower tempera tures, than any other diatomic molecular gas.

The most prevalent hydrogen nucleus is the proton, a elementary fermion, with unit mass and charge, in nuclear units Roughly 1 in 5000 hydrogen nuclei, however, are deuterons, com binations of one proton and one neutron, having unit charge an mass 2. As mentioned earlier, deuterons behave like bosons witl respect to each other. Thus there are three different kinds of diatomi hydrogen possible, each with a different internal wave-functio symmetry: H_2, HD, and D_2, if we call the proton H and th deuteron D. The molecule H_2, with its pair of protons, must have a wave function which is antisymmetric with respect to interchange of its two nuclei. The wave function of D_2, on the other hand, must be symmetric with respect to interchange of its two deuterons. No specific symmetry is required of the wave function for HD in regard to interchange of its dissimilar nuclei.

As mentioned in Chapter 22, the electrons in a diatomic molecule surround the two nuclei, their motion and their negative charge binding the positively charged nuclei into a stable molecule. The lighter electrons move rapidly enough so the electron cloud accommodates itself to the slower motion of the relatively massive nuclei. Thus, in Eq. (21-1) we split the molecular energy into four parts; the energy of translation of the molecule as a whole; the energy of rotation of the two nuclei about their center of mass; that of vibration of the nuclei, along a line between them, about their equilibrium distance; and the energy of the particular state of the electron cloud surrounding the nuclei.

This also is the division of the molecular wave function into factors. There is a factor ψ_{el} corresponding to the electronic motions and spins; this must be antisymmetric for interchange of any electron with any other. The factor related to nuclear vibration does not need to satisfy any symmetry requirements. But the factor related to the rotation of the nuclei about their center of mass, times the nuclear spin factor, must have symmetry appropriate to the pair of nuclei that constitute the molecule. Finally the factor for the molecular translation as a whole is combined with those of the other molecules of the gas to form the symmetric combination required by the fact that homonuclear diatomic molecules behave as bosons do. However, even with H_2, the temperature for which the substance is a gas is high enough so that there is negligible difference between the translational properties of a BE gas and a MB gas, of the sort discussed in Chapters 21 and 22.

The factors which exhibit the nuclear statistics in a measurable way are the rotational and the nuclear spin factors. The angle factors in the molecular wave function were illustrated in Figure 27-2. Here, as with the earlier, simplified example, the orientation of the line joining the nuclei is given by the angles ϑ and φ. The angle factor in the wave function is thus a function of ϑ and φ, having two quantum numbers, l and m, the first determining the total angular momentum of the molecule and the second the component of this angular momentum along the polar axis.

As indicated prior to Eq. (22-1), the energy of rotation of the molecule depends only on the quantum number l, being equal to $(\hbar^2/2I)l(l + 1)$, where $I = [M_1 M_2 r_{12}^2/(M_1 + M_2)]$ is the moment of inertia of the molecule about a line through the center of mass perpendicular to the internuclear axis. There is only one wave function for $l = 0$, the constant factor 1, independent of ϑ or φ. For $l = 1$ there are three different rotational wave functions: $\cos \vartheta$, $\sin \vartheta e^{i\varphi}$, and $\sin \vartheta e^{-i\varphi}$ (with values of m equal to 0, 1, and -1, respectively) as indicated in Figure 27-2. For $l = 2$ there are five different wave functions (i.e., five different quantum states, for $m = -2, -1, 0, 1, 2$) and so on. For our purposes we need only report that all angle factors for which l is an odd integer reverse their sign when the direction of $\mathbf{r}_{12}$ is reversed (i.e., if the spatial positions of the two nuclei are interchanged, if the molecule is homonuclear). Thus the nuclear spatial factor is *antisymmetric* for nuclear spatial interchange when l is *odd*; when l is *even* the factor is *symmetric* in the spatial coordinates (when the molecule is homonuclear).

Para- and Orthohydrogen

Now let us examine the H_2 molecule in detail. Since the nuclei are protons, which are fermions, the molecular wave function must be antisymmetric for interchange of the two nuclei. If the space part is symmetric ($l = 0, 2, 4,...$) then the nuclear spin part must be antisymmetric [i.e., must be the singlet factor of Eq. (27-11)]. If the space part is antisymmetric ($l = 1, 3, 5,...$) then the nuclear spin factor must be symmetric [i.e., one of the three triplet factors of Eq. (27-11)]. Therefore the partition-function sum of Eq. (22-1) cannot be the correct form for the rotational factor for H_2; there must be a different weighting factor distinguishing the odd values of l from the even values.

Since each quantum state has equal a priori probability, the triplet states must be three times as numerous as the singlet states. Thus the antisymmetry of the nuclear part of the H_2 wave function has coupled the nuclear spin orientations to the evenness or oddness of the rotational quantum number l. This has a measurable effect on the intensities of the rotational band spectrum of the molecule and also influences the relative prevalence of the even and odd l states in the partition-function sum. If thermal equilibrium is established between nuclear spin orientations, the sum must have the form

$$z_{\text{rot}}^{eq} = z_{\text{rot}}^{e} + 3z_{\text{rot}}^{o}$$

$$z_{\text{rot}}^{e} = \sum_{l=0,2,4,\ldots} (2l + 1) \exp[-\theta_{\text{rot}}l(l + 1)/T] \qquad (27\text{-}18)$$

$$z_{\text{rot}}^{o} = \sum_{l=1,3,5,\ldots} (2l + 1) \exp[-\theta_{\text{rot}}l(l + 1)/T]$$

The terms for l even can occur only when the molecular nuclei are in the singlet state (which has multiplicity of 1) and the terms for l odd can occur only when they are in the triplet state (which has multiplicity of 3).

When T is considerably larger than θ_{rot} this makes no difference in the results. The sum z_{rot}^{eq} can be evaluated by changing from summation to integration, as with Eq. (22-2). It turns out to be equal to $(2T/\theta_{\text{rot}})$, so the rotational part of the heat capacity, when the nuclear spins are in thermal equilibrium, is

$$C_V^{\text{rot}} = NkT\frac{\partial^2}{\partial T^2}[T \ln z_{\text{rot}}^{eq}] \to Nk \qquad T \gg \theta_{\text{rot}} \qquad (27\text{-}19)$$

which is identical with Eq. (22-2). However, the curve for smaller values of T is *not* the same as that shown in Figure 22-1, for the simpler partition function of Eq. (22-1). The C_v for Eq. (27-18) has a somewhat higher peak at a rather higher temperature, about twice θ_{rot}, as compared with the C_v of Figure 22-1, which has a peak at about θ_{rot}.

Unfortunately the measured rotational heat capacity of H_2 gas is not like either curve; it has no peak, it just rises monotonically from nearly zero at about 50°K (well above the boiling point at 25°K) to nearly Nk at room temperatures, as shown in the solid curve of Figure 27-3a. Since θ_{rot} for H_2 is about 85° the curve predicted by Eq. (27-19) would be that shown in Figure 27-3a, marked "equilibrium," and the curve predicted by Eq. (22-1) would be that of Figure 22-1, with a maximum at about $T = 75$°K.

What is wrong with the C_v obtained from Eq. (27-18) is that thermal equilibrium of nuclear spin orientations is not easy to attain at low temperatures. The proton magnetic moments (which are coupled to the proton spins) are small compared to electronic moments (in inverse ratio of their masses) and they are well protected against outside influences by the electron cloud. Consequently at low temperatures it takes several days for spin equilibrium of the nuclei to be reached. During the time usually taken to measure heat capacity the nuclear spins must be considered to be fixed in metastable equilibrium. Those H_2 molecules which are in the singlet state

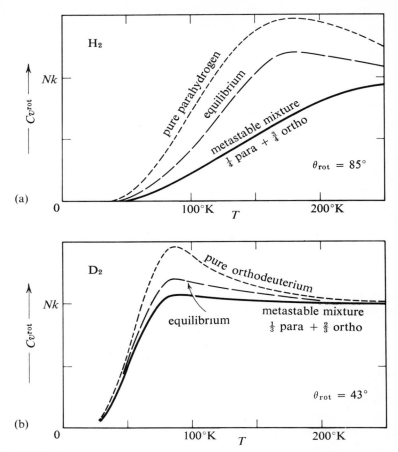

FIGURE 27-3

Heat capacities of hydrogen and deuterium gas under different equilibrium conditions.

will stay in the singlet state and those in the triplet state will stay triplet.

In other words the gas, at low temperatures, acts as though it were a mixture of two different gases, with relative concentrations which do not change with temperature. One gas, which is called *parahydrogen*, is H_2 in the singlet state, and thus (for H_2) has only *even* rotational states ($l = 0, 2, 4,...$). The other gas, called *orthohydrogen*, is H_2 in the triplet state and thus (for H_2) has only *odd* rotational states ($l = 1, 3, 5, ...$). If the gas has been quickly cooled down from room temperature, where there is thermal equilibrium between the two varieties and thus a ratio of 1 to 3 in relative amounts, the cold gas will be a mixture of $\frac{1}{4}N$ parahydrogen molecules and $\frac{3}{4}N$ orthohydrogen molecules. If the heat capacity is measured in a time short compared to the time for nuclear spin orientation to reach thermal equilibrium, the gas will act as though it is a permanent mixture of the two. Therefore its partition function will be the *product* of the partition functions of each gas [see the discussion between Eqs. (19-7) and (19-8)],

$$Z_{\text{rot}} = (z_{\text{rot}}^e)^{(1/4)N}(z_{\text{rot}}^o)^{(3/4)N}$$

and thus

$$C_V^{\text{rot}} = NkT\left[\frac{1}{4}\frac{\partial^2}{\partial T^2}(T\ln z_{\text{rot}}^e) + \frac{3}{4}\frac{\partial^2}{\partial T^2}(T\ln z_{\text{rot}}^o)\right] \qquad (27\text{-}20)$$

(metastable mixture)

where z_{rot}^e and z_{rot}^o are the partition functions, given in Eq. (27-18), for pure parahydrogen and for pure orthohydrogen. The formula (27-20) is for a $\frac{1}{4}$ and $\frac{3}{4}$ mixture.

This formula fits the measured values very closely; the solid line of Figure 27-3a is the heat capacity predicted by Eq. (27-20); the measured values deviate from the curve by only a few per cent. The check constitutes a comprehensive verification of the long chain of reasoning outlined in this chapter.

But there is more to come. The introduction of a catalyst, such as activated carbon, can speed up thermal equilibrium between nuclear spin states at low temperatures. Since the lowest allowed state of orthohydrogen is the $l = 1$ state, which has a higher energy than the lowest state of parahydrogen ($l = 0$), at temperatures considerably below θ_{rot} the equilibrium ratio of orthohydrogen to parahydrogen would be proportional to the small quantity $[3z_{\text{rot}}^o/z_{\text{rot}}^e] \to 9e^{-2\theta_{\text{rot}}/T}$. By leaving the H_2 gas at low temperatures in contact with the catalyst, we can obtain almost pure parahydrogen. If we then remove the catalyst and measure the heat capacity we shall be measuring the capacity of parahydrogen,

$$C_V^{\text{rot}} = NkT\frac{\partial^2}{\partial T^2}(T\ln z_{\text{rot}}^e) \qquad \text{(pure parahydrogen)} \qquad (27\text{-}21)$$

This is shown as the dotted curve of Figure 27-3a. Again the measured values check the theory surprisingly well.

Finally, if we leave the H_2 in contact with the catalyst while the heat capacity is being measured, and if the measurements are made slowly enough, we should be able to measure the equilibrium value of C_V^{rot}, given in Eq. (27-19) and shown as the dashed curve of Figure 27-3a. Again the agreement with experiment is satisfactory, though the difficulty of measurement precludes the accuracy of check obtained in the other two situations.

Ortho- and Paradeuterium

For the heteronuclear molecule HD there are no symmetry requirements, no para or ortho states and the heat capacity should follow the formula of Eq. (22-2), shown in Figure 22-1, with $\theta_{rot} = 64°K$. Thus the peak in the C_v^{rot} curve should come at about $50°K$, considerably below those for either H_2 or D_2. The measurements bear this out in detail.

For the homonuclear molecule D_2 the nuclear factor in the wave function should be symmetric in interchange of the two deuterons, which should behave like bosons. Therefore for the *even* rotational states ($l = 0, 2, 4,...$) the spin factor should also be *symmetric*. For the *odd* rotational states ($l = 1, 3, 5,...$) the spin factor must be *antisymmetric*.

As was noted earlier, a deuteron has a spin 1, so its spin should have three possible states: the one with the factor α representing a component $+1$, the state with a factor γ representing a component 0, and the function with a factor β representing the -1 component possibility. The possible antisymmetric combinations of spin factors for two deuterons are threefold,

$$[\alpha(1)\gamma(2) - \alpha(2)\gamma(1)]/\sqrt{2}, \qquad [\alpha(1)\beta(2) - \alpha(2)\beta(1)]/\sqrt{2},$$

$$\text{and} \qquad [\gamma(1)\beta(2) - \gamma(2)\beta(1)]/\sqrt{2}$$

which corresponds to a triplet. These *antisymmetric spin* factors go with the *odd rotational* factors to make up the *paradeuterium* states of D_2. There are six possible *symmetric spin* factors,

$$\alpha(1)\alpha(2), \qquad [\alpha(1)\gamma(2) + \alpha(2)\gamma(1)]/\sqrt{2}, \qquad \gamma(1)\gamma(2)$$

$$[\gamma(1)\beta(2) + \gamma(2)\beta(1)]/\sqrt{2} \qquad \text{and} \qquad \beta(1)\beta(2),$$

$$\text{and} \qquad [\alpha(1)\beta(2) + \alpha(2)\beta(1)]/\sqrt{2}$$

representing a singlet and a quintet. These go with the *even rotational* factors to make up the *orthodeuterium* states of D_2.

Therefore the equilibrium partition function for D_2 is

$$z_{\text{rot}}^{eq} = 6z_{\text{rot}}^{e} + 3z_{\text{rot}}^{o} \tag{27-22}$$

which is to be compared with the formula of Eq. (27-18) for H_2. The curve, for $\theta_{\text{rot}} = 43°K$, is plotted as the dashed curve of Figure 27-3b. The heat capacity for the stable form at low temperatures, pure orthodeuterium, is the same as that for pure parahydrogen, except for the change in temperature scale; it is shown as the dotted curve of Figure 27-3b. And finally the metastable mixture of $\frac{2}{3}$ orthodeuterium plus $\frac{1}{3}$ paradeuterium, obtained by cooling deuterium gas without the presence of a catalyst, is given by the formula

$$C_V^{\text{rot}} = NkT\left[\frac{2}{3}\frac{\partial^2}{\partial T^2}(T\ln z_{\text{rot}}^e) + \frac{1}{3}\frac{\partial^2}{\partial T^2}(T\ln z_{\text{rot}}^o)\right] \tag{27-23}$$

(metastable mixture)

and is shown, for $\theta_{\text{rot}} = 43°K$, as the solid line of Figure 27-3b. All of these curves are verified by experimental data, taken under the different conditions, appropriate for the three cases.

Thus the effects of quantum statistics turn up in odd corners of the field, at low temperatures and for substances a part of which can stay gaslike to low-enough temperatures for the effects of degeneracy to become evident. For the great majority of substances and over the majority of the range of temperature and density, classical statistical mechanics is valid, and the calculations using the canonical ensemble of Chapters 19 through 22 quite accurately portray the observed results. The situations where quantum statistics must be used to achieve concordance with experiment are in the minority (luckily; otherwise our computational difficulties would be much greater). But, when they are all considered, these exceptional situations add up to exhibit an impressive demonstration of the fundamental correctness of quantum statistics. They are some of the aspects of reality, mentioned at the beginning of Chapter 16, the explanation of which is the sole justification for the theoretical superstructure of statistical mechanics discussed in these last ten chapters.

PROBLEMS

Chapter 2

2-1. The amount of heat ΔQ added to a unit mass of some homogeneous material raises its temperature by an amount ΔT, where $\Delta Q = c_m \Delta T$, c_m being the heat capacity of the material per unit mass. A temperature gradient in the material will produce a heat flow $\mathbf{J}$ heat units per second per unit perpendicular to the flow. The relationship is $\mathbf{J} = -\kappa_T \operatorname{grad} T$, where κ_T is the thermal conductivity of the material. Use the equation of continuity for fluid flow to relate the divergence of $\mathbf{J}$ to the time rate of change of Q and therefore of T. From this obtain the differential equation for the temperature distribution in the material, as a function of x, y, z, and t, in terms of c_m, κ_T, and ρ, the density of the material.

2-2. Suppose that all the atoms in a gas are moving with the same speed v, but that their directions of motion are at random.

(*a*) Average over directions of incidence to compute the mean number of atoms striking an element of wall area dA per second (in terms of N, V, v, and dA) and the mean momentum per second imparted to dA.

(*b*) Suppose, instead, that the number of atoms having speeds between v and $v + dv$ is $2N[1 - (v/v_m)](dv/v_m)$ for $v < v_m$ (the directions still at random). Calculate for this case the mean number per second striking dA and the mean momentum imparted per second, in terms of N, V, v_m, and dA. Show that Eq. (2-4) holds for both of these cases.

2-3(*a*) For a perfect gas of point particles the mean kinetic energy per particle $\langle \text{K.E.} \rangle_{\text{tran}}$ is a function $f(T)$ of the temperature T. Thus any change in temperature will produce a change in the internal energy U of the gas. But an addition of heat ΔQ produces an increase ΔT in temperature; by definition the relationship is $\Delta Q = C \Delta T$, where C is the heat capacity of the gas. Therefore, for the gas of Eq. (2-1), an addition of heat is equivalent to an increase in internal energy of the gas. If ΔQ is expressed in the same energy units (joules) as U, what is the equation relating c_m, the heat capacity per unit mass of the gas, to N_m, the number of molecules per unit mass and f', the slope of $f(T)$?

(b) When there is a temperature gradient in a substance there cannot be thermodynamic equilibrium because heat is flowing (see Problem 2-1). There can be mechanical equilibrium, however, if the pressure is uniform throughout the material. Suppose the material is a perfect gas of point particles, and suppose the temperature gradient is small enough so that Eq. (2-4) holds at each point in the gas, with $\langle K.E. \rangle_{tran} = f(T)$. What must be the value of the gradient of the density ρ of the gas, in terms of the gradient of T, of ρ, c_m and U, for there to be mechanical equilibrium?

Chapter 3

3-1. The coefficient of thermal expansion β and the compressibility κ of a substance are defined in terms of partial derivatives

$$\beta = (1/V)(\partial V/\partial T)_P \qquad \kappa = -(1/V)(\partial V/\partial P)_T$$

(a) Show that

$$(\partial\beta/\partial P)_T = -(\partial\kappa/\partial T)_P \text{ and that } (\beta/\kappa) = (\partial P/\partial T)_V$$

for any substance.

(b) It is found experimentally that, for a given gas,

$$\beta = \frac{RV^2(V - nb)}{RTV^3 - 2an(V - nb)^2}$$

$$\kappa = \frac{V^2(V - nb)^2}{nRTV^3 - 2an^2(V - nb)^2}$$

where a and b are constants, and also that the gas behaves like a perfect gas for large values of T and V. Find the equation of state of the gas.

3-2. A gas obeys equation of state (3-4). Show that for just one critical state, specified by the values T_c and V_c, both $(\partial P/\partial V)_T$ and $(\partial^2 P/\partial V^2)_T$ are zero. Write the equation of state giving P/P_c in terms of $T \: T_c$ and $V \: V_c$. Plot three curves for P/P_c as a function of $V \: V_c$, one for $T = \frac{1}{2}T_c$, one for $T = T_c$, and one for $T = 2T_c$. What happens physically when the equation indicates three allowed values of V for a single P and T?

3-3(a) By considering the charging of a parallel-plate condenser, calculate the work done by a homogeneous, isotropic dielectric in terms of the volume v of the dielectric and the uniform electric field $\mathfrak{E}$ applied, when the electric polarization of the dielectric changes by $d\mathfrak{P}$.

(b) It is found experimentally that

$$(\partial\mathfrak{E}/\partial\mathfrak{P})_T = [T/(AT + B)]$$
$$\text{and} \qquad (\partial\mathfrak{E}/\partial T)_{\mathfrak{P}} = [B\mathfrak{P}/(AT + B)^2]$$

Compute $(\partial\mathfrak{P}/\partial T)_{\mathfrak{E}}$ and, by integration, obtain the equation of state, $\mathfrak{P}$ as a function of $\mathfrak{E}$ and T.

3-4. Express the van der Waals equation of state (3-4) in the form of the virial equation (3-5). Find $B(T)$ and $C(T)$ in terms of a, b, and T.

Chapter 4

4-1. A gas with van der Waals' equation of state (3-4) has an internal energy

$$U = \tfrac{3}{2}nRT - (an^2/V) + U_0$$

Compute C_V and C_P as functions of V and T and compute T as a function of V for an adiabatic expansion.

4-2. An ideal gas for which $C_V = \tfrac{5}{2}nR$ is taken from point a to point b in Figure P-1, along three paths, acb, adb, and ab, where $P_2 = 2P_1$, $V_2 = 2V_1$.

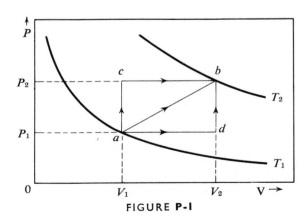

FIGURE **P-I**

(*a*) Compute the heat supplied to the gas, in terms of n, R, and T_1, in each of the three processes.

(*b*) What is the heat capacity of the gas, in terms of R, for the process *ab*?

4-3. A paramagnetic solid, obeying Eqs. (3-6), (3-8), and (4-8) and having a heat capacity $C_{v\mathfrak{M}} = nAT^3$, is magnetized iso-thermally (at constant volume) at temperature T_0 from $\mathfrak{M} = 0$ to a maximum magnetic field of $\mathfrak{H}_m$. How much heat must be lost? It is then demagnetized adiabatically (at constant volume) to $\mathfrak{M} = 0$ again. Compute the final temperature T_1 of the solid, in terms of $\mathfrak{H}_m$, T_0, A, and D. How do you explain away the fact that, if $\mathfrak{H}_m$ is large enough or T_0 small enough, the formula you have obtained predicts that T_1 should be negative?

4-4. Derive equations for $(\partial U/\partial T)_P$ and $(\partial U/\partial P)_T$ analogous to Eqs. (4-4) and (4-6). Obtain an expression for $(\partial H/\partial P)_T$. For a perfect gas, with $C_P = \frac{5}{2}nR$ and $(C_P - C_V)/(\partial V/\partial T)_P = P$, integrate the partials of H to obtain the enthalpy.

4-5. Figure P-2 shows a thermally isolated cylinder, divided into two parts by a thermally insulating, frictionless piston. Each side contains n moles of a perfect gas of point atoms. Initially both sides have the same temperature; heat is then supplied slowly and reversibly to the left side until its pressure has increased to $(243P_0/32)$.

(*a*) How much work was done on the gas on the right?

(*b*) What is the final temperature on the right?

(*c*) What is the final temperature on the left?

(*d*) How much heat was supplied to the gas on the left?

4-6. The ratio C_P/C_V for air is equal to 1.4. Air is com-pressed from room temperature and pressure, in a diesel engine compression, to 1/15 of its original volume. Assuming the com-pression is adiabatic, what is the final temperature of the gas?

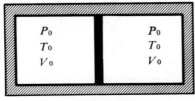

FIGURE **P-2**

4-7.   Find the relationship between T and V for the reversible, adiabatic compression of a gas obeying van der Waals equation (3-4) and having a heat capacity at constant volume C_v which is independent of T and of V.

4-8.   A gas, for which $(\partial U/\partial V)_T = 0$, satisfies the virial equation of state (3-5), with B and C independent of T.

(a) How much heat must be supplied to expand the gas, in a quasistatic, isothermal process, from volume V_1 to volume V_2?

(b) Express the difference $(C_P - C_V)$ between the heat capacities of this gas in terms of T and V.

Chapter 5

5-1.   Compute ΔQ_{12} and ΔQ_{43}, for a Carnot's cycle using a perfect gas of point particles, in terms of nR and T_h and T_c. Using the perfect-gas scale of temperature, show that $\Delta W_{23} = -\Delta W_{41}$. Show that the efficiency of the cycle is $(T_h - T_c)/T_h$ and thus prove that the perfect-gas temperature scale coincides with the thermodynamic scale of Eq. (5-5).

5-2.   A magnetic material, satisfying Eqs. (4-8) and (3-8) has a constant heat capacity, $C_{V\mathfrak{M}} = C$. It is carried around a Carnot cycle shown in Figure P-3, $\mathfrak{M}$ being reduced isothermally from $\mathfrak{M}_1$ to $\mathfrak{M}_2$ at T_h, then reduced adiabatically from $\mathfrak{M}_2$ to $\mathfrak{M}_3$, when it has temperature T_c, then remagnetized isothermally at T_c to $\mathfrak{M}_4$, and thence adiabatically back to T_h and $\mathfrak{M}_1$.

(a) Express $\mathfrak{M}_3$ in terms of T_h, T_c, D, C, and $\mathfrak{M}_2$ and relate $\mathfrak{M}_4$ similarly with $\mathfrak{M}_1$.

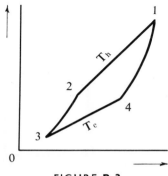

FIGURE **P-3**

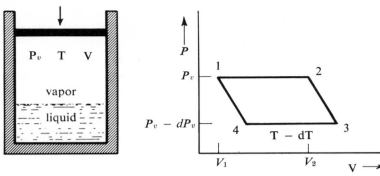

FIGURE **P-4**

(b) How much heat is absorbed in process 1-2? How much given off in process 3-4?

(c) How much magnetic energy dW is given up by the material in each of the four processes? Show that $dW_{23} = -dW_{41}$.

(d) Show that the efficiency of the cycle heat-magnetic energy is $(T_h - T_c)/T_h$.

5-3. When a mole of liquid is evaporated at constant temperature T and vapor pressure $P_v(T)$, the heat absorbed in the process is called the latent heat L_v of vaporization. A Carnot cycle is run as shown in Figure P-4, going isothermally from 1 to 2, evaporating n moles of liquid and changing volume from V_1 to V_2, then cooling adiabatically to $T - dT$, $P_V - dP_V$ by evaporating an additional small amount of liquid, then recondensing the n moles at $T - dT$, from V_3 to V_4, and thence adiabatically to P_V, T again.

(a) Show that $V_2 - V_1 = V_g - V_l$ where V_g is the volume occupied by n moles of the vapor, and V_l the volume of n moles of the liquid.

(b) Find the efficiency of the cycle, in terms of dP_V, $V_g - V_l$, and nL_V.

(c) If this cycle is to have the same efficiency as any Carnot cycle, this efficiency must be equal to $(T_h - T_c)/T_h = dT/T$. Equating the two expressions for efficiency, obtain an equation for the rate of change dP_V/dT of the vapor pressure with temperature in terms of $V_g - V_l$, n, L_V, and T.

5-4. In the tropics the water near the surface is considerably warmer than the deep water. Would an engine operating between these two levels violate the second law?

5-5. Electromagnetic radiation in an evacuated vessel of volume V, at equilibrium with the walls at temperature T, behaves like a gas of photons, having internal energy U and pressure P: $U = aVT^4$, $P = \frac{1}{3}aT^4$, where a is Stefan's constant [see Eq. (7-8)].

(a) Find the relation between P and V for an adiabatic process and the relationship between heat absorbed and change of volume in an isothermal process, for this gas.

(b) Plot the closed curve, on the $P - V$ plane, for a Carnot cycle.

(c) Compute the heat absorbed and the work done in each part of the cycle and prove that the efficiency is $[(T_h - T_c)/T_h]$.

5-6. Prove that it is impossible for two quasistatic adiabatics to intersect. (*Hint*: show that the second law is violated if they do.)

5-7. Suppose we desire to use a temperature scale τ such that the efficiency deficit function Ψ of Eq. (5-2) is a subtractive function of τ, $\Psi(\tau_h - \tau_c)$.

(a) Show that the thermodynamic temperature T is an exponential function of τ.

(b) If we wish τ to be zero at the melting-point of ice and to be 100 at the boiling point of water, what is the exact functional relationship between T and τ? What value of τ corresponds to $T = 0$?

(c) What is the equation of state and the heat capacities C_v and C_p for a perfect gas of point atoms in terms of the τ scale?

5-8. An ideal gas, satisfying Eqs. (4-7) and (4-12) is carried around the cycle shown in Figure P-5; 1-2 is at constant volume, 2-3 is adiabatic, 3-1 is at constant pressure, V_3 is $8V_1$, and n moles of the gas are used.

(a) What is the heat input, the heat output, and the efficiency of the cycle, in terms of P_1, V_1, n, and R?

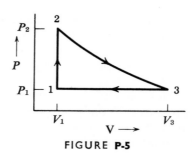

FIGURE P-5

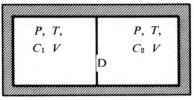

FIGURE P-6

(b) Compare this efficiency with the efficiency of a Carnot cycle operating between the same extremes of temperature.

Chapter 6

6-1. An amount of perfect gas of one kind is in the volume $C_1 V$ of Figure P-6 at temperature T and pressure P, separated by an impervious diaphragm D from a perfect gas of another kind, in volume $C_2 V$ and at the same pressure and temperature ($C_1 + C_2 = 1$). The volume V is isolated thermally. What is the entropy of the combination? Diaphragm D is then ruptured and the two gases mix spontaneously, ending at temperature T, partial pressure $C_1 P$ of the first gas, $C_2 P$ of the second gas, all in volume V. What is the entropy now? Devise a pair of processes, using semipermeable membranes (one of which will pass gas 1 but not 2, the other which will pass 2 but not 1), which will take the system from the initial to the final state reversibly and thus verify the change in entropy. What is the situation if gas 1 is the same as gas 2?

6-2. Two identical solids, each of heat capacity C_V (independent of T), one at temperature $T + t$, the other at temperature $T - t$, may be brought to common temperature T by two different processes:

(a) The two bodies are placed in thermal contact, insulated thermally from the rest of the universe and allowed to reach T spontaneously. What is the change of entropy of the bodies and of the universe caused by this process?

(b) First a reversible heat engine, with infinitesimal cycles, is operated between the two bodies, extracting work and eventually bringing the two to common temperature. Show that this common temperature is not T, but $\sqrt{(T^2 - t^2)}$. What is the work produced and what is the entropy change in this part of process b? Then heat is added reversibly to bring the temperature of the two bodies to T. What is the entropy change of the bodies during the whole of reversible process b? What is the change in entropy of the universe?

6-3. Show that $T\,dS = C_v\,dT + T(\partial P/\partial T)_V\,dV$, and $T\,dS = C_v(\partial T/\partial P)_v\,dP + C_P(\partial T/\partial V)_P\,dV$.

6-4. A paramagnetic solid, obeying Eqs. (3-6), (3-8), and (4-8), has a heat capacity $C_{p0}(T)$ (at zero magnetic field) dependent solely on temperature. First, show that

$$T\,dS = C_P\,dT - T(\partial V/\partial T)_P\,dP + T(\partial \mathfrak{M}/\partial T)_{\mathfrak{H}}\,d\mathfrak{H}$$

and, analogous to Eq. (8-13), that $(\partial C_{P\mathfrak{H}}/\partial \mathfrak{H})_{TP} = T(\partial^2 \mathfrak{M}/\partial T^2)_{\mathfrak{H}}$. From this, show that

$$S = \int_0^T (C_{p0}/T)\,dT - \tfrac{1}{2}nD(\mathfrak{H}/T)^2 - \beta V_0 P$$

and thence obtain energy U as a function of T, P, and $\mathfrak{H}$. Obtain S as a function of T, V, and $\mathfrak{M}$ and thence obtain U as a function of T, V, and $\mathfrak{M}$.

6-5. Express the entropy and the volume of a perfect gas of point atoms in terms of its pressure and temperature. Then integrate the Gibbs-Duhem equation (6-6) to obtain the chemical potential μ of a mole of the gas.

6-6. At very low temperatures the entropy, magnetization, and internal energy of the atomic magnets in a paramagnetic crystal are fairly accurately represented by the equations

$$S_m = (\partial \mathfrak{G}/\partial T)_{\mathfrak{H}}; \qquad \mathfrak{M} = (\partial \mathfrak{G}/\partial \mathfrak{H})_T; \qquad U_m = TS_m + \mathfrak{H}\mathfrak{M} - \mathfrak{G}$$

where

$$\mathfrak{G} = -nR_m T \ln[1 + 256e^{-8T_m/T} \cosh^8(\beta T_m \mathfrak{H}/T)]$$

and where R_m, β and $T_m = 0.2°\text{K}$ are constants.

(*a*) Compute $(\partial \mathfrak{H}/\partial T)_{\mathfrak{M}}$ and thence find $(\partial U_m/\partial \mathfrak{M})_T$ as a function of $\mathfrak{H}$ and T. Plot it as a function of T for $0 < T < 5T_m$ for $\mathfrak{H} = 0$ and for $\beta\mathfrak{H} = 2$. What does this predict regarding the relation between heat absorbed and rate of gain of U_m in an isothermal process?

(*b*) Plot S_m as a function of T for $0 < T < 5T_m$ for $\mathfrak{H} = 0$ and $\beta\mathfrak{H} = 2$. Use these curves to recapitulate the arguments regarding the third law of thermodynamics. Roughly, what temperature can be reached by starting at $T = 1°\text{K}$, $\mathfrak{H} = 0$, increasing to $\beta\mathfrak{H} = 2$ isothermally and then reducing to $\mathfrak{H} = 0$ isentropically? How much further could you go in two more such steps? How much in eight more steps?

Chapter 7

7-1. A gas obeys the virial equation of state (3-5) with B and C independent of T and its heat capacity at constant volume $C_v = \frac{3}{2}nR$. Compute the Joule–Thomson coefficient $(\partial T/\partial P)_H$ as a function of T and V. Does this gas have an inversion point?

7-2. The Stirling cycle consists of two isothermal processes, at T_h and T_c and two processes at constant volume, as is shown in Figure P-7. The characteristics of the cycle depend on how the heat is transferred to and from the engine during the constant-volume parts of the cycle.

(a) Consider first the case where only two heat reservoirs are available, at temperatures T_h and T_c. Find the entropy change of the universe for section 41 of the cycle.

(b) Show that this change is always positive.

(c) Find the efficiency of the cycle for this case, assuming that the working substance is a perfect gas of point atoms.

(d) Now suppose that a continuous set of heat reservoirs, for every temperature between T_c and T_h, is available, so that 23 and 41 can be traversed reversibly. What now is the efficiency of the cycle?

(e) In view of your answers to (c) and (d), explain the detrimental effect of the increase in entropy in the former case.

7-3. A very approximate set of formulas for the entropy and density of the normal fluid in helium II is $s_n = 1600$ joules/kg deg (which thus equals the entropy per unit mass of the whole fluid s_λ at the transition point) and $\rho_n = \rho(T/T_\lambda)^4$, where $\rho = \rho_n + \rho_s = 144$ kg/m^3 is the density of the whole fluid and where $T_\lambda = 2.2\,°$K is the temperature of the transition point.

(a) Enough heat is supplied to the coil H of Figure 7-2 to

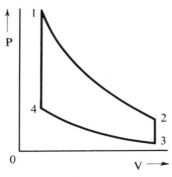

FIGURE **P-7**

make the temperature $1.2°K$, whereas the temperature outside the vessel is $1.1°K$. What is the pressure difference between the inside and outside of the vessel, in atmospheres?

(b) Using the improved formula $(\rho_s \rho_n s_n^2 T / \rho^2 c)^{1/2}$, plot the velocity of second sound versus T for the range $\frac{1}{2}T_\lambda < T < T_\lambda$.

7-4. Show that, when temperature T and applied magnetic field $\mathfrak{H}$ are the independent variables, the heat capacity for constant $\mathfrak{H}$ is $C_\mathfrak{H} = (\partial H_m / \partial T)$, where $H_m = U_m - \mathfrak{H}\mathfrak{M}$, and thus that H_m is the magnetic heat content of the material at constant applied field.

(a) Find $C_\mathfrak{H}$ for a material obeying Curie's law and having a heat capacity for constant magnetization $C_\mathfrak{M} = [nR_m T^2 / (\theta^2 + T^2)]$. What is the entropy of this material? Can this formula be correct down the $T = 0$?

(b) Find $C_\mathfrak{H}$ for the material of Problem 6-6. Plot it, as a function of T for $0 < T < 5T_m$, for $\beta\mathfrak{H} = 0$ and for $\beta\mathfrak{H} = 2$. Show that for T much greater than both T_m and $\beta\mathfrak{H}T_m$ this material obeys Curie's law and that its $C_\mathfrak{H}$ is inversely proportional to T^2.

Chapter 8

8-1. The tension J in a plastic rod, stretched to a length L, at temperature T, is $J = A(T)(L - L_0)$, where L_0 is independent of T. Its heat capacity at constant L is $C_L = C(T) - M(T)N(L)$, where $N = 0$ when $L = L_0$ and where $MN \ll C$ over the range of interest of T and L.

(a) Show that $M(T)N(L) = \frac{1}{2}TA''(T)(L - L_0)^2$, where the primes on A indicate differentiation with respect to T. What is the heat capacity of the rod at constant tension?

(b) Compute the entropy of the rod, as function of T and L, if $S \to 0$ as $T \to 0$. What does this require of function $A(T)$?

(c) Find the rate of change $(\partial T / \partial L)_s$ of the temperature of the rod for adiabatic stretching. If the temperature rises when the rod is stretched, does the entropy increase or decrease as the rod is stretched? What does this indicate about the change in the amount of disorder of the molecules of the rod on stretching?

(d) Compute the internal energy U of the rod and then compute $F = U - TS$ as a function of T and L. Calculate the partials of F with respect to T and L and show that they equal $-S$ and J, respectively.

8-2. A gas obeys the van der Waals' equation of state (3-4) and has heat capacity at constant volume $C_V = \frac{3}{2}nR$. Write the equation of state in terms of the quantities $t = T/T_c$, $p = P/P_c$, and

$v = V/V_c$, where $T_c = 8a/27Rb$, $P_c = a/27b^2$, $V_c = 3nb$ (see Problem 3-2). Calculate T_cS/P_cV_c in terms of t and v, likewise F/P_cV_c and G/P_cV_c. For $t = \frac{1}{2}$ plot p as a function of v from the equation of state. Then, for the same value of t, calculate and plot G/P_cV_c as a function of p, by graphically finding v for each value of p from the plot, and then computing G/P_cV_c for this value of v (remember that for some values of p there are three allowed values of v). The curve for G/P_cV_c crosses itself. What is the physical significance of this?

8-3. The work done by a homogeneous, isotropic dielectric for a small change of electric polarization is $-V\mathfrak{E}\,d\mathfrak{P}$, where $\mathfrak{E}$ is the applied electric field strength and $\mathfrak{P}$ is the polarization, if the field is uniform.

(a) Find the Maxwell relations for this case. Assume constant V.

(b) Find the heat supplied to the dielectric when a parallel-plate condenser is charged isothermally to a final field strength $\mathfrak{E}_0$. The dielectric obeys the law $\mathfrak{P} = [a + (b/T)]\mathfrak{E}$. Interpret your result in terms of entropy and disorder.

8-4. The Gibbs function of a nonideal gas is

$$G = nRT \ln P + P[nb - (na/RT)] + f(T)$$

where a and b are constants and f is a function of T only. Find the equation of state and show that it agrees with the van der Waals equation of state if second-order terms in a and b are neglected. Compute the heat capacity at constant pressure.

8-5. Find the magnetic Gibbs function and the magnetic Helmholtz function for the material of Problem 6-6. Calculate expressions for the partials $(\partial T/\partial \mathfrak{H})_s$ and $(\partial T/\partial \mathfrak{H})_U$ as functions of T and $\mathfrak{H}$.

8-6. The equation of state and the heat capacity of a gas are

$$P = RT[(n/V) + (n/V)^2 B(T)];$$

$$C_v = \tfrac{3}{2}nRT - (n^2R/V)\frac{d}{dT}(T^2B')$$

where $B'(T) = (dB/dT)$ and where the second term in the brackets is small compared to the first term for the useful range of T and V.

(a) Show that the second term in the expression for C_v must be present if the equations are to be self-consistent. Compute C_P.

(b) Compute S and U and show that $(\partial U/\partial V)_T$ is not zero.

(c) Compute F, H, and μ, all as functions of T and V.

8-7. The Helmholtz function for a homogeneous, isotropic solid is

$$F = (1/2\kappa_0 V_0)(V - V_0)^2$$

$$+ 3nRT \ln\left\{1 - \exp\left(-\frac{\theta_0}{T}\right)[1 + \alpha(V_0 - V)]\right\}$$

where R, κ_0, V_0, α, and θ_0 are independent of T and V.

(a) Compute the equation of state and thence find the compressibility $\kappa = -(1/V)(\partial V/\partial P)_T$ and the coefficient of thermal expansion $\beta = (1/V)(\partial V/\partial T)_P$.

(b) Compute the entropy S and the internal energy U of the solid. Plot $(S/3nR)$ as a function of (T/θ_0) from 0 to 2, for $\alpha(V_0 - V) = 0$ and $= 0.5$. Is this set of formulas consistent with the third law?

(c) Calculate the expressions for $(\partial T/\partial V)_U$ and $(\partial T/\partial V)_S$.

(d) Compute the heat capacities C_v and C_P. What are their limiting formulas for $T \ll \theta_0$ and $T \gg \theta_0$?

8-8. Express the entropy of the gas of Eq. (8-22) in terms of n, U, and V, rather than n, T, and V. Then show that the entropy parameters $F_0 \equiv (\partial S/\partial U)_{Vn} = (1/T)$, $F_v \equiv (\partial S/\partial V)_{Un} = (P/T)$, and $F_n \equiv (\partial S/\partial n)_{UV} = (\mu/T)$.

8-9. Use Eqs. (8-48) to show that, when electric current J_e flows through a junction between a wire of material A to one of material B, all kept at constant temperature T, heat is evolved.

8-10(a) Show that the enthalpy $H = G - T(\partial G/\partial T)_P$, where G is the Gibbs function. Thus show that q in Eq. (8-38) would be zero if the thermo-osmotic process were reversible.

(b) The gas diffusing through the porous partition is a perfect gas of point particles. From Eqs. (8-42) devise an experiment to measure q indirectly without having to measure Q or κ or grad P. What is the formula for q in terms of the quantities measured?

Chapter 9

9-1. Assume that near the triple point the latent heats L_m and L_v are independent of P, that the vapor has the equation of state of a perfect gas, that the volume of a mole of solid or liquid is negligible compared to its vapor volume, and that the difference $V_l - V_s$ is positive, independent of P or T and is small compared to nL_m/T. Using these assumptions, integrate the three Clausius-Clapeyron equations for the vapor-pressure, sublimation-pressure,

and melting-point curves. Sketch the form of these curves on the
P-T plane.

9-2.   The heat of fusion of ice at its normal melting point is
3.3×10^5 joules/kg and the specific volume of ice is greater than
the specific volume of water at this point by 9×10^{-5} m^3/kg. The
value of $(1/V)(\partial V/\partial T)_P$ for ice is 16×10^{-5} per degree and the value
of $-(1/V) \times (\partial V/\partial P)_T$ is 12×10^{-11} m^2/newton.

(a) Ice at $-2°C$ and at atmospheric pressure is compressed
isothermally. Find the pressure at which the ice starts to melt.

(b) Ice at $-2°C$ and atmospheric pressure is kept in a container at constant volume and the temperature is gradually increased.
Find the temperature at which the ice begins to melt.

(c) Ice at $-2°C$ and atmospheric pressure is compressed
adiabatically. At what pressure will the ice melt?

9-3.   Obtain an expression for the fraction $(\Delta U/nL)$ of the
latent heat of evaporation, which goes into a change of internal
energy in the change of phase, in terms of P_v, T, and (dP_v/dT).
What is the value of this fraction for the transition from water to
ice at $0°C$?

9-4.   The specific heat of ice, near the melting point, is
2100 joules/kg deg and that of water is 4200 joules/kg deg. How
much does the heat of fusion of ice change with an increase of
pressure of 10 atm?

9-5.   A mass m of water, originally at temperature T_h, is
placed inside an electric refrigerator. The refrigerator operates between the water and the room, at T_h, which may be considered as
a heat reservoir. The water is frozen and the ice is cooled further
to a temperature T_0. The specific heats c_w and c_i of water and ice
may be considered as a constant over this temperature range. Find
the minimum amount of electrical energy that must be supplied to
the refrigerator for this process. What is the numerical value of
this minimal energy, in kilowatt-hours, if $m = 1$ kg, $T_h = 27°C$,
and $T_0 = -23°C$?

9-6.   The temperature of a long vertical column of a certain
material is everywhere equal. When this temperature is T_0 the
material below a certain point in the column is found to be solid;
that above this point is liquid. When the temperature is reduced by
$0.2°C$ the solid–liquid interface is observed to shift upward by
0.5 m. The heat of fusion of the material is 10,000 joules/kg and the
density of the liquid is 1000 kg/m^3. What is the density of the
solid? (*Hint*: the pressure at the original interface point is not
changed by the temperature change.)

9-7. Show that $(\partial T_c/\partial P)_{\mathfrak{H}_c} = (T_c/nL_c)(V_n - V_s)$ for a super-conductor, where T_c is the transition temperature at pressure P, and where V_n and V_s are the volumes of the normal and the super-conducting phases.

9-8. The magnetic field $\mathfrak{H}_c$ for superconductive transition, for a particular material, depends on T as follows:

$$\mathfrak{H}_c = \mathfrak{H}_0[1 - (1 - \alpha)(T/T_0)^2 - \alpha(T/T_0)^3]$$

Its heat capacity, in the superconductive phase, at very low temperatures, goes to zero quadratically with T,

$$C_s = \beta_2 V(T/T_0)^2 + \beta_3 V(T/T_0)^3 + \cdots$$

Show that the heat capacity of the normal phase (for $\mathfrak{H} > \mathfrak{H}_c$) at low temperatures must depend on T as follows:

$$C_n = \gamma_1 V(T/T_0) + \gamma_2 V(T/T_0)^2 + \cdots$$

Determine the values of γ_1 and γ_2 in terms of μ_0, T_0, $\mathfrak{H}_0$, α, and β_2.

Chapter 10

10-1. The observed equilibrium constant $K(T)$ for a gaseous reaction is usually expressed in the series form

$$\ln K(T) = (Q/T) + A \ln T + B + CT + \cdots$$

Show that Q is the heat evolved in the reaction at zero temperature.

10-2. The coefficients in the expansion for $\ln K$ (see Problem 10-1) for the iodine dissociation reaction $2I \rightarrow I_2$ are $Q = 17400$, $A = 0.75$, $B = 1$, and C, D, etc., are negligible, where T is in °K and the dimensions of K are $(atm)^{-1}$. What fraction of the iodine is dissociated at 1 atm and 500°K? At what rate is this fraction changing with T?

10-3. The equilibrium constant for the reaction $2 SO_3 \leftrightarrow 2 SO_2 + O_2$ has the value 0.29 atm at $T = 1000$°K. If one mole of SO_2 and 2 moles of O_2 are introduced into a vessel and maintained at $P = 4$ atm and $T = 1000$, find the number of moles of SO_3 present at equilibrium.

Chapter II

11-1. Calculate the mean value $\langle n \rangle$ and variance σ_n^2 of n for the following discrete probability distributions:

Geometric $\qquad P_n = (1 - \alpha)\alpha^n$

Binomial $\qquad P_n = [N!/n!(N - n)!]p^n(1 - p)^{N-n}$

Poisson $\qquad P_n = (N^n/n!)e^{-N}$

11-2. Calculate the mean value $\langle x \rangle$ and variance σ_x^2 of x for the continuous distributions, having the following probability densities:

Exponential $\qquad f(x) = (1/\lambda)e^{-x/\lambda}$

Erlang $\qquad f(x) = (4x/\lambda^2)e^{-2x/\lambda}$

11-3. The probability that a certain trial (throw of a die or drop of a bomb, for example) is a success is p for every trial. Show that the probability that m successes are achieved in n trials is

$$P_m(n) = \frac{n!}{m!(n - m)!}p^m(1 - p)^{n-m} \qquad \text{(this is the binomial distribution)}$$

Find the average number $\bar{m}$ of successes in n trials, the mean-square $(\bar{m}^2)$ and the standard deviation Δm of successes in n trials.

11-4. The probability of finding n photons of a given frequency in an enclosure that is in thermal equilibrium with its walls is $P_n = (1 - \alpha)\alpha^n$, where $\alpha(0 < \alpha < 1)$ is a function of temperature, volume of the enclosure, and the frequency of the photons. What is the mean number $\bar{n}$ of photons of this frequency? What is the fractional deviation $\Delta n/\bar{n}$ of this number from the mean? Express this fractional deviation in terms of $\bar{n}$, the mean number. For what limiting value of $\bar{n}$ does the fractional deviation tend to zero?

11-5(a) Show that the number of different ways in which M *distinguishable* objects can be placed in C numbered boxes, with *no restriction* on the number of objects per box, is C^M.

(b) Show that the number of different ways in which M distinguishable objects can be placed in C numbered boxes, with *no more than one* object per box, is $[C!/(C - M)!]$, where $M \leq C$.

(c) Show that the number of ways in which M *indistinguishable* objects can be placed in C numbered boxes, with *no more than one* object per box is $[C!/M!(C - M)!]$, where $M \leq C$.

(d) Show that the number of ways in which M indistinguishable objects can be placed in C numbered boxes, with *no restriction* on the number of objects per box, is $[(C + M - 1)!/M!(C - 1)!]$.

(e) Show that the number of ways in which M indistinguishable objects can be placed in C numbered boxes, with *at least* one object per box, is $[(M - 1)!/(C - 1)!(M - C)!]$, where $(M \geq C)$.

11-6. Aircraft arrive at an airport at times that are more or less randomly distributed in time. Suppose the probability density that *the next* arrival occurs at a time t *after the previous* arrival is $a(t)$.

(a) Show that the probability that *no* arrivals occur in a time t *after the last one* is

$$A_0(t) = \int_t^\infty a(x)\,dx$$

where $A_0(0) = 1$, and that

$$T = \int_0^\infty ta(t)\,dt = \int_0^\infty A_0(t)\,dt$$

is the mean time between arrivals. Compute $A_0(t)$ for an exponential distribution, $a(t) = (1/T)e^{-t/T}$ and for an Erlang distribution $a = (4t/T^2)e^{-2t/T}$.

(b) Show that the probability that no arrivals occur in a time t *chosen at random* is

$$U_0(t) = (1/T)\int_0^\infty dx \int_t^\infty a(x + y)\,dy = (1/T)\int_t^\infty A_0(y)\,dy$$

Compute $U_0(t)$ for the exponential and the Erlang arrival distributions, and compare them with the postarrival probabilities obtained in part (a). Why does $U_0 = A_0$ in one case and $U_0 \neq A_0$ in the other?

Chapter 12

12-1. A molecule in a gas collides from time to time with another molecule. These collisions are at random in time, with an average interval τ, the mean free time. Show that, starting at time $t = 0$ (not an instant of collision) the probability that the molecule has not yet had its *next* collision at time t is $e^{-t/\tau}$. What is the expected time to this *next* collision? Show also that the probability that its *previous* collision (the last one it had before time $t = 0$) was earlier than time $-T$ is $e^{-T/\tau}$. What is the mean time of this *previous* collision? Does this mean that the average time interval between collisions is 2τ? Explain the paradox.

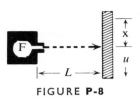

FIGURE P-8

12-2. In interstellar space, the preponderant material is atomic hydrogen, the mean density being about 1 hydrogen atom per cc. What is the probability of finding no atom in a given cc? Of finding 3 atoms? How many H atoms cross into a given cc, through one of its 1-cm^2 faces, per second, if the temperature is 1°K? If T is 1000°K?

12-3. A closed furnace F (Figure P-8) in an evacuated chamber contains sodium vapor heated to 1000°K. What is the mean speed $\bar{v}$ of the vapor atoms? At $t = 0$ an aperture is opened in the wall of the furnace, allowing a collimated stream of atoms to shoot out into the vacuum. The aperture is closed again at $t = \tau$. At a distance L from the aperture, a plate is moving with velocity u, perpendicular to the atom stream, so that the stream deposits its sodium atoms along a line on the plate; the position of the stream that strikes at time t hits the line at a point a distance $X = ut$ from its beginning. Obtain a formula for the density of deposition of sodium as function of X along the line, assuming that $\tau \ll (L/\bar{v})$, and find the value of X for which this density is maximum. Sketch a curve of the density versus X.

12-4. The momentum distribution in a gas, which has particle density (N/V), is

$$f(\mathbf{p}) = (2\pi mkT)^{-3/2} e^{-p^2/2mkT}(1 + \varepsilon \cos \alpha)$$

where $\varepsilon \ll 1$ and where α is the angle between $\mathbf{p}$ and the x axis.

(a) Compute the mean drift velocity $\mathbf{U}$ of the gas.

(b) How many atoms per second are crossing, in the positive x direction, a unit area of the y–z plane? How many are crossing in the negative direction? How are these quantities related to U_x?

12-5. Most conduction electrons in a metal are kept from leaving the metal by a sudden rise in electric potential energy, at the surface of the metal, of an amount eW_0, where W_0 is the electric potential difference between the inside and the outside of the metal. Show that if the conduction electrons inside the metal are assumed

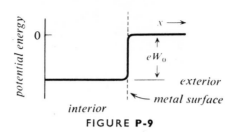

FIGURE **P-9**

to have a Maxwell distribution of velocity, there will be a thermionic emission current of electrons from the surface of a metal at temperature T that is proportional to $\sqrt{T} \exp(-eW_0/kT)$ (see Figure P-9). What is the velocity distribution of these electrons just outside the surface? [The measured thermionic current is proportional to $T^2 \exp(-e\phi/kT)$, where $\phi < W_0$; see Problem 26-2.]

12-6. A gas of molecules with a Maxwell distribution of velocity at temperature T is in a container having a piston of area A, which is moving outward with a velocity u (small compared to $\langle v \rangle$), expanding the gas adiabatically (Figure P-10). Show that, because of the motion of the piston, each molecule that strikes the piston with velocity $\mathbf{v}$ at an angle of incidence θ rebounds with a loss of kinetic energy of an amount $2mvu \cos \theta$ ($u \ll v$). Show that consequently the gas loses energy, per second, by an amount $-dU = PAu = P\,dV$, where dV is the increase in volume of the container per second.

12-7. Helium atoms have a collision cross section approximately equal to $2 \times 10^{-16}\ \mathrm{cm^2}$. In helium gas at standard conditions (1 atm pressure, 0°C), assuming a Maxwell distribution,

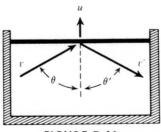

FIGURE **P-10**

what is the mean speed of the atoms? What is their mean distance apart? What is the mean free path? The mean free time?

12-8. The Doppler formula for the observed frequency f from a source moving with velocity v_x along the line of sight to the observer is

$$f = f_0[1 + (v_x/c)]$$

where f_0 is the frequency radiated when the source is at rest and c is the velocity of light.

(a) What is the distribution in frequency of a particular spectrum line, radiated from a gas at temperature T?

(b) What is the variance $\langle (f - f_0)^2 \rangle$ of this radiated frequency? The square root of the variance is called the *breadth* of the line.

(c) Atomic hydrogen and atomic oxygen are both present in a hot gas. How much broader will the hydrogen lines be, compared to oxygen lines of roughly the same frequency?

Chapter 13

13-1. A particle, moving vertically under the influence of gravity, periodically strikes and rebounds elastically from a horizontal plane. Sketch its phase-space trajectory between bounces. If particle 1 is released at $t = 0$ a distance h above the plane, with no initial velocity, and if particle 2 is released under similar conditions at time Δt, how far apart in phase space are the two particles when particle 2 is released? How far apart are they when particle 1 just reaches the horizontal plane? Assume that Δt is small compared to the time between bounces of each particle. If the two particles are released simultaneously, one at height h the other at height $h + \Delta h$, how far apart in phase space are they when the first particle hits the plane?

13-2. Use the Maxwell–Boltzmann distribution to show that, if the atmosphere is at uniform temperature, the density ρ and pressure P a distance z above the ground is $\exp(-mgz/kT)$ times ρ_0 and P_0, respectively (where g is the acceleration of gravity). Express ρ_0 and P_0 in terms of g, T, and M_a, the total mass of gas above a unit ground area. Obtain this same expression from the perfect gas law, $P = \rho kT/m$ and the equation $dP = -\rho g\, dz$ giving the fall-off of pressure with height (assuming T is constant). Find the corresponding expressions for ρ and P in terms of z if the temperature varies with pressure in the way an adiabatic expansion does, i.e., $P = (\rho/C)^\gamma$, $T = (D\rho)^{\gamma-1}$, where $\gamma = (C_p/C_v)$ [see Eqs. (4-12)].

13-3. The collision cross section of an air molecule for an electron is about 10^{-19} m². At what pressure will 90 per cent of the electrons emitted from a cathode reach an anode 20 cm away?

13-4. A gas-filled tube is whirled about one end with an angular velocity ω. Find the expression for the equilibrium density of the gas as a function of the distance r from the end of the tube.

13-5(*a*) The phase-space distribution for the N atoms in a gas is

$$f(\mathbf{r}, \mathbf{p}) = (1 + \gamma x)(2\pi mkT)^{-3/2}e^{-p^2/2mkT}$$

where the origin of position coordinates is at the center of a rectangular box of volume V, occupied by the gas. What is the pressure in the gas? Is the gas in mechanical equilibrium if no forces are acting? What sort of force would have to be applied for the gas to be in mechanical equilibrium? Assume that γx is smaller than 1 everywhere inside the box.

(*b*) The phase-space distribution is changed to

$$f(\mathbf{r}, \mathbf{p}) = [2\pi mkT(x)]^{-3/2}e^{-p^2/2mkT(x)}, \quad T(x) = T_0(1 - \gamma x)$$

what is the pressure now? Is the gas in mechanical equilibrium?

(*c*) Finally, the distribution is changed to

$$f(\mathbf{r}, \mathbf{p}) = (1 + \gamma x)[2\pi mkT(x)]^{-3/2}e^{-p^2/2mkT(x)}$$

where $T = T_0(1 - \gamma x)$ as before. To the first order in the small quantity γ is the gas in mechanical equilibrium? Is it in thermal equilibrium?

13-6. The atomic magnets in a paramagnetic crystal are not completely free to align themselves along the applied magnetic field; there is a small energy of binding (γ per magnet, where γ/k is a fraction of a degree K) to the crystal lattice. In the bound state the magnets cancel their polarization in pairs. In the unbound state the magnets can align themselves parallel or antiparallel to the field (i.e., $J = \frac{1}{2}$ for each magnet). The behavior can be approximated by a model having five different quantum states for each pair of magnets:

0—both magnets unbound, energy $\varepsilon_0 = 0$
1—unbound, both magnets parallel to field,
$\quad \varepsilon_1 = 2\gamma - 2 \cdot \frac{1}{2}m_B\mathcal{B}$
2—unbound, magnet 1 parallel, magnet 2 antiparallel,
$\quad \varepsilon_2 = 2\gamma$

3—unbound, magnet 1 antiparallel, magnet 2 parallel, $\varepsilon_3 = 2\gamma$

4—unbound, both magnets antiparallel, $\varepsilon_4 = 2\gamma + m_B\mathfrak{B}$

Show that the magnetization of the crystal is

$$\mathfrak{M} = \mu_0 N_m m_B \tanh\left(\frac{m_B\mathfrak{B}}{2kT}\right)\frac{X^2}{1 + X^2};$$

$$X = 2e^{-\gamma/kT}\cosh\left(\frac{m_B\mathfrak{B}}{2kT}\right)$$

where m_B is the Bohr magneton, N_m is the total number of atomic magnets, and $\mu_\sigma = (\mathfrak{B}/\mathfrak{H})$. Sketch or describe the dependence of $\mathfrak{M}$ on $\mathfrak{H}$ when $kT \gg \gamma$ and when $kT \ll \gamma$. Over what range of T and $\mathfrak{H}$ is Curie's law valid?

Chapter 14

14-1. A tube of length $L = 2\,\text{m}$ and of cross section $A = 10^{-4}\,\text{m}^2$ contains CO_2 at normal conditions of pressure and temperature (under these conditions the diffusion constant D for CO_2 is about $10^{-5}\,\text{m}^2/\text{sec}$). Half the CO_2 contains radioactive carbon, initially at full concentration at the left-hand end, zero concentration at the right-hand end, the concentration varying linearly in between. What is the value of t_c for CO_2 under these conditions? Initially, how many radioactive molecules per second cross the mid-point cross section from left to right? [Use Eqs. (14-15).] How many cross from right to left? Compute the difference and show that it checks with the net flow, calculated from the diffusion equation (net flow) $= -D(dn/dx)$.

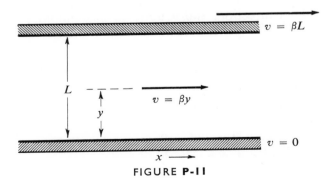

FIGURE P-11

14-2. A gas is confined between two parallel plates, one moving with respect to the other, so that there is a flow shear in the gas, the mean gas velocity a distance y from the stationary plate being βy in the x direction (Figure P-11). Show that the zero-order velocity distribution in the gas is

$$(1/2\pi mkT)^{3/2} \exp\{-(1/2mkT)[(p_x - m\beta y)^2 + p_y^2 + p_z^2]\}$$
$$= f_0(p_x - m\beta y, p_y, p_z)$$

Use Eq. (14-7) to compute f to the first order of approximation. Show that the mean rate of transport of x momentum across a unit area perpendicular to the y axis is

$$(N/V)(t_c\beta/m) \int \int \int p_y^2 f_0 \, dp_x \, dp_y \, dp_z = (N/V)t_c\beta kT,$$

which equals the viscous drag of the gas per unit area of the plate, which equals the gas viscosity η times β, the velocity gradient. Express η in terms of T and λ (mean free path) and show that the diffusion constant D of Eq. (14-16) is equal to (η/ρ), where ρ is the density of the gas.

14-3. A vessel containing air at standard conditions is radiated with x-rays, so that about 10^{-10} of its molecules are ionized. A uniform electric field of 10^4 volts/meter is applied. What is the initial net flux of electrons? Of ions? (See Problem 13-3 for the cross section for electrons; the cross section for the ions is four times this. Why?) What is the ratio between drift velocity and mean thermal velocity for the electrons? For the ions?

14-4. The free electrons in a partly ionized gas (plasma) have a density (N_e/V) and an equilibrium distribution

$$f_0(p) = (1/V)(2\pi mkT)^{-3/2}e^{-p^2/2mkT}.$$

An oscillating electric field $\mathfrak{E} = \mathfrak{E}_0 e^{j\omega t}$ is applied, where $\mathfrak{E}_0$ is pointed in the x direction, and is uniform throughout V.

(a) Show that the Boltzmann equation for the resulting distribution is

$$f + t_c(\partial f/\partial t) = f_0 - e\varepsilon_0 t_c e^{j\omega t}(\partial f_0/\partial p_x)$$

and thus show that f must have the form $f_0(p) + g(\mathbf{p})e^{j\omega t}$.

(b) By equating coefficients of the time-dependent and the time-independent terms, show that

$$g(\mathbf{p}) = \left(\frac{e\mathfrak{E}_0\lambda}{kT}\right)\frac{1 - j\omega t_c}{1 + (\omega t_c)^2}\left(\frac{p_x}{m}\right)\langle 1/v\rangle f_0(p)$$

where we have assumed that $t_c = \lambda\langle 1/v\rangle$.

(c) Compute the drift velocity V_x. Show that when $\omega t_c \ll 1$ it is in phase with $\mathfrak{E}$ and corresponds to Eq. (14-12). Show that when $\omega t_c \gg 1$ it is out of phase with $\mathfrak{E}$ and corresponds to the separate motion of each electron in an oscillating field.

(d) Compute the joule heat production per unit volume. Explain its dependence on frequency.

14-5. A paramagnetic gas is subjected to an oscillating magnetic field $\mathfrak{B} = \mathfrak{B}_0 \cos \omega t (\mathfrak{B} = \mu_0 \mathfrak{H})$. The atomic magnets have quantized directions, either being parallel to the field, with moment $\frac{1}{2} m_B$ and energy $\frac{1}{2} m_B \mathfrak{B}$, or being antiparallel, with moment $-\frac{1}{2} m_B$ and energy $+\frac{1}{2} m_B \mathfrak{B}$, where m_B is the Bohr magneton.

(a) Use the Boltzmann equation to find the first-order distribution function f for orientation of the atomic magnets, corrected to allow for the time variation of $\mathfrak{B}$. Assume that $\frac{1}{2} m_B \mathfrak{B} \ll kT$, expand the zero-order Maxwell–Boltzmann distribution f_0 in powers of $(m_B \mathfrak{B}/2kT)$, and discard all powers higher than the first.

(b) Calculate the magnetization $\mathfrak{M} = \mu_0 N_M \langle \text{moment} \rangle$ of the gas as a function of time, to the first order in $(m_B \mathfrak{B}/2kT)$.

(c) Compute the energy gained by the gas $\int \mathfrak{H} \, d\mathfrak{M}$ (and therefore lost by the oscillator) per second, to the first order.

14-6. A plasma is subjected to an electric field $\mathfrak{E}_x$ in the x direction and a magnetic field $\mathfrak{B}_z$ in the z direction, both fields being independent of position and time.

(a) Show that the Boltzmann equation for the distribution in momentum of the electrons in the plasma is

$$f = f_0 - et_c \left[\mathfrak{E}_x \left(\frac{\partial f}{\partial p_x} \right) + \mathfrak{B}_z \frac{p_y}{m} \left(\frac{\partial f}{\partial p_x} \right) - \mathfrak{B}_z \frac{p_x}{m} \left(\frac{\partial f}{\partial p_y} \right) \right]$$

(b) Compute f to the second order in t_c by setting $f = f_0 + t_c f_1 + t_c^2 f_2$ and equating coefficients of powers of t_c up to the second. Show that if f_0 is the Maxwell distribution, $f_1 = (e \mathfrak{E}_x p_x / mkT) \times f_0$ and

$$f_2 = [-(e^2 \mathfrak{E}_x^2 / mkT) + (e^2 \mathfrak{E}_x^2 p_x^2 / m^2 k^2 T^2)$$
$$- (e^2 \mathfrak{E}_x \mathfrak{B}_z p_y / m^2 kT)] f_0$$

Why does the magnetic field have no first-order effect on the distribution?

(c) Compute the components U_x and U_y of the drift velocity. What is the physical significance of the drift in the y direction?

14-7. It is suggested that the conduction electrons in a metal might be considered to be a gas in thermal equilibrium with the atomic ions, which are held together in the crystal lattice. The electron collisions would be mainly with the ions and the t_c in the Boltzmann equation would then be the time between collisions with the ions. In Chapter 26 we see that this model is not a very good approximation, but it is useful to see what it predicts.

(a) Use the formulas of this chapter to compute the electric conductivity ($I/\mathfrak{E}$) of silver (mol.wt. = 108, density 10,500 kg/m^3, density of ions = 6×10^{28}/m^3) in terms of e, N_e/V, λ, m, k, T, and $\langle 1/v \rangle$. The electric conductivity of silver at room temperatures is 7×10^7 mhos/m. If we use the Maxwell distribution to compute $\langle 1/v \rangle$, what does this predict for the mean free path λ of the electrons in silver? How does this compare with the distance between the ions? How does the formula predict the conductivity should vary with T? The actual conductivity is proportional to T^{-1} at room temperatures and to T^{-4} at lower temperatures (see page 376).

(b) Write out the formula for the heat conductivity of the electrons, assuming they have a Maxwell distribution, in terms of the same quantities as in (a). Show that the ratio between thermal and electric conductivity, for this model, divided by T, is just $(5k^2/2e^2)$, independent of T, λ, or $\langle 1/v \rangle$. The measured value of this quantity, for silver, for $250°K < T < 500°K$, is 2.3×10^{-8}. What does the model predict? Below 150°K the measured value of this ratio decreases rapidly.

(c) Use the formulas of this chapter to evaluate the thermo-electric power, of Eqs. (8-47) et seq., for silver.

14-8. Suppose the mean free path λ of the electrons is not constant, but is proportional to the electron's velocity v. How then would the conductivity depend on T? Does this suggest an answer to some of the discrepancies of Problem 14-7?

Chapter 15

15-1. A solid cylinder of mass M is suspended from its center by a fine elastic fiber so that its axis is vertical. A rotation of the cylinder through an angle θ from equilibrium requires a torque $K\theta$ to twist the fiber. When suspended in a gas-filled container at temperature T, the cylinder exhibits rotational fluctuation due to Brownian motion. What is the standard deviation ($\Delta\theta$) of the amplitude of rotation and what is the standard deviation ($\Delta\omega$) of its angular velocity? What would these values be if the container were evacuated?

15-2. An inductance L connected across the terminals of a capacitance C has electromagnetic energy of oscillation

$$U = (Q^2/2C) + (LI^2/2), \qquad I = (dQ/dt)$$

Q being the charge on the capacitance plates. The circuit is in thermal equilibrium with its surroundings at temperature T. There is a thermal noise current and a fluctuating charge in the capacitor.

(a) Calculate $\langle Q^2 \rangle$ and $\langle I^2 \rangle$.

(b) Assuming $C = 10^{-12}$ farad, what is the rms voltage across the capacitor plates at $T = 300°K$?

15-3. A piston slides without friction in a cylinder of internal cross section A, and encloses a perfect gas of N molecules. The force which pushes the piston downward has a magnitude $F = P_0 A$ no matter what the displacement of the piston. Also assume that F is large enough so that we can neglect the effect of the mass of the piston.

(a) What is the equilibrium distance x_0 of the piston from the bottom of the cylinder, if the temperature of the gas is T?

(b) What additional force is required to displace the piston a distance x (small compared to x_0) from equilibrium? Express this force in terms of x, x_0, N, and T. Calculate the potential energy of the displacement in terms of the same quantities.

(c) The piston does not stay in its equilibrium position; it oscillates up and down because of the fluctuations of pressure of the gas of the cylinder. What is the relative probability that the piston is displaced from equilibrium between x and $x = dx$?

(d) What is the mean-square amplitude $\langle x^2 \rangle$ of its displacement from equilibrium? Express the result in terms of x_0 and N.

15-4. Show that if the Hamiltonian energy of a molecule depends on a generalized coordinate q or momentum p in such a way that $H \to \infty$ as p or $q \to \pm\infty$, it is possible to generalize the theorem on equipartition of energy to

$$\left(q \, \frac{\partial H}{\partial q} \right)_{av} = \left(p \, \frac{\partial H}{\partial p} \right)_{av} = kT$$

Verify that this reduces to ordinary equipartition when H has a quadratic dependence on q or p. If H has the relativistic dependence on the momentum

$$H = c\sqrt{(p_x^2 + p_y^2 + p_z^2) + m^2 c^2}$$

show that

$$(c^2 p_x^2/H)_{av} = \cdots = (c^2 p_z^2/H)_{av} = kT$$

15-5. Analyze the thermal oscillations of electromagneti waves along a conducting wire of length L. In this case of one dimensional standing waves, the nth wave will have the form $\cos(\pi n x/L)e^{i\omega t}$, where $\omega = 2\pi f = \pi n c/L$, c being the wave velocity and f the frequency of the nth standing wave. Show that the number of allowed frequencies between f and $f + df$ is $(2L/c)\,df$, and thus if every standing wave has a mean energy kT, the energy content of the waves with frequencies between f and $f + df$ is $(2LkT/c)\,d$. If the wire is part of a transmission line, which is terminated b its characteristic impedance, all the energy will be delivered to the impedance in a time $(2L/c)$. Show, consequently, that the power delivered to the terminal impedance, the thermal noise power, i the frequency band df at frequency f, is $kT\,df$ if every wave is degree of freedom and if every degree of freedom carries a mea thermal energy kT. This formula corresponds with experiment a frequencies less than about 10^{10} cycles per second or so, but cannot be correct for all frequencies, clear out to $f \to \infty$, or els each wire would contain an infinite amount of electromagneti energy. (See Problem 25-5 for the answer.)

15-6. Observations of the Brownian motion of a spherica particle of radius 4.4×10^{-7} m in water at 300°K, which ha viscosity $\eta = 10^{-3}$ newton-sec/m^2 were made every 4 sec. Th displacements in the x direction, $\delta = x(t) - x(t - 4)$ are tabulate for 25 readings, as follows:

−5.8	+3.1	−1.0	−2.0	−1.5
+3.4	−0.2	+2.6	−1.9	+1.8
−1.8	−3.5	+0.3	−2.2	−0.2
−0.5	+1.3	+0.4	−0.4	+2.5
+0.5	+0.3	+0.6	+1.5	+1.9

Compute the mean value of δ and its variance. How close is thi distribution to the normal distribution of Eq. (11-23)? Use Eq (15-11) to compute Avogadro's number N_0 from the data, assumin that $R = kN_0$ is known.

Chapter 18

18-1. A gas of N point particles, with negligible (but nc zero) collision interactions, enclosed in a container of volume V has a total energy U. Show that the system point for the gas ma be anywhere on a surface in phase space which encloses a volum $[V^{3N/2}(2\pi m U)^{3N/2}/(3N/2)!]$. For an ensemble of these systems t represent an equilibrium state, how must the system points of the ensemble be distributed over this surface?

18-2. A harmonic oscillator has a Hamiltonian energy H related to its momentum p and displacement q by the equation

$$p^2 + (m\omega q)^2 = 2mH$$

When $H = U$, a constant energy, sketch the path of the system point in two-dimensional phase space. What volume of phase space does it enclose? In the case of N similar harmonic oscillators, which have the total energy E given by

$$\sum_{j=1}^{N} p_j^2 + \sum_{j=1}^{N} (M\omega q_i)^2 = 2MU$$

with additional coupling terms, too small to be included, but large enough to ensure equipartition of energy, what is the nature of the path traversed by the system point? Show that the volume of phase space "enclosed" by this path is $(1/N!)(2\pi U/\omega)^N$.

18-3. Use the final result of Problem 18-2 to show that the entropy of N distinguishable harmonic oscillators, according to the microcanonical ensemble is

$$S = Nk[1 + \ln(kT/\hbar\omega)]$$

18-4. A linear array of N particles is spaced equally along a straight line. Suppose each particle has two possible states, which may be designated as states A and B, and that we have adjusted the energy origin so the A state has energy $-\beta$ and the B state has energy $+\beta$. Suppose also that there is interaction between nearest neighbors in the linear array, so that every pair with like states (AA or BB) contributes 0 to the energy and every pair with unlike states (AB or BA) contributes α. (This is a one-dimensional model of the magnets in a ferromagnetic or antiferromagnetic material, where state A is orientation parallel to the field and B is antiparallel orientation, with $\beta = m\mathfrak{B}$; the energy α is the difference between the coupling energy between neighboring antiparallel magnets and that between neighboring parallel magnets. It could also be a model for the mixing of two substances in a binary alloy; state A is the presence of one sort of atom, state B the presence of the other sort.)

(*a*) Let m be the number of particles in state A (so that $N - m$ particles are in state B). These A particles can be placed anywhere along the line; they may be all at the beginning of the line, and all the B states may be at the end, or there may be a group

of A's at the beginning and at the end, with all the B's grouped i
the middle, and so on. Suppose the A's are grouped in n group
$(n \leq m)$, each group comprised one or more A's, separated from
the next A group by a group of one or more B's. Show that the
number of B groups is either $n - 1$ or n or $n + 1$ and that the
number of unlike pairs (divisions between A groups and B group:
is either $2n - 2$ or $2n - 1$ or $2n$, respectively. Show that, when N
n, and m are large enough, the energy of the linear array is $U = -N$
$+ 2\beta m + 2\alpha n$, where $m \geq n$.

(b) Use the results of Problem 11-5e to show that the numbe
of different ways in which the m states A may be arranged amon
the n different A groups, with at least one particle in each group
is $[m!/n!(m - n)!]$ and the number of different ways in which the
$(N - m)$ B states are arranged within the n B groups is

$$[(N - m)!/n!(N - m - n)!].$$

(c) Show that the entropy of a microcanonical ensembl
having a linear array of m particles in state A and $(N - m)$ particle
in state B and having the A states in n separated groups is

$$S = kT[m \ln m + (N - m) \ln(N - m) - 2n \ln n$$

$$- (m - n) \ln(m - n) - (N - m - n) \ln(N - m - n)$$

(d) Show that application of Eq. (18-1) leads to the pair o
equations

$$e^{2\beta/kT} = \frac{(N - m)(m - n)}{(N - m - n)m}; \qquad e^{2\alpha/kT} = \frac{(N - m - n)(m - n)}{n^2}$$

which must be solved simultaneously to find equilibrium values o
n and m and thus of U and S, for a specified T.

(e) If $\beta = 0$ (zero applied magnetic field, if the system is a
linear array of magnets) show that the equilibrium value of m i
$\frac{1}{2}N$, and those of n, and S are

$$n = \frac{(N/2)}{e^{\alpha/kT} + 1}; \qquad S = NkT \ln\left[2 \cosh\left(\frac{\alpha}{2kT}\right)\right]$$

$$- \frac{1}{2}\alpha N \tanh\left(\frac{\alpha}{2kT}\right)$$

What is the most likely ordering of the states along the line wher
$kT \gg \alpha$? What is the most likely ordering for $kT \ll \alpha$, when α i
positive (binding between unlike pairs stronger than binding be
tween like pairs)? When α is negative?

Chapter 19

19-1. A system consists of three distinguishable molecules at rest, each of which has a quantized magnetic moment, which can have its z component $+M$, 0, or $-M$. Show that there are 27 different possible states of the system; list them all, giving the total z component M_{zi} of the magnetic moment for each. Compute the entropy $S = -k\Sigma f_i \times \ln(f_i)$ of the system for the following a priori probabilities f_i:

(a) All 27 states are equally likely (no knowledge of the state of the system).

(b) Each state is equally likely for which the z component M_z of the total magnetic moment is zero; $f_i = 0$ for other states (we know that $M_z = 0$).

(c) Each state is equally likely for which $M_z = M$; $f_i = 0$ for all other states (we know that $M_z = M$).

(d) Each state is equally likely for which $M_z = 3M$; $f_i = 0$ for all other states (we know that $M_z = 3M$).

(e) The distribution for which S is maximum, subject to the requirements that $\Sigma f_i = 1$ and the mean component $\Sigma f_i M_{zi}$ is equal to γM. Show that for this distribution

$$f_i = \exp[(3M - M_{zi})\alpha]/(1 + x + x^2)^3$$

where $x = e^{\alpha M}$ (α being the Lagrange multiplier) and where the value of x (thus of α) is determined by the equation $\gamma = 3(1 - x^2)/(1 + x + x^2)$. Compute x and S for $\gamma = 3$, $\gamma = 1$, and $\gamma = 0$. Compare with a, b, c, and d.

19-2. Suppose we wish to set up an ensemble for a system depending on the thermal variables T, S and the mechanical variables Y, X [see Eqs. (8-17) et seq.], and the system is quantized so we can write down the allowed values E_ν of the system's energy and the allowed values X_ν of the mechanical variable X, for each quantum state ν of the system. We prepare the ensemble as follows: all systems in the ensemble have N particles, all have been brought to thermal equilibrium at temperature T, and they all have been brought to mechanical equilibrium at a specified value of the *intensive* variable Y.

(a) Show that the distribution function f must satisfy the following requirements: $S = -k\Sigma f_\nu \ln f_\nu$ is minimum, subject to

$$\sum f_\nu = 1, \qquad \sum f_\nu E_\nu = U \qquad \text{and} \qquad \sum f_\nu X_\nu = X$$

where X is the mean value of X_ν, averaged over the ensemble,

satisfying the equation of state for the specified values of Y and T
(b) Show that the solution of these requirements is

$$f_v = (1/Z_y)e^{(YX_v - E_v)/kT}; \qquad Z_y = \sum_v e^{(YX_v - E_v)/kT}$$

$$G_y = U - TS - YX = -kT \ln Z_y,$$

$$S = -(\partial G_y/\partial T)_Y, \qquad X = -(\partial G_y/\partial Y)_T$$

where Y and T are the independent variables.

19-3. A system contains N particles, confined in a box o
volume V at temperature T. Each particle has one quantum stat
with zero energy, has another one with energy $k\theta$, two more wit
energy $k\theta + kT_0(V_0/V)$, three more with energy $k\theta + 2kT_0(V_0/V)$
and so on, there being $(n + 1)$ states having energy $k\theta + nkT_0(V_0/V)$
The lowest state is a bound state, with binding energy $k\theta$; th
unbound states have energies which spread apart as the volume
decreases.

(a) Use the canonical ensemble to show that the Helmholtz
function and the pressure of the system are

$$F = -NkT \ln \ 1\left\{+ \frac{e^{\theta/T}}{(1 - e^{-T_0V_0/TV})^2}\right\}$$

$$P = \frac{2NkT_0(V_0/V^2)}{(e^{T_0V_0/TV} - 1)[1 + e^{\theta/T}(1 - e^{-T_0V_0/TV})^2]}$$

(b) Plot (PV_0/NkT_0) as a function of (V/V_0) for $T = T_0$ and
for $e^{\theta/T_0} = 2$, over the range $0 \leq V \leq 3V_0$. What are the limiting.
approximate formulas for F and P for $V \to 0$, for $V \to \infty$?

(c) What is the meaning of the sudden drop in P as V i
made less than $\frac{1}{2}V_0$? Is the system stable in this region? Does this
imply a phase change, from a bound phase to an unbound one
If so, how would you determine the pressure at which this phase
change occurs, for a given temperature?

(d) Plot the Gibbs function

$$(G/NkT_0) = (F/NkT_0) + (PV/NkT_0)$$

as a function of (V/V_0) for $T = T_0$ and $e^{\theta/T_0} = 2$. Combine this
plot, graphically, with the plot of (PV_0/NkT_0) to obtain a plot of
(G/NkT_0) as a function of (PV_0/NkT_0), instead of (V/V_0). Explain
the behavior of the curve.

Chapter 20

20-1. Show that a reasonable approximation to the value of ω_m, of Eq. (20-13), when expressed as a function of the velocities c_t and c_l of transverse and longitudinal waves, respectively, is

$$\omega_m = (18\pi^2 N/V)^{1/3}[(1/c_l^3) + (2/c_t^3)]^{-1/3}$$

The elastic constants for aluminum are: density 2700 kg/m³, bulk modulus $\beta = (1/\kappa) = 7.5 \times 10^5$ newtons/m², shear modulus $\alpha = 3.3 \times 10^5$. The two velocities are related to the moduli by the equations $c_l^2 = [(\beta + \frac{4}{3}\alpha)/\rho]$ and $c_t^2 = [\alpha/\rho]$. Compute the Debye temperature of aluminum from these data and compare with the value 400° which is obtained by fitting the Debye curve to the measured specific heats.

20-2. For the solid described by Eq. (20-16) show that $P = [(V_0 - V)/\kappa V_0] + (\gamma U_D/V)$, where $U = [(V_0 - V)^2/2\kappa V_0] + U_D$ and $\gamma = (-V/\theta) \times (d\theta/dV)$. Thence show that, for any temperature, if γ is independent of temperature, the thermal expansion coefficient β is related to γ by the formula

$$\beta = (1/V)(\partial V/\partial T)_P = \kappa(\partial P/\partial T)_V = (\kappa\gamma C_v/V)$$

Constant γ is called the Grüneisen constant.

20-3. When a crystal is stretched along one of its axes, the frequency of lattice vibrations in the direction of the stretch are changed, becoming $\omega_s = \omega_0[1 + (\alpha/2L_0^2)(L - L_0)^2]$, where L_0 is the equilibrium length. Use the Einstein model, with energy

$$E_v = (QA/2L_0)(L - L_0)^2 + \sum_{i=1}^{2N} \hbar\omega_0 n_i + \sum_{j=1}^{N} \hbar\omega_s n_j$$

to set up a canonical ensemble for this case (Q is Young's modulus of elasticity, A is the cross section of the crystal perpendicular to the tension, so that AL_0 is the crystal volume). Calculate the equation of state for elongation, $J = (\partial F/\partial L)_T$, for temperatures larger than $(\hbar\omega_0/k)$. If α is positive does the temperature rise or fall when the crystal is stretched adiabatically?

20-4. Set up the partition function of Eq. (20-19) for the one-dimensional array of magnets discussed in Problem 18-4. Let $E_v = 2\alpha n$ and let $\mathfrak{M}_v = (2m - N)\beta$, where $\beta = \frac{1}{2}\mu_0 \mathfrak{m}_B \mathfrak{H}$.

(*a*) Show that the partition function is

$$Z_m = \sum_{m,n} \frac{(N-m)!\,m!}{(n!)^2(m-n)!(N-m-n)!}$$

$$\times \exp\left[\frac{(2m-N)\beta - 2\alpha n}{kT}\right] = \sum_{m,n} e^{f(m,n)}$$

Write out the approximate expression for the exponent $f(m, n)$ using Stirling's approximation for the factorial functions.

(*b*) Expand the function f in a double-power series by setting $m = \frac{1}{2}N + x$ and $n = \frac{1}{4}N + y$ and using the Taylor series

$$f(m, n) = f\left(\frac{N}{2}, \frac{N}{4}\right) + xf_m\left(\frac{N}{2}, \frac{N}{4}\right)$$

$$+ yf_n\left(\frac{N}{2}, \frac{N}{4}\right) + \tfrac{1}{2}x^2f_{mm}\left(\frac{N}{2}, \frac{N}{4}\right)$$

$$+ xyf_{mn}\left(\frac{N}{2}, \frac{N}{4}\right) + \tfrac{1}{2}y^2f_{nn}\left(\frac{N}{2}, \frac{N}{4}\right) + \cdots$$

where the subscripts indicate partial derivatives. Then change from summation over m, n to integration with respect to dm and dn, obtaining a second-order expression, useful for $kT \gg \alpha$ and β,

$$Z_m \simeq 2^N \exp[-(N\alpha/2kT) + \tfrac{1}{2}N(\alpha/2kT)^2 + \tfrac{1}{2}N(\beta/kT)^2]$$

Compute the entropy and the magnetization to this approximation.

Chapter 21

21-1. Each of the N particles (distinguishable particles), in a box of volume V has two possible energies, 0 or E, where E is inversely proportional to the volume ($E = b/V$, where $b = $ constant). What is the equation of state of this system?

21-2. A system consists of a box of volume V and a variable number of indistinguishable (MB) particles each of mass m. Each particle can be "created" by the expenditure of energy γ; once created it becomes a member of a perfect gas of point particles within the volume V. The allowed energies of the system are therefore $n\gamma$ plus the kinetic energies of n particles inside V, for $n = 0$, 1, 2, ... Show that the Helmholtz function for this system (canonical ensemble) is

$$F = kT \ln\left[\sum_{n=0}^{\infty} (V^n \bar{X}^n/n!)\right] = -kTV\bar{X}$$

where $X = (2\pi mk/T/h^2)^{3/2}e^{-\gamma/kT}$. Calculate the probability that n particles are present in the box and thence obtain an expression for $\bar{N}$, the mean number of free particles present as a function of γ, T, and V. Also calculate S, C_v, and P from F and express these quantities as functions of $\bar{N}$, T, and V.

21-3. A gas consists of a mixture of $\frac{1}{2}N$ molecules of mass M_1 and $\frac{1}{2}N$ molecules of mass M_2, in a volume V at temperature T. Set up the partition function, assuming that the molecular rotation and vibration can be neglected and that the molecules of each kind are indistinguishable from each other (but distinguishable from those of the other kind). Calculate the entropy, internal energy, and pressure of the mixture. Compare the formula for S with that of Eq. (6-15).

Chapter 22

22-1. What fraction of the molecules of H_2 gas are in the first excited rotational state ($l = 1$) at $20°K$, at $100°K$, and at $5000°K$? What are the corresponding fractions for O_2 gas? What fraction of the molecules of H_2 gas are in the first excited vibrational states ($n = 1$) at $20°K$ and $5000°K$? What are the corresponding fractions for O_2 gas?

22-2. Plot the heat capacity C_v, in units of Nk for O_2 gas, from 100 to $5000°K$.

22-3. The solid of Eqs. (20-14) sublimes at low pressure, at a sublimation temperature T_s which is large compared to θ, the resulting vapor being a perfect diatomic gas, with properties given by Eqs. (21-14) and (22-2) (where $\theta_{rot} \ll T_s \ll \theta_{vib}$). Show that the equation relating T_s and the sublimation pressure P_s is approximately

$$G_s = V_0 P_s + \tfrac{3}{2}Nk\theta + 3NkT_s \ln(\theta/T_s) - NkT_s$$

$$= G_g \simeq NkT_s \ln(P_s V_0 T_s^{7/2}/N_0 k\theta^{5/2}) - NkT_s$$

where the equation

$$N_0 = V_0(4\pi I ek\theta/h^2)(2\pi mk\theta/h^2)^{3/2}$$

defines the constant N_0. Since $V_0 \ll V_g = NkT_s/P_s$ and $\theta \ll T_s$, show that this reduces to

$$P_s \simeq N_0 k\sqrt{T\theta}/V_0$$

Also show that the latent heat of sublimation is

$$L_s = T_s(S_g - S_s) \simeq \tfrac{1}{2}NkT_s$$

Chapter 23

23-1. At very high temperatures a gas is composed entirel of atoms of mass m. In a container of volume V its grand partitio function (MB statistics) is

$$\mathfrak{z} = \exp\left[Ve^{\mu_a/kT}x^{3/2}\right]$$

where $x = (2\pi mkT/h^2)$, and μ_a is the chemical potential of th atom. At lower temperatures some atoms combine in pairs to forr diatomic molecules, each of mass $2m$, chemical potential μ_m, an binding energy γ. Assume that these molecules have no rotation; or vibrational states, so each molecule has energy $-\gamma$ plus i translational, kinetic energy.

(a) Write out the expression for the grand partition functio (MB statistics) for the mixture of atoms and molecules, as a functio of V, x, γ, μ_a, μ_m, and kT. From this, obtain the grand potenti; $\Omega = -PV$ and, by differentiation, obtain the expressions for $\bar{N}$ and $\bar{N}_m$, the mean number of atoms and molecules in the systen

(b) Write out the expression for the Gibbs function $G = \mu_a \bar{N}$ $+ \mu_m \bar{N}_m$ in terms of kT, $\bar{N}_a$, $\bar{N}_m$, γ, x, and V. For thermodynami equilibrium at constant P and T, what must be the ratio betwee μ_m and μ_a?

(c) From this ratio compute the value of $\bar{N}_m$ as a functio of $\bar{N}_a$, V, x, and (γ/kT). Over what range of temperatures is $\bar{N}_m \gg \bar{N}_a$ What is the equation of state (in terms of $\bar{N} = \bar{N}_a + 2\bar{N}_m$) in thi range of temperature? What is the equation of state in terms of I in the range where $\bar{N}_m \ll \bar{N}_a$?

23-2. Work out the grand canonical ensemble for a gas o point atoms, each with spin magnetic moment, which can hav magnetic energy $+\frac{1}{2}\mu\mathfrak{H}$ or $-\frac{1}{2}\mu\mathfrak{H}$ in a magnetic field $\mathfrak{H}$ in additio; to its kinetic energy. Obtain the expression for $\bar{N}$ and expression for Ω, μ, S, U, $C_{V\mathfrak{M}}$, and the equation of state, in terms of $\bar{N}$, T and $\mathfrak{H}$. How much heat is given off by the gas when the magneti; field is reduced from $\mathfrak{H}$ to zero isothermally, at constant volume?

Chapter 24

24-1. A system consists of three particles, each of which has three possible quantum states, with energy 0, $2E$, or $5E$, respec tively. Write out the complete expression for the partition functior Z for this system: (a) if the particles are distinguishable; (b) if th particles obey Maxwell–Boltzmann statistics; (c) if they obey

Einstein–Bose statistics; (d) if they obey Fermi–Dirac statistics. Calculate the entropy of the system in each of these cases.

24-2. The particles of Problem 19-3 were distinguishable particles. Work out the grand canonical ensemble for particles having the same energy levels but different statistical behavior.

(a) Suppose the particles are MB particles. Compute the grand potential, the mean number $\bar{N}$, and the Helmholtz function $\Omega + \mu\bar{N}$. Show that when F is expressed as a function of $\bar{N}$, T, and V, it differs from the F of Problem 19-3 by the term $\bar{N}kT(\ln \bar{N} - 1)$. Why?

(b) Suppose the particles are bosons. Show that

$$\Omega = kT \ln[1 - e^{x/T}] + kT \sum_{n=0}^{\infty} (n + 1)$$

$$\times \ln\left\{1 - \exp\left[\frac{x - \theta - n(T_0 V_0/V)}{T}\right]\right\}$$

$$\bar{N} = [e^{-x/T} - 1]^{-1} + \sum_{n=0}^{\infty} (n + 1)$$

$$\times \left\{\exp\left[\frac{n(T_0 V_0/V) + \theta - x}{T}\right] - 1\right\}^{-1}$$

where $x = (\mu/k)$. Show that, when $T < \theta$ and $\bar{N} \gg 1$, the quantity $x \simeq -T[\bar{N} - e^{-\theta/T}(1 - e^{-V_0 T_0})^{-2}]^{-1}$, that $N_b = \bar{N} - N_f$ of the particles are in the bound state, and $N_f \simeq e^{-\theta/T}(1 - e^{-V_0 T_0/VT})^{-2}$ are in the upper, unbound states, and that $P \simeq (N_f kT/V)$ and $G = kx\bar{N}$. Is there a phase transition in this range of T?

(c) Suppose the particles are fermions. Write out the expressions for Ω and $\bar{N}$ as functions of T, V, and $x = \mu/k$. By using the equations

$$\sum_{n=0}^{n_0} (n + 1) \simeq \tfrac{1}{2}n_0^2 \qquad \text{and} \qquad \sum_{n=0}^{n_0} n(n + 1) \simeq \tfrac{1}{3}n_0^3$$

show that, when $T < \theta$ and $N \gg 1$, $x \simeq (T_0 V_0/V)(2\bar{N})^{1/2}$ and $P \simeq \tfrac{1}{3}\bar{N}xk = \tfrac{1}{3}G$; compare with the P of b. Is there a phase transition in this range of T?

Chapter 25

25-1. The maximum intensity per unit frequency interval, in the sun's spectrum, occurs at a wavelength of 5000 Å. What is the surface temperature of the sun?

25-2. Show that, for Einstein–Bose particles (bosons)

$$S = -k \sum_i \left[\bar{n}_i \ln(\bar{n}_i) - (1 + \bar{n}_i) \ln(1 + \bar{n}_i) \right]$$

25-3. It has been reported that the fission bomb produces a temperature of a million °K. Assuming this to be true over a sphere 10 cm in diameter: (a) What is the radiant-energy density inside the sphere? (b) What is the rate of radiation from the surface? (c) What is the radiant flux density 1 km away? (d) What is the wavelength of maximum energy per unit frequency interval?

25-4. The Planck distribution can be obtained by considering each standing electromagnetic wave in a rectangular enclosure $(L_x L_y L_z)$ as a degree of freedom, with coordinate Q_v proportional to the amplitude of the electric vector, with momentum P_v proportional to the amplitude of the magnetic vector, and with a field energy, corresponding to a term in the Hamiltonian, equal to $2\pi c^2 P_v^2 + (\omega_v^2/8\pi c^2)Q_v^2$, where c is the velocity of light and where the allowed frequency of the vth standing wave is given by

$$\omega_v^2 = \pi^2 c^2 [(k_v/L_x)^2 + (m_v/L_y)^2 + (n_v/L_z)^2]$$

(because of polarization, there are two different waves for each trio k_v, m_v, n_v). Use the methods of Eqs. (20-4) to (20-11) to prove that the average energy contained in those standing waves with frequencies between ω and $\omega + d\omega$ is $dE = (\hbar/\pi^2 c^3)\omega^3 \, d\omega/(e^{\hbar\omega/kT} - 1)$. Compare this derivation with the one dealing with photons, which produced Eq. (25-3).

25-5. Work out the correct form for the distribution-in-frequency of the thermal energy of electromagnetic waves along a wire, discussed in Problem 15-5. Prove that the energy content of the waves with frequencies between f and $f + df$ is

$$[2Lhf \, df/c(e^{hf/kT} - 1)]$$

and that the energy between f and $f + df$ which is delivered to a matching terminal impedance is $[hf \, df/(e^{hf/kT} - 1)]$. Show that this agrees with the results of Problem 15-5 for low frequencies. Above what frequency (for $T = 300°K$) is the equipartition result in error? What is the total energy delivered per second, over all frequencies.

25-6. A container of volume V has N short-range attractive centers (potential wells) fixed in position within the container. There are also bosons within the container. Each particle can either be bound to an attractive center, with an energy $-\gamma$ (one level per center), or can be a free boson, with energy equal to its kinetic energy, E. Use the analysis of this chapter to show that the equation relating the mean number $\bar{N}$ of bosons to their chemical potential μ is

$$\bar{N} = \frac{N}{e^{(\gamma - \mu)/kT} - 1} + N_0 f_{1/2}\left(\frac{-\mu}{kT}\right); \qquad N_0 = gV\left(\frac{2\pi m k T}{h^2}\right)^{3/2}$$

Draw curves for $-\mu/kT$ as a function of N_0/N for $\bar{N}/N = 1$ and for $\gamma/kT = 0.1$ and 1.0, using Table 25-1. Draw the corresponding curves for PV/NkT.

25-7. Suppose the particles of Problem 25-6 are MB particles instead of bosons. Calculate the partition function Z for a canonical ensemble and compare it with the Z for Problem 21-2.

Chapter 26

26-1. Show that, for Fermi–Dirac particles (fermions),

$$S = -\sum_i \left[\bar{n}_i \ln(\bar{n}_i) + (1 - \bar{n}_i) \ln(1 - \bar{n}_i)\right]$$

26-2. The conduction electrons of Problem 12-5 are, of course, fermions. Show that, for FD statistics, the thermionic emission current from the metal surface at temperature T is proportional to $T^2 \exp(-e\phi/kT)$, where $\phi = W_0 - \mu \simeq W_0 - \mu_0$ is called the thermionic work function of the surface.

26-3. The container and N attractive centers of Problem 25-6 have N fermions, instead of bosons, in the system. By using Eqs. (26-2) and (26-3) show that the equation relating μ and T and V is

$$\bar{N} = \frac{N}{e^{(\gamma - \mu)/kT} + 1} + N_0 \eta(\mu/kT), \qquad N_0 \text{ as in Problem 25-6}$$

Plot μ/kT as function of N_0/N for $\gamma/kT = 0.1$ and 1.0, using Table 26-1. Draw the corresponding curves for PV/NkT.

26-4. A simple model for a nucleus is that of a Fermi gas, held together by a harmonic oscillator potential. The allowed energies of each nucleon are $\varepsilon_n = (n + \frac{3}{2})\hbar\omega$ ($n = 0, 1, 2, ...$). The nth quantum level has $g_n = 2(n + 2)(n + 1)$ distinct states. Then, start with the expression for the grand potential

$$\Omega = -kT\sum_n g_n \ln\left[1 + e^{(\mu - \varepsilon_n)/kT}\right]$$

(a) Calculate the expression for the average number of particles $\bar{N}$.

(b) Show that this expression is a sum of terms $\bar{n}_n$. Sketch the plot of $\bar{n}_n$ as a function of ε_n, when $kT \ll \mu$.

(c) The energy-level spacing $\hbar\omega$ for a given nucleus is 10 Mev at $T = 0$. Its Fermi energy $\mu_F = kT_F$ turns out to equal 50 Mev. How many nucleons are there in the nucleus?

(d) For $0 < T \ll T_F$, the internal energy is

$$U(T) \simeq \bar{N}kT_F[\tfrac{3}{5} + (\pi^2/4)(T/T_F)^2 + \cdots]$$

The nucleus of part c absorbs 50 Mev from an incident proton. What is its resultant temperature (give kT in Mev)?

(e) What energy must an incident proton give up to the nucleus to transform it into a MB gas?

26-5. The equilibrium distribution function for conduction electrons in a metal is not the Maxwell distribution, but

$$f_0 = (2/Nh^3)[e^{(\varepsilon-\mu)/kT} + 1]^{-1}$$

where $\varepsilon = (p_x^2 + p_y^2 + p_z^2)/2m$ is the kinetic energy of one of the N electrons and $f_0 \, dp_x \, dp_y \, dp_z = mf_0 \, d\beta \sin \alpha \, d\alpha\sqrt{(2m\varepsilon)} \, d\varepsilon$ is the probability that the electron's position is in a unit volume and its momentum is in the element $dp_x \, dp_y \, dp_z$ at p.

(a) Evaluate the upper limit μ_F of the Fermi distribution at $T = 0$ in terms of h, m, and the number (N/V) of conduction electrons per unit volume.

(b) A uniform electric field $\mathfrak{E}$ is applied to the metal in the x direction. Solve the Boltzmann equation

$$f + t_c(\partial f/\partial t) \simeq f_0 - t_c[(\mathbf{p}/m) \cdot \mathrm{grad}_r f_0 + \mathbf{F} \cdot \mathrm{grad}_p f_0]$$

to obtain the first-order approximation for f in terms of t_c, f_0, $(\partial f_0/\partial p_x) = (p_x/m)(\partial f_0/\partial\varepsilon)$, and the various constants.

(c) Write out the integral expression for the resultant current I in the metal; integrate over the angles α and β and obtain the current as an integral over ε only, with the integrand containing $(\partial f_0/\partial\varepsilon)$.

$$\left[\iint p_x \, d\beta \sin \alpha \, d\alpha = 0; \qquad \iint p_x^2 \, d\beta \sin \alpha \, d\alpha = (8\pi/3)m\varepsilon. \right]$$

(d) Show that when $kT \ll \mu_F$,

$$\int_0^\infty F(\varepsilon)(\partial f_0/\partial\varepsilon) \, d\varepsilon \simeq -(2/Nh^3)F(\mu_F)$$

where F is any smoothly varying function of ε. Use this to evaluate the integral for the current and thus obtain I in terms of $N, e, \mathfrak{E}$, t_c, and m. Compare this with the MB formula (14-13).

Chapter 27

27-1. Calculate the heat capacity of D_2 as a function of T/θ_{rot} from 0 to 1.

27-2. The Schrödinger equation for a one-dimensional harmonic oscillator is

$$H\psi \equiv -\frac{\hbar^2}{2m}\frac{\partial^2\psi}{\partial x^2} + \tfrac{1}{2}m\omega^2 x^2\psi = E\psi$$

Its allowed energies and corresponding wave functions are

$$E_n = \hbar\omega(n + \tfrac{1}{2})$$

$$\psi_n(x) = \sqrt{m\omega/2^n n!h\sqrt{\pi}}\ H_n(x\sqrt{m\omega/\hbar})\exp(-m\omega x^2/2\hbar)$$

where $H_0(z) = 1$, $H_1(z) = 2z$, $H_2(z) = 4z^2 - 2$, $H_3(z) = 8z^3 - 12z$, etc.

Two identical, one-dimensional oscillators thus have a Schrödinger equation

$$H(x, y)\Psi \equiv -\frac{\hbar^2}{2m}\left(\frac{\partial^2}{\partial x^2} + \frac{\partial^2}{\partial y^2}\right)\Psi + \tfrac{1}{2}m\omega^2(x^2 + y^2)\Psi = E\Psi$$

where x is the displacement of the first particle from equilibrium and y that of the second.

(a) Show that allowed solutions of this equation, for the energy $\hbar\omega(n + \tfrac{1}{2})$, may be written either as linear combinations of the products $\psi_m(x)\psi_{n-m}(y)$ for different values of m between 0 and n, or else as linear combinations of the products

$$\psi_m\left(\frac{x+y}{\sqrt{2}}\right)\psi_{n-m}\left(\frac{x-y}{\sqrt{2}}\right)$$

(b) Express the solutions

$$\psi_m\left(\frac{x+y}{\sqrt{2}}\right)\psi_0\left(\frac{x-y}{\sqrt{2}}\right), \qquad \psi_m\left(\frac{x+y}{\sqrt{2}}\right)\psi_1\left(\frac{x-y}{\sqrt{2}}\right)$$

and

$$\psi_m\left(\frac{x+y}{\sqrt{2}}\right)\psi_2\left(\frac{x-y}{\sqrt{2}}\right)$$

for $m = 0, 1$, and 2 as linear combinations of the solutions $\psi_m(x)\psi_n(y)$ for $m, n = 0, 1, 2$.

(c) Which of these solutions are appropriate if the two particles are bosons? Which if they are fermions?

(d) Suppose the potential energy has an interparticle repulsive term $-\frac{1}{2}m\kappa^2(x - y)^2$ (where $\kappa^2 < \omega^2$) in addition to the term $\frac{1}{2}m\omega^2(x^2 + y^2)$. Show that, in this case, the allowed energies for bosons differ from those for fermions. Which lie higher and why?

REFERENCES

The texts listed below have been found useful to the writer of this volume. They represent alternative approaches to various subjects treated here, or more complete discussions of the material.

E. Fermi, "Thermodynamics," Prentice-Hall, Englewood, N.J., 1937, is a short, readable discussion of the basic concepts.

W. P. Allis and M. A. Herlin, "Thermodynamics and Statistical Mechanics," McGraw-Hill, New York, 1952, presents some alternative approaches.

F. W. Sears, "Introduction to Thermodynamics, the Kinetic Theory of Gases, and Statistical Mechanics," Addison-Wesley, Reading, Mass., 1953, also provides some other points of view.

H. B. Callen, "Thermodynamics," Wiley, New York, 1960, is a "postulational" development of the subject.

Charles Kittel, "Elementary Statistical Physics," Wiley, New York, 1958, contains short dissertations on a number of aspects of thermodynamics and statistical mechanics.

J. C. Slater, "Introduction to Chemical Physics," McGraw-Hill, New York, 1939, has a more complete treatment of the application of statistical mechanics to physical chemistry.

L. D. Landau and E. M. Lifchitz, "Statistical Physics," Addison-Wesley, Reading, Mass., 1958, includes a thorough discussion of the quantum aspects of statistical mechanics.

Allen L. King, "Thermophysics," Freeman, San Francisco, 1962, gives a number of examples, taken from modern physics.

T. L. Hill, "Statistical Mechanics," McGraw-Hill, New York, 1956, treats the Maxwell–Boltzmann statistics in considerable detail, including the modern theories of liquids and solids.

J. D. Fast, "Entropy," McGraw-Hill, New York, 1962, is a readable discussion of the entropy concept in thermal physics.

CONSTANTS

Gas constant $R = 1.988$ kg-cal/mole-deg
$= 8.317 \times 10^3$ joules/mole-deg
Avogadro's number N_0 (No. per kg mole) $= 6.025 \times 10^{26}$
Number of molecules per m^3 at standard conditions $= 2.687 \times 10^{25}$
Standard conditions: $T = 0°C = 273.16°K$, $P = 1$ atm
Volume of perfect gas at standard conditions $= 22.42$ m^3/mole
Dielectric constant of vacuum $\varepsilon_0 = \frac{1}{9} \times 10^{-9}$ farad/m
Electronic charge $e = 1.602 \times 10^{-19}$ coulomb
Electron mass $m = 9.109 \times 10^{-31}$ kg
Proton mass $= 1.672 \times 10^{-27}$ kg
Planck's constant $h = 6.625 \times 10^{-34}$ joule-sec
$\hbar = (h/2\pi) = 1.054 \times 10^{-34}$
Boltzmann's constant $k = 1.380 \times 10^{-23}$ joule/°K
Acceleration of gravity $g = 9.8$ m/sec^2
Velocity of light $c = 2.998 \times 10^8$ m/sec

1 atm	$= 1.013 \times 10^5$ newtons/m^2
1 cm Hg	$= 1333$ newtons/m^2
1 newton	$= 10^5$ dynes
1 joule	$= 10^7$ ergs
1 electronvolt (ev)	$= 1.602 \times 10^{-19}$ joules
	$= k(11600°K)$
Velocity of 1 ev electron	$= 5.93 \times 10^5$ m/sec
1 kg-cal	$= 4182$ joules
Ratio of proton mass to electron mass	$= 1836$

Stefan's constant $a = (\pi^2 k^4/15\hbar^3 c^3) = 7.56 \times 10^{-16}$ joule/m^3(°K)^4
Bohr magneton $\mathfrak{m}_B = (e\hbar/2m) = 0.927 \times 10^{-23}$ amp-m^2 for electrons
$= 5.05 \times 10^{-27}$ amp-m^2 for protons

$\bar{v} = \sqrt{(8kT/\pi m)} = 1.96 \times 10^4$ m/sec for electrons at $T = 1°K$
$= 146$ m/sec for protons at $T = 1°K$

$h^2 N_0^{2/3}/2\pi mk = 3961°K - m^2$ for electrons, $= 2.157°K - m^2$ for protons

$h^3 N_0/(2\pi mkT)^{3\,2} = V_b/T^{3\,2}$;

$V_b \begin{cases} = 2.49 \times 10^5 \text{ m}^3 \text{ for electrons} \\ = 3.17 \text{ m}^3 \text{ for protons [see Eq. (26-2)]} \end{cases}$

GLOSSARY

Symbols used in several chapters are listed here with the numbers of the pages on which they are defined or discussed.

a	van der Waals' constant, 28, 316	h_i	scale factor, 188
a	Stefan's constant, 82, 351	H	enthalpy, 79, 95
A	area, 16	H	Hamiltonian function, 188, 244
b	van der Waals' constant, 28, 316	$\mathfrak{H}$	magnetic intensity, 21, 196
$\mathfrak{B}$	magnetic induction, 21, 193	I	amount of information, 252
c	specific heat, 19	J	tension, 30
C	heat capacity, 19, 101	$\mathbf{J}$	flux density, 108
d	perfect differential, 32	k	Boltzmann's constant, 27, 166, 252
$đ$	imperfect differential, 33		
D	Curie's constant, 30, 195	L	latent heat, 121
D	diffusion constant, 214, 234	m	mass of particle, 14
e	= 2.7183, nat. log. base	M	mass of molecule, 304
e	charge on electron, 210	$\mathfrak{m}_B$	Bohr magneton, 197, 295
E	energy of system, 248	$\mathfrak{M}$	magnetization, 21, 195
$\mathfrak{E}$	electric intensity, 210	n	number of moles, 19
f	distribution function, 163, 243	n_j	occupation number, 332
F	Helmholtz' function, 97, 261	N	number of particles, 14, 243
		N_0	Avogadro's number, 27, 446
F_n	entropy parameter, 106	$\mathbf{p}$	momentum, 162, 244
$\mathfrak{F}$	Faraday's constant, 145	P	pressure, 14, 24
g	multiplicity of state, 279, 353	P_i	partial pressure, 141
		$\mathfrak{P}$	magnetic polarization, 21
G	Gibbs' function, 98, 293	q	coordinate, 188, 244
h	Planck's constant, 82, 248	Q	heat, 10
$\hbar$	= $h/2\pi$	Q	collision function, 184, 200

r	internuclear distance, 312	μ_0	permeability of vacuum, 21
$\mathbf{r}$	position vector, 182	v	quantum number, 248
R	gas constant, 25, 446	v_i	stoichiometric coefficient, 138
S	entropy, 59, 251	π	= 3.1416
t	time, 182	ρ	density, 44, 228
t_c	relaxation time, 204	σ	electric conductivity, 116
T	Temperature, 11, 26, 55	σ	collision cross section, 173, 201
U	internal energy, 15, 35, 65	σ_n	standard deviation of n, 152
$\mathbf{v}$	velocity, 15	σ_n^2	is variance of n, 158, 165
V	volume, 14	τ	mean free time, 173, 206
$\mathbf{V}$	drift velocity, 163, 211	ϕ	number of degrees of freedom, 243
W	work, 35	$\phi(q)$	potential energy, 188
x, y, z	coordinates	χ	magnetic susceptibility, 21, 196
Z	normalizing constant, 189	χ_i	concentration, 141
Z	partition function, 275	ψ	particle wave function, 379
$\mathfrak{Z}$	grand partition function, 328	Ψ	efficiency debit, 53
		Ψ	wave function, 379
α	Lagrange multiplier, 256, 274	ω	= 2π (frequency), 225, 260
β	thermal expansion coefficient, 30, 130, 405	ω_j	oscillator constant, 284
		Ω	mobility, 212
γ	C_p/C_v, 43	Ω	grand potential, 99, 327
ε	particle energy, 278, 302, 332	$\simeq$	approximately equal to
η	heat efficiency, 50	$\langle\,\rangle$	average value, 15, 168
η	viscosity, 210	∂	partial derivative, 31
θ	Debye temperature, 289	$\ln$	natural logarithm, 69
Θ	temperature, 25, 53	$n!$	factorial function, 160, 266
κ	compressibility, 30, 131, 405		
λ	mean free path, 173		
μ	chemical potential, 22, 138, 215, 327		

INDEX